Study Guide

to accompany

COLLEGE PHYSICS

Study Guide

to accompany

Paul A. Tipler
COLLEGE PHYSICS

Granvil C. Kyker, Jr.
Rose-Hulman Institute of Technology

Worth Publishers, Inc.

For the first, and best, Study Guide I ever encountered—my father

Study Guide
to accompany
COLLEGE PHYSICS by Paul A. Tipler
By Granvil C. Kyker, Jr.

Copyright © 1987 by Worth Publishers, Inc.

Printed in the United States of America
ISBN: 0-87901-269-2
First printing, September 1987

Worth Publishers, Inc.
33 Irving Place
New York, New York 10003

Contents

To the Student

How Do You Study Physics?

Of course there isn't a single answer to that one; you have to discover and develop the methods that work best for you. The important thing is that you find out what system does the most for you. In this *Study Guide,* which has been written to be used with Paul A. Tipler's *College Physics,* I've put together several kinds of material to try to help you do this. Note that there are complementary review materials at the end of each chapter in Tipler's text.

The chapters of this *Study Guide* supplement the corresponding chapters in Tipler. Each contains the following:

I. *Key Ideas* A brief discussion of what seem to me to be the basic physical concepts discussed in the text chapter.

II. *Numbers and Key Equations* A list, for quick review, of the new physical quantities, constants, and conversion factors encountered in the chapter as well as the important basic equations that have been discussed in it.

III. *Possible Pitfalls* These are points that, in my experience as a teacher, are easy for students to misunderstand or become confused about.

IV. *Questions and Answers* These are questions that require some physical reasoning, based on

material discussed in the chapter, to answer. Most are conceptual, although a few involve simple calculations. Complete answers are given for each question.

V. *Problems and Solutions* The problems are drawn from the material of the chapter. They range in difficulty from quite easy to pretty hard; they are meant not only to help you review but also to develop your problem-solving skills. Each problem includes detailed directions for solution, and detailed solutions of the odd-numbered (usually the more difficult) ones are given at the end of the section.

The single most important resource you have in studying physics is your instructor. You will have questions, and you should always assume that he or she is ready to answer them, in or out of class. First, though, do *your* part by pinpointing just what your question *is.* If you can narrow down your difficulty to a single point at which you start to get lost, it will save you and your instructor a lot of time and frustration, and it will usually get you a clearer and more useful answer. Often, too, in the process of doing this, you'll find that you've cleared up the difficulty for yourself.

Identifying just where your problem areas are is one of the ways in which this *Study Guide* can help you study physics. Start working through each chapter, reviewing the Key Ideas and Possible Pitfalls sections as soon as you've finished reading the corresponding chapter in Tipler. Review these again, together with the Summary and Review material at the end of the text chapter, *after* you've studied them in detail in the text, gone over your lecture notes, and done your homework assignments. Together with the material in the text, they provide a quick and convenient review of the chapter.

The most important equations that appear in the chapter are given, too, together with whatever new *numbers*—constants, conversion factors, or whatever—have appeared in the chapter. The equations *aren't* necessarily there to be memorized. As a rule, physics students spend too much time memorizing equations. If you have studied the text chapter in detail, worked a lot of problems, talked the basic ideas over with your fellow students and your instructor, and gone back to review, you'll find that you *know* the few most central equations. If you haven't done this, committing a long list of equations to memory without understanding their meanings and limitations is useless.

There is no substitute for the hard work of reading, analyzing, outlining, applying, rereading, and reviewing. I can't absorb everything in a piece of scientific writing in one reading, and you shouldn't expect to. In particular, don't try to use what's in this *Study Guide* as any kind of substitute for concentrated study of your text.

The questions in Section IV of each chapter require verbal answers. Most can be dealt with in a couple of sentences. Although they don't involve any but the most elementary calculations, for many of them you'll have to think a little beyond exactly what's said in the text chapter and perhaps put together a couple of ideas in a slightly different way.

Probably the most demanding, and I think the most rewarding, part of your study of physics is learning how to apply the ideas you're encountering to the solution of specific problems. At some point, you are likely to think, "I understand the theory but I just can't do the problems." Sorry, but it just isn't so. Unless and until the physical

ideas and mathematical equations you're studying become tools that are yours to apply at will to specific situations, you haven't really learned them. I think there are at least two major aspects to learning to solve problems: skill and drill. By "drill" I mean going through a lot of problems that involve the very direct application of a particular idea until you start to feel familiar with the way that idea fits into physical situations. One of the outstanding features of Tipler's text is the large set of exercises at the end of each chapter; they are meant for just this purpose. *Do a lot of these!* Almost certainly you should do *more* than you'll be required to do for homework. When you come across a particular point that isn't clear to you, then do as many exercises relating to that point as you can find time for. The answers to most of the odd-numbered exercises and problems are in the back of the book.

At the same time, however, you need to be tackling problems that go beyond the direct application of a single concept. It is precisely as you develop skill in bringing different ideas to bear on a particular situation and causing them to interact that they start to become your own. As you find out that you *can* deal with more complex problems, you may even start to enjoy the process. I can't tell you exactly how to do this; there isn't any universal problem-solving method. But I can suggest a pretty good plan of attack for most problems:

1. *Identify what you know.* The statement of the problem describes a physical situation and probably makes some simplifying assumptions. A numerical problem also provides you with specific values for some of the variables involved. Write out a list of the information you're given.

2. *Identify your goal.* Get entirely clear in your mind what it is that you are supposed to prove, calculate, or explain, and write out a clear statement of this.

3. *Choose your weapons.* What laws, principles, or relationships apply to this problem? All physical laws have some conditions or limitations under which they apply. Does the law you are trying to use really fit? Do you have to make any assumptions beyond those that you were given?

4. *Proceed logically.* You should know how you are going to get your answer and why you will choose each intermediate step. Write out your solution step by step, indicating what principle or law you are using in each one.

5. *Consider your result.* You may know what the result should be (as in a proof), or the answer may be given. If you don't know the correct answer, look for something you can *check* your answer against. Sometimes you can do this by working backwards from your answer to check that you get the given data. Also, ask yourself if your answer is *reasonable*. If the problem is about the speed of a runner and your answer says she was doing 10^{12} m/s, you probably made a mistake somewhere!

The examples in Part V of each chapter in this *Study Guide* are designed to help you develop your problem-solving skills. Most of these are not very easy. But with each problem, there is a set of suggestions as to how to carry out the solution. Try to solve the problem at this point. If you have trouble, the solutions to the odd-numbered problems are presented in detail at the end of the section. Study the solution and then look back at the suggestions to see how they point you to the solution. Then try one of the even-numbered problems; the detailed solutions to these aren't given, but most of them use the same ideas as the preceding odd-numbered problem.

One bit of advice: Be sure, throughout your physics course, to keep the forest in view behind all those trees. Physics is not a huge collection of unrelated formulas and calculations and numbers; there are really only a few themes that turn up in all sorts of places and in all sorts of disguises. There *is* an overall continuity to the various ideas you will find yourself working with; if you can keep these interrelationships in mind, it will make all the details a good deal more intelligible and a lot more fun.

Acknowledgments

I want to thank Paul Tipler for suggesting this project to me, for producing a text that has been a pleasure to work with, and for reading much of this *Study Guide* in manuscript and making many helpful suggestions. I am deeply indebted to June Fox for her confidence in me, to Anne Vinnicombe, who has worked long and hard to make this a better book than it could otherwise have been, and to many others at Worth Publishers for their ready and expert support. I am grateful to all those who read this version and the previous one and whose constructive suggestions helped me clarify my thinking on many points, and to my colleagues at Rose-Hulman Institute of Technology for their understanding of the demands this project has made on me. Above all, I must express my gratitude to my wife, Penny, for the support that has made this effort possible.

Granvil C. Kyker, Jr.
Terre Haute, Indiana
June, 1987

CHAPTER 1

Introduction

I. Key Ideas

Standard Units of Measurement When we measure a physical quantity, we are comparing it to some agreed-upon standard *unit*. A value for a physical quantity is meaningless unless the units of measurement are given (or understood). Scientists use a system of units called the international system (SI). In it, the standard units of length, mass, and time are the *metre*, the *kilogram*, and the *second*, respectively. (This is also referred to as the *mks system*—for *metre*, *kilogram*, and *second*.) SI contains other standard units which we will encounter later; for now, in studying mechanics, these three are all we need.

Other Unit Systems There are other systems of units in use defined in terms of the standard system. The *cgs system* (centimetre, *gram*, and second) is based on decimal multiples of the SI units. Both the mks and the cgs system are metric systems—decimal systems based on the metre. In everyday matters the *U.S. customary system* (foot, pound, and second) is still used. It is not a decimal system and is thus much less convenient to use.

Dimensions The *dimensions* of a physical quantity express what kind of quantity it is—whether it is a length, time, mass, etc. For example, the *dimensions* of velocity are length per unit time. The corresponding *units* might be miles per hour or metres per second. Whenever we add or subtract quantities, they must have the same dimensions. Also both sides of an equation must have the same dimensions. Checking that dimensions are in fact the same is often a useful way of checking for mistakes in setting up equations.

Scientific Notation Very often in physics we find ourselves dealing with very large or very small numbers. This is much simpler to do if the numbers are written in scientific notation, that is, as a number between 1 and 10 multiplied by the appropriate power of ten. An example is 1.67×10^5 for the number 167,400. When numbers are multiplied (or divided), the powers of 10 are added (or subtracted).

Significant Figures The quantities we deal with in physics are not pure numbers, but (in principle or in fact) results of measurements. They are not known to unlimited precision. The number of digits of a quantity that we write is, by implication, an expression of its *uncertainty*; we do not carry digits that we know are meaningless, that is, of no *significance*. The number of significant digits in a product or quotient can't be larger than the least number of significant digits in any of the factors.

II. Numbers and Key Equations

Numbers

$$1 \text{ foot (ft)} = 0.3048 \text{ m}$$

$$1 \text{ mile (mi)} = 1.61 \times 10^3 \text{ m}$$

$$1 \text{ litre (L)} = 1 \times 10^{-3} \text{ m}^3$$

Key Equations

There are no equations to review for this chapter.

III.   Possible Pitfalls

Remember that physical quantities are meaningless without units! If you try to work a problem using whatever numbers are given, and then you guess what the units of the answer are, you are in trouble. The best way to make sure you come out with the right answer, expressed in the right units, is to convert *all* the quantities given into a *consistent* system of units (preferably SI) *before* plugging them into an equation.

You will be working a lot of problems in this course and will almost certainly want to use a calculator. But *don't* trust it too far. It will do what you tell it to do, whether that makes sense or not. Check it by making a rough estimate *before* you start punching buttons. And when it gives you an answer of ten digits, only three of which are meaningful, don't mindlessly write down all ten. To do so is to claim a precision for your answer that it does not really possess.

Be careful about zeros when you think about how many significant digits there are in a number. Some zeros are only there to place the decimal point. For example, there is no way to tell how many of the zeros in a number like 16,400,000 are significant. (That's one of the advantages of scientific notation—the difference in the number of significant figures between 1.64×10^7 and 1.6400×10^7 is clear.)

It is possible to get all tangled up in a problem when there are several unit conversions to do. If you remember to do all conversions by *multiplying by a conversion factor whose magnitude is 1*, as discussed in the text, you will avoid a lot of errors.

Be careful when *adding* or *subtracting* numbers expressed in scientific notation. The power of ten on each number is a multiplier; so before you can add or subtract such numbers, you must express them as multiples of the *same* power of ten.

IV.   Questions and Answers

Questions

1. If you use a calculator to divide 3411 by 62.0, you will get something like 55.016129. (Exactly what you get will depend on your calculator.) Of course, you know that all those decimal places aren't significant. How *should* you write the answer?

2. If two physical quantities are to be added, must they have the same dimensions? The same units? What if they are to be divided? If your answers are different, explain why.

3. How else might you express the quantity 2×10^3 mockingbirds?

4. Is it possible to define a system of units in which length is not one of the fundamental quantities?

5. What properties should an object, system, or process have for it to be a useful standard of measurement of a physical quantity such as length or time?

6. Following the rules for determining the number of significant figures in a quantity allows us to estimate the precision of that quantity from the way it's expressed. Consider the number 61,000. What is the least number of significant figures this might have? The greatest number? If the same number is expressed as 6.10 x 10^4, how many significant figures does it have?

7. Acceleration has dimensions L/T^2 (or LT^{-2}), where L is length and T is time. What are the SI units of acceleration? What are its units in the U.S. customary system?

8. Acceleration, including the acceleration g due to gravity of a falling body, has dimensions L/T^2, where L is length and T is time; those of velocity are L/T. Suppose while taking a test you remember that the velocity of an object that has fallen a distance h under gravity is either $v = (2gh)^2$ or $v^2 = 2gh$—but you can't recall which. Which one *must* it be?

Answers

1. The number 3411 has four significant figures, and 62.0 has three (if it had only two, it should be written 62). The answer should be given to three figures, the least number of significant figures in the factors. So it should be 55.0.

2. To be added or subtracted, quantities *must* have the same dimensions. Otherwise, you are adding things of different kinds—like adding doughnuts to armchairs. When you actually come to add or subtract the numbers, they must also be expressed in the same units. Feet and metres both measure length, but adding them doesn't give a meaningful answer. When quantities are multiplied or divided, their dimensions and units need not be the same. In such cases the units are treated as another algebraic quantity: 2.5 m is "2.5" times "1 metre."

3. 2 kilomockingbird. (Sorry about that.)

4. Certainly it is possible. We might take time and velocity as the "fundamental" units, and define length to be velocity times time. (This might be pretty clumsy sometimes, however.) There has to be a unit of length, but it doesn't have to be one of those chosen as "fundamental."

5. Plainly it should be invariable, that is, not subject to change with time and, as nearly as may be, independent of extraneous conditions. Comparison of measurements should be relatively straightforward and possible at different places without carting the standard around. Convenience of use is more important for secondary than for primary standards.

6. The number 61,000 can have anywhere from two to five significant figures as we do not know how many of the zeros are significant. The number 6.10×10^4, however, because of the way it is written has only three significant figures. This is another advantage of scientific notation.

7. The units for acceleration are m/s^2 in SI and ft/s^2 in the U.S. customary system.

8. The expression $2gh$ has dimensions

$$(L/T^2)(L) = (L^2/T^2) = (L/T)^2$$

that is, velocity squared. The second version must be the correct one.

V. Problems and Solutions

Problems

1. In the following equation, x, v, and t are distance, velocity, and time and A, B, and C are constants. What are the dimensions of A, B, and C?

$$x = Avt + Bv \sin(Ct)$$

How to Solve It
- Remember that both sides of an equation must have the same dimensions.

- Remember that to add or subtract two quantities, they must have the same dimensions.

- In multiplying or dividing quantities, the dimensions are treated as any algebraic quantity would be.

- Mathematical functions such as the sine and cosine and their arguments must be pure numbers, that is, dimensionless.

2. In the following equation, x, v, and t are distance, velocity, and time expressed in SI units; A, B, and C are constants. What are the SI units of A, B, and C?

$$v = Ax[x + B \cos(Ct)]$$

How to Solve It

- Remember that both sides of an equation must have the same units. What are the SI units of velocity?

- Mathematical functions such as the sine and cosine and their arguments must be pure numbers, that is, dimensionless.

3. A stack of one hundred new $100 bills is 1.4 cm thick, and U.S. currency measures 6.6 × 15.8 cm. How much money in new $100 bills can you fit into an attache case that is 4 cm deep, 40 cm long, and 30 cm wide?

How to Solve It

- How much volume does each bill occupy?

- Assume that the bills fill the entire volume of the case, that is, that you jam bills into the corners, etc., so that you don't waste any space.

4. You hire a printer to print concert tickets. He delivers them in circular rolls labeled as 1000 tickets each. You want to check this without counting thousands of tickets. You decide to do it by measuring the diameter of the rolls. If the tickets are 2 in long and 0.22 mm thick and are rolled on a core 3 cm in diameter, what should be the diameter of a roll of 1000 tickets?

How to Solve It

- Don't forget to convert to the same units before you plug numbers in!

- If the tickets are rolled tight, the side area of the roll will be the number of tickets multiplied by the area of the edge of a ticket.

- What is the side area of the roll? Don't forget to subtract for the area of the 3 cm core.

5. The United States is about the only country left that uses the units feet, miles, gallons, etc. You see some car specifications in a German magazine which give fuel efficiency as 7.6 kilometres per kilogram of fuel. If a mile is 1.602 km and a gallon is 3.785 litres, and if a litre of gasoline has a mass of 1.75 kg, what is the car's fuel efficiency in miles per gallon?

How to Solve It

- Remember that every conversion factor can be expressed as a number equal to 1.

- You can treat the mass of a litre of gasoline as a conversion factor in this case, although strictly speaking it's a separate physical quantity called the density of gasoline.

- Multiply the given fuel efficiency by the appropriate conversion factors and the density of gasoline in such a way as to cancel units and leave the result in miles per gallon.

6. A furlong is one eighth of a mile and a fortnight is two weeks. Of what physical quantity is a furlong per fortnight a unit? What is the SI unit for this quantity? Find the factor for converting from furlong per fortnight to the corresponding SI unit.

How to Solve It
- What physical quantity is equal to distance per unit time?

- Remember that every conversion factor can be expressed as a number equal to 1.

- Multiply 1 furlong/fortnight by the appropriate conversion factors in such a way as to cancel units and leave the result in SI units.

7. In a certain experiment, light is found to take 37.1 microseconds (μs) to traverse a measured distance of 11.12 km. Calculate the speed of light from these data. Express your answer in SI units and in scientific notation, with the appropriate number of significant figures.

How to Solve It
- Note how many significant figures there are in each quantity and decide how many significant figures the result will have.

- Write each quantity in SI units and scientific notation before you divide them.

8. The time it takes a heavy object to fall to the ground from an initial height h is given by

$$t = [2h/g]^{1/2}$$

where g is the acceleration due to gravity. If you drop a brick from the top of a building which is 90.0 m high and find it takes 4.4 s to fall, what value do you calculate for the acceleration due to gravity? Express your answer in scientific notation, with the appropriate number of significant figures.

How to Solve It
- Note how many significant figures there are in each quantity and decide how many significant figures the result must have.

- Write each quantity in scientific notation before you divide them.

- What effect do you suppose squaring a number has on the number of significant figures?

Solutions

1.   Each of the two terms that are added on the right-hand side of the equation must have the same dimensions as x on the left-hand side of the equation. Thus, Avt and Bv (since the sine function is dimensionless) must each have dimensions of length. Since vt has dimensions $(L/T)\,(T) = L$, A is a *dimensionless number.* The argument Ct of the exponential is dimensionless also, so C must have dimensions of T^{-1}.

2. A is in 1/metres-second; B is in metres; C is in 1/seconds.

3.   The volume occupied by each bill is

$$V = L \times W \times H = (6.6 \text{ cm})(15.8 \text{ cm}) \left(\frac{1.4}{100} \text{ cm}\right) = 1.46 \text{ cm}^3$$

and the volume available in the briefcase is

$$(30 \text{ cm})(40 \text{ cm})(4 \text{ cm}) = 4.8 \times 10^3 \text{ m}^3$$

So if the briefcase is filled "solid," it can hold

$$\frac{4.8 \times 10^3}{1.46} = 3300 \text{ bills or } \$330{,}000$$

4.   The roll should be 12.3 cm in diameter.

5.   Multiply by "1" for each conversion:

$$7.6 \text{ km/kg} \left(\frac{1 \text{ mi}}{1.602 \text{ km}}\right) \left(\frac{1.75 \text{ kg}}{1 \text{ L}}\right) \left(\frac{3.785 \text{ L}}{1 \text{ gal}}\right) = 31.4 \text{ mi/gal}$$

6.   These are the units of velocity. The SI units of velocity are metres per second. 1 furlong/fortnight = 1.663×10^{-4} m/s.

7.   The travel time is 37.1×10^{-6} s = 3.71×10^{-5} s. The distance traveled is 11.12×10^3 m = 1.112×10^4 m. The speed is thus

$$c = \frac{x}{t} = \frac{1.112 \times 10^4 \text{ m}}{3.71 \times 10^{-5} \text{ s}} = 0.2997 \times 10^9 \text{ m/s}$$

$$= 3.00 \times 10^8 \text{ m/s}$$

The answer is shown to three significant figures because that's all we had for t. Note, however, that it's a good idea to carry the extra figures through the intermediate steps in any calculation to avoid the accumulation of roundoff errors.

8.   9.7 m/s^2 from these data.

CHAPTER 2

Motion in One Dimension

I. Key Ideas

Motion of a Particle in One Dimension To begin with, we simplify our discussion by considering the *motion of a single particle*. We can apply what we learn to anything small enough (in a given situation) so that we can locate it by specifying the position of a single point in space. Also, for the time being, we will consider only *motion in one dimension*. If we define an x axis along the direction of motion with its origin O at some reference point, we can specify an object's position by a single number x. The sign of x indicates to which side of the origin O the object is. We describe the object's motion by saying how its position changes with time.

Speed, Displacement, and Velocity The *average speed* of an object is the distance it has traveled without regard to direction divided by the time taken. If we define the *displacement* $\Delta x = x_2 - x_1$ of an object as the change in its position from time t_1 to time t_2, then the *average velocity* of the object is its displacement divided by the time interval $\Delta t = t_2 - t_1$. Velocity has *direction* (in one dimension, a sign) as well as *magnitude*. Note that the average velocity depends only on the initial and final positions of the object, not on any details of what happens in between.

Instantaneous Velocity and Acceleration What we usually mean when we say "velocity" is instantaneous velocity. *Instantaneous velocity* is defined to be the average velocity over a very small time interval, or more exactly, it is the limit of the average velocity as the time interval goes to zero. If a moving object's instantaneous velocity is not constant, it is said to be *accelerating*. (The acceleration of an object is important to us as it is the quantity directly related to the forces that act on the object.) The *average acceleration* over any time interval is the change in velocity over that interval divided by the interval. The *instantaneous acceleration* is the average acceleration over a very small time interval.

Graphs Some of these quantities are easier to understand if we represent them graphically. If we draw a graph of *position as a function of time* to represent an object's motion, then the average velocity for a particular time interval $\Delta t = t_2 - t_1$ is the slope of a straight line drawn between two points (x_1, t_1) and (x_2, t_2) on the graph. We defined the instantaneous velocity at a particular time t_1 as the average velocity over a very small time interval. On the graph, it is the slope of a line drawn tangent to the graph at that point (x_1, t_1). In just the same way, from the graph of *velocity as a function of time* you can find the acceleration: the average acceleration for a particular time interval is the slope of a straight line drawn between those two points on a *v*-versus-*t* graph, and the instantaneous acceleration at a particular time is the slope of a line drawn tangent to the graph at that point. If an object's velocity is constant, the *x*-versus-*t* graph is a straight line; if its acceleration is constant, the *v*-versus-*t* graph is linear.

Constant Acceleration In many real-life situations, objects move with constant (or nearly constant) acceleration, so this is an important special case of accelerated motion. An example is *free-fall:* an object moving near the earth and acted on by the earth's gravity (and free of other forces) falls with an acceleration (downward) of 9.81 m/s². If we know the position and velocity of an object moving with constant acceleration at one particular time, we can write down equations that give its position and velocity at any other time.

II. Numbers and Key Equations

Numbers

Acceleration Due to Gravity

$$g = 9.81 \text{ m/s}^2 = 32.2 \text{ ft/s}^2$$

Key Equations

Velocity

$$v_{av} = \frac{\Delta x}{\Delta t} = \frac{x_2 - x_1}{t_2 - t_1} \qquad v = \text{limit}_{\Delta t \to 0} \frac{\Delta x}{\Delta t}$$

Acceleration

$$a_{av} = \frac{\Delta v}{\Delta t} \qquad a = \text{limit}_{\Delta t \to 0} \frac{\Delta v}{\Delta t}$$

Motion with constant acceleration

$$v = v_0 + at$$

$$x = x_0 + v_0 t + \tfrac{1}{2} at^2$$

$$v^2 = v_0^2 + 2a \, \Delta x = v_0^2 + 2a \, (x - x_0)$$

$$v_{av} = \tfrac{1}{2} (v_0 + v)$$

III. Possible Pitfalls

A lot of what we do in physics depends on representing various quantities by algebraic variables. We use *t* for the time of an event, for example. If this sort of thing is new to you, try not to confuse the *value of* a quantity with a *change in*, or *increment of*, that quantity. For instance, 10:30 A.M. on Tuesday is a specific point in time, but 10.5 hours is a time interval. Similarly, *x* is the position of a particle at some instant, *but* the distance it has gone in some time interval, its displacement, is $\Delta x = x_2 - x_1$.

Signs are likely to cause you trouble until you get used to doing kinematics problems. The important thing to remember is that the signs of all the quantities in a problem have to be *consistent*. If you're doing a free-fall problem, for instance, and you measure position upward from the ground, then the acceleration is negative *g* because you've chosen to call the upward direction positive and freely falling objects accelerate downward. If you're consistent, the sign of the answer will tell you which way something is going.

Speed is the magnitude of velocity, but average speed is not necessarily the magnitude of average velocity. Average speed is defined as the total distance traveled divided by the time interval. Velocity and acceleration can be either positive or negative. In everyday usage we say "deceleration" for a decrease in speed; don't let this confuse you. Properly, "acceleration" means a change in velocity of either sign.

Acceleration and velocity are separate quantities and vary independently. Don't confuse them! It's perfectly possible for a particle to have zero velocity and nonzero acceleration or vice versa; or to have positive velocity and negative acceleration or vice versa.

In the last part of text Chapter 2, the equations for motion with constant acceleration are developed. You need to know these and how to use them. But don't forget that they *only* apply to this special case! You can't use them unless you know that the acceleration is constant (or can assume it's constant as a good approximation).

The equations for motion with constant acceleration are worked out *with reference to* the initial conditions of the moving object—where it was (x_0) and how fast it was moving (v_0) at time $t = 0$, where $t = 0$ can be any time you like. In working a particular problem, it's often convenient to choose the time $t = 0$ such that $x_0 = 0$ and/or $v_0 = 0$. In any case, be sure you have all the quantities defined in a *consistent* way.

IV. Questions and Answers

Questions

1. A car of ordinary size is driven down the road at 65 km/h. Briefly state under what circumstances it is and is not appropriate to consider the car as a particle for the purpose of studying its motion.

2. Figure 2-1 is a graph of the position of a particle as a function of time. (*a*) Over what interval(s) is the particle's velocity apparently zero? (*b*) Over what interval(s) is the particle's velocity nonzero but apparently constant? (*c*) Over what interval(s) is the particle's average velocity less than zero? (*d*) Over what interval(s) is the particle's acceleration apparently zero? (*e*) Is the acceleration positive or negative at point D? (*f*) Identify a point or points at which the particle's acceleration is clearly not constant.

Figure 2-1

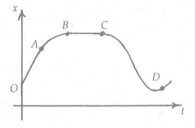

3. At some instant, a car's velocity is 15 m/s; one second later it is 11 m/s. If the car's acceleration is constant, what is its average velocity over this interval? How far does it go? Could you answer these questions without knowing that the acceleration is constant?

4. Can a particle whose velocity is zero have nonzero acceleration? Can a particle's speed be changing if its acceleration is zero?

5. The motion of a particle is represented by the graph in Figure 2.2. Over the interval from $t = 0$ to $t = 12$ s, are the average velocity and the average speed the same?

Figure 2-2

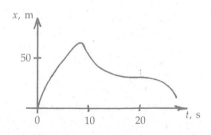

6. A world-class sprinter can run 100 m in 10 s, but cannot run 30 m in 3 s. Why not? Neither can he run 400 m in 40 s. Why not?

7. A high diver leaps 2.5 m straight up from a springboard, then falls past the edge of the board to the water 6 m below it. Is there a point at which her velocity is zero? Is there a point at which her acceleration is zero?

Answers

1.   If the motion of a single point through space tells you all you want to know about the car's motion, then it's a particle. This is probably the case if, say, you want to study its motion as it proceeds down a straight road for an hour. On the other hand, if the car is skidding and turning over after a collision, then clearly for the purposes of studying its motion you can't treat the car as a particle.

2.   (*a*) From *B* to *C*, the position of the particle isn't changing in time, so $v = 0$. (*b*) From the origin *O* to point *A*, the graph is a straight line, so the velocity is constant. (*c*) From *C* to *D*, the value of *x* decreases, so the average velocity is less than zero. (*d*) If the particle's acceleration is zero, the velocity will be constant and the graph, a straight line. This is true, for example, from the origin *O* to point *A* and from *B* to *C*. (*e*) The acceleration is positive at *D*. (*f*) Since from *C* to *D* the acceleration changes from negative to positive, it's clearly not constant at points in that interval.

3.   If the acceleration is constant, the average velocity over the interval is halfway between that at the beginning and that at the end of the interval. Thus, the car's average velocity here is 13 m/s, and it goes 13 m in the one-second interval. However, if the acceleration were not constant, we could not calculate v_{av} or the distance the car traveled without knowing just how the velocity changed over the interval.

4.   A particle's velocity can be zero at an instant no matter what its acceleration is. For example, a ball thrown up in the air has zero velocity at the topmost point of its flight, but its acceleration is not zero. If the acceleration of an object is zero, on the other hand, its speed (and its velocity) is constant.

5.   No. As shown on the graph, the particle reverses direction so that the displacement will be less than the distance traveled, so average speed will be greater than average velocity.

6.   The sprinter takes longer to run the 30 m because he needs more than 3 s to reach his top speed; thus his average velocity is lower over the shorter distance. He takes longer to run 400 m because his body can't sustain his top speed for that long.

7.   The diver's velocity is zero at the top point of her leap—at that instant she has ceased to go up (*v* positive) and is about to go down (*v* negative). Since she is in free fall (acceleration equal to *g*, downward), her acceleration is never zero.

V. Problems and Solutions

Problems

1. A car travels 40 km along a straight road at a speed of 86 km/h, and then goes 40 km farther at a speed of 50 km/h. What is the car's average velocity for the entire trip?

How to Solve It
- Average velocity is the displacement divided by the time interval.

- The total displacement is the sum of the two distances, and the total time interval is the sum of the two time intervals.

- In *each* of the two intervals the car was moving at constant speed.

2. You're bicycling across central Oklahoma on a perfectly straight road. You started at 8:30 A.M. and you've covered 21 miles by 11:15 A.M. when your chain breaks. You haven't any spare parts with you, so you have to walk the bike back to the last town at a speed of 2.6 mi/h. If you get there at 1:30 P.M., what was (*a*) your displacement, (*b*) your average velocity, and (*c*) your average speed for the whole trip?

How to Solve It
- Displacement is *net* change in position, never mind what happened along the way. What's the difference between your initial and your final positions? Don't forget that you changed directions!

- Average velocity is the displacement divided by the time interval.

- The average speed is the *distance gone*, without regard to direction, divided by the time interval.

3. Starting at a time we'll call $t = 0$, a man walks east from his office to McDonald's, has lunch, and then walks to his bank. His trip is shown on the graph of position versus time in Figure 2-3. (*a*) What was his average velocity from his office to the bank? (*b*) How much time did he spend at lunch? (*c*) How far is it from McDonald's to the bank? (*d*) At what point in the trip was he walking fastest?

How to Solve It
- On an *x*-versus-*t* graph, the average velocity between any two points is the slope of a straight line drawn between those points.

- Where is McDonald's? What's happening on the graph while the man is having his lunch?

- The instantaneous velocity at a point is the slope of a line drawn tangent to the curve at that point.

Figure 2-3

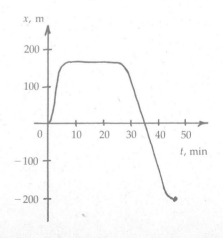

4. A woman drives west from Nashville toward Memphis on I-40. (Assume that the highway is straight.) At Bucksnort, Tennessee she develops car trouble and has to be towed back to Nashville for repairs before she can continue her journey, finally arriving at Memphis at midnight. The entire trip is described on the position-versus-time graph in Figure 2-4. (*a*) What was her average velocity before her car broke down? (*b*) What was it for the whole trip? (*c*) Assume that, without the breakdown, she would have continued at her initial (average) velocity. How much time did the breakdown cost her? (*d*) In what time interval was her velocity the greatest?

Figure 2-4

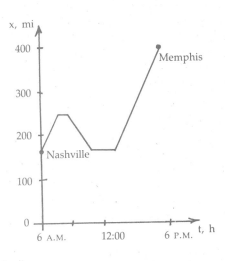

How to Solve It
- Average velocity is the displacement divided by the time interval, so the average velocity times the time interval is equal to the displacement.

- The (instantaneous) velocity is greatest where the slope of a line drawn tangent to the graph is steepest.

- No, I didn't make up Bucksnort. It's there.

5. A ball rolls down a hill, starting from rest at the top of the hill at time $t = 0$, picking up speed as it goes. Suppose that its velocity is given by the equation

$$v = (0.06 \text{ m/s}^3) \, t^2$$

(*a*) What is the ball's average acceleration over the interval from $t = 0$ to $t = 2$ s? (*b*) What is its average acceleration from $t = 2$ to $t = 4$ s? (*c*) Is the ball moving with constant acceleration?

How to Solve It
- Average acceleration is the change in (instantaneous) velocity over a time interval divided by the length of the interval.

- The equation stated in the problem gives you the (instantaneous) velocity at any time t.

- If the ball's acceleration were constant, what would you expect to get for the average acceleration in the two different time intervals?

6. The table at right shows acceleration data for an automobile (a 1987 Whizbang) from a standing start. From these data, (*a*) calculate the average acceleration of the car over the interval from $t = 1$ s to to $t = 3$ s. (*b*) Draw a graph of these data and determine the car's instantaneous acceleration at $t = 2$ s.

How to Solve It
- Average acceleration is the change in (instantaneous) velocity over a time interval divided by the length of the interval.

t, s	v, m/s
0	0
1	12.3
2	17.7
3	21.5
4	24.6
5	27.3
6	29.9

- On a graph of velocity versus time, the acceleration at a particular time is the slope of a line drawn tangent to the curve at that point.

7.   Al sees Bob drive past his house at 20 m/s and realizes he must catch Bob to get the day's physics assignment. Al jumps in his car and starts it—this takes 15 s—and takes off after Bob. Suppose Al accelerates at a *uniform* rate of 1.4 m/s², and Bob continues driving at a constant speed of 20 m/s. *Where* and *when* does Al catch up to Bob?

How to Solve It
- Write an equation for Al's motion, that is, one that gives his position x as a function of time. Then write an equation for Bob's motion.

- At the particular time t when Al catches up with Bob, the two equations will give the same value of x!

- Now you know *when* Al catches up to Bob; use either equation to calculate *where* this happens.

8.   A car is traveling at 18 m/s along a straight level road. Suddenly the driver sees a truck 30 m ahead in her lane, moving in the same direction at 8 m/s. If it takes the driver 0.2 s to hit her brakes and she brakes thereafter with a constant acceleration, what must her acceleration be if she is to avoid a collision?

How to Solve It
- Write the equation for the truck's motion, which is really simple. Write the equations for motion for the position and velocity of the car.

- Calculate the time at which the car and the truck have the same velocity. The unknown acceleration will appear in your result; that's okay.

- Where are car and truck when their velocities are the same? Remember if the driver succeeded in avoiding a collision, her car must still be behind the truck at this time.

9.   A world-class sprinter can run the 100 m sprint in 10.0 s. Assume that he starts from rest and runs with uniform acceleration for the first 50 m and thereafter runs at his top (constant) speed. (*a*) What is his acceleration in the first part of the run? (*b*) Assuming his acceleration and top speed to be the same as in part (*a*), how long would it take him to run 200 m?

How to Solve It
- There's nothing subtle about the physics here, but you need to give a little thought to the algebra!

- There are two time intervals—from 0 to some time t_1, during which the sprinter is accelerating, and from t_1 to $t = 10$ s.

- First find his velocity at the end of the first interval, in terms of t_1. This is his top speed; he continues at this speed for a time $\Delta t = 10$ s $- t_1$, during which he covers the second 50 m.

- Once you've solved for t_1, you can calculate his acceleration in the first 50 m.

10. A toy rocket starting from rest moves straight upward with an acceleration three times that of gravity, until it is 300 m above the ground. At that point its engine quits. How long after it was fired does it hit the ground? Neglect air resistance.

How to Solve It
- First write an equation for the *powered part* of the flight. How long does this part take?

- When the engine quits, what is the rocket's upward speed?

- You know the rocket's height at the moment the engine quits, and you want to find out how long (starting from this point) it's in the air.

11. A man standing on a cliff, 50 m above the level of the ground below, throws a stone straight up in the air. The stone falls back past the edge of the cliff and strikes the ground 5 s after it was thrown. With what initial velocity did the man throw the stone?

How to Solve It
- The stone is in free-fall, so you know how to write an equation of motion for it.

- Notice that you don't need to solve for any intermediate quantities such as how high the stone went.

- If you write an equation for the stone's position at time $t = 5$ s, you know everything in the equation except the initial velocity.

12. The speed of sound in air is about 340 m/s. If you drop a stone into a well and hear the splash 3.80 s later, how deep is the well?

How to Solve It
- Use the free-fall equations to get the time the stone took to fall, in terms of the depth.

- Don't forget the time the sound takes to get back to you.

- These two times add up to 3.80 s.

Solutions

1. In the first leg of the trip the car covers 40 km at a speed of 86 km/h in time

$$\Delta t_1 = \frac{\Delta x}{v_1} = \frac{40 \text{ km}}{86 \text{ km/h}} = 0.465 \text{ h}$$

The second leg takes

$$\Delta t_2 = \frac{\Delta x}{v_2} = \frac{40 \text{ km}}{50 \text{ km/h}} = 0.80 \text{ h}$$

The car has gone 80 km in a total of 0.465 + 0.80 = 1.265 hours, and so

$$v_{av} = \frac{\Delta x}{\Delta t} = \frac{80 \text{ km}}{1.265 \text{ h}} = \underline{63.2 \text{ km/h}}$$

2. (a) 15.2 mi   (b) 3.04 mi/h   (c) 5.36 mi/h

3. (a) The man got to the bank, at $x = -200$ m (see Figure 2-3), 46 min after he started. His average velocity was therefore

$$v_{av} = \frac{\Delta x}{\Delta t} = \frac{-200 \text{ m}}{46 \text{ min}} \times \frac{1 \text{ min}}{60 \text{ s}} = \underline{-0.072 \text{ m/s}}$$

(b) He is at rest (presumably at McDonald's) from $t = 6$ min to $t = 26$ min; so his lunch must have taken 20 min.

(c) McDonald's is therefore at $x = 170$ m, and the distance from there to the bank is

$$|-200 - 170| = \underline{370 \text{ m}}$$

(d) His velocity is the slope of the graph, so the greatest speed corresponds to the steepest slope. It looks like this occurs at $t \approx 4$ min; presumably he was in a hurry to get his lunch.

4. (a) 50 mi/h   (b) 21.4 mi/h   (c) 6.4 h   (d) $t = 1$ to 5 P.M.

5. (a) To get the velocity at time $t = 2$ s, we plug into the equation given:

$$v_{2 \text{ s}} = (0.06 \text{ m/s}^3)(2 \text{ s})^2 = 0.24 \text{ m/s}$$

In the same way, we find $v_0 = 0$ and $v_{4 \text{ s}} = 0.96$ m/s. The average acceleration over an interval is the change in velocity divided by the time interval, so

$$a_{0\text{-}2 \text{ s}} = \frac{\Delta v}{\Delta t} = \frac{0.24 \text{ m/s} - 0}{2 \text{ s}} = \underline{0.12 \text{ m/s}^2}$$

(b) And likewise $a_{2\text{-}4 \text{ s}} = \underline{0.36 \text{ m/s}^2}$.

(c) Since the acceleration is not the same in the two time intervals, clearly it is <u>not</u> constant.

6. (a) <u>4.6 m/s²</u> (b) You should get about <u>4.0 m/s²</u>.

7. Let the instant at which Al sees Bob pass be $t = 0$. Fifteen seconds later Al starts out with a constant acceleration of 1.4 m/s². Since he starts from rest, his position at any time t is therefore given by

$$x_1 = x_0 + v_0 t + \frac{1}{2} at^2 = \frac{1}{2} at^2 = \frac{1}{2} (1.4 \text{ m/s}^2)(t - 15 \text{ s})^2$$

[The way we have defined t, Al has been moving for a time $(t - 15 \text{ s})$.] Bob's position at any time t is $x_2 = v_2 t = (20 \text{ m/s})(t)$. Al catches up with him at some time t, at which point $x_1 = x_2$. Substituting into this equation, we obtain:

$$(0.7)(t - 15)^2 = 0.7 t^2 - 21 t + 157.5 = 20 t$$

or

$$t^2 - 58.6 t + 225 = 0$$

This is a quadratic equation and can be solved using the quadratic formula:

$$t = \frac{58.6 \pm \sqrt{58.6^2 - (4)(225)}}{2} = \underline{54.5 \text{ s}} \text{ (or 4.1 s)}$$

Clearly the second root isn't meaningful here as Al hasn't even started moving at $t = 4.1$ s. Thus, Al catches up 54.5 s after he sees Bob pass by. At this time they are at

$$x_1 = x_2 = v_2 t = (20 \text{ m/s})(54.5 \text{ s}) = \underline{1090 \text{ m}}$$

Notice that this isn't a particularly reasonable problem: it assumes that Al's car accelerates at a constant rate from 0 to 200 km/h!

8. The car's acceleration must be negative with a magnitude of at least <u>1.79 m/s²</u>.

9. (a) The sprint is in two parts, with $t_1 + t_2 = t = 10.0$ s. In the first part, the runner moves with constant acceleration a, so

$$50 \text{ m} = \frac{1}{2} (a)(t_1)^2$$

At the end of this part, his velocity is $v_1 = at_1$. He continues at this speed for the remaining 50 m, so

$$50 \text{ m} = v_1 t_2 = (at_1)(10 \text{ s} - t_1) = 10at_1 - at_1^2 = (10 \text{ s})v_1 - at_1^2$$

So, substituting 100 m for at_1^2 in the preceding equation, we obtain

$$50 \text{ m} = (10 \text{ s})v_1 - 100 \text{ m}$$

or $v_1 = 15 \text{ m/s}$

So $t_2 = \dfrac{x}{v_1} = \dfrac{50 \text{ m}}{15 \text{ m/s}} = 3.33 \text{ s}$

and $t_1 = t - t_2 = 10 \text{ s} - 3.33 \text{ s} = 6.67 \text{ s}$

and finally

$$a = \frac{v_1}{t_1} = \frac{15 \text{ m/s}}{6.67 \text{ s}} = \underline{2.25 \text{ m/s}^2}.$$

(b) Assuming his acceleration and top speed to be the same as in part (a), he will take 10 s for the first 100 m and so

$$\frac{x}{v_1} = \frac{100 \text{ m}}{15 \text{ m/s}} = 6.67 \text{ s}$$

for the second 100 m for a total time of 10.0 s + 6.67 s = $\underline{16.67 \text{ s}}$. Nobody runs the 200 m that fast, of course, which suggests that our model of the sprints is an oversimplified one.

10. The rocket is in the air for $\underline{33.7 \text{ s}}$.

11. Let's take the ground below the cliff to be height $x = 0$ and the values of height to be positive upward. At $t = 0$, the stone is at $x_0 = 50$ m and its acceleration $a = -g = -9.81$ m/s^2. At $t = 5$ s, the stone is at $x = 0$. Notice that the stone's acceleration is negative because we chose to treat upward as positive. Thus in the equation

$$x = x_0 + v_0 t + \frac{1}{2} at^2$$

we know everything except v_0, so we can solve for that:

$$0 = (50 \text{ m}) + v_0 (5 \text{ s}) + \frac{1}{2} (-9.81 \text{ m/s}^2)(5 \text{ s})^2$$

which gives

$$v_0 (5 \text{ s}) = 72.6 \text{ m}$$

or $v_0 = \underline{14.5 \text{ m/s}}$

12. The well is $\underline{63.7 \text{ m}}$ deep.

CHAPTER 3

Motion in Two and Three Dimensions

I. Key Ideas

Vectors Many physical quantities must be described by a magnitude *and* a direction in space. For example, in motion that is not confined to a straight line, the directions of the displacement, velocity, and acceleration, as well as their magnitudes, must be specified. Such quantities are called *vectors*.

Displacement The displacement vector is the quantity that gives the straight-line distance and direction from one point in space to another as illustrated in Figure 3-1. Notice that it does *not* depend on the actual path followed from one point to another.

Figure 3-1

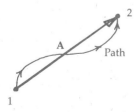

Vector Addition and Subtraction We can use displacement to see graphically what is meant by adding two vector quantities. In Figure 3-2a, the

sum of displacement **A** (from point 1 to point 2) and displacement **B** (from point 2 to point 3) is clearly the displacement from point 1 to point 3—vector **C** in the drawing. Figure 3-2a defines vector addition and shows how it can be done graphically, using a ruler and a protractor. Subtracting **B** is just like adding −**B**, which is equal to **B** in magnitude but opposite in direction (see Figure 3-2b).

Figure 3-2

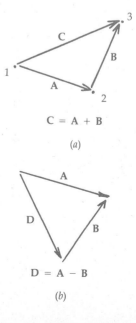

19

Vector Components It is often convenient, instead, to represent vectors in terms of their components (projections) along the x and y coordinate axes. Vector addition and subtraction are then performed analytically by adding and subtracting their corresponding components.

Velocity and Acceleration for Two- and Three-dimensional Motion The definitions of velocity and acceleration in terms of displacement and velocity are the same as they were for one-dimensional motion; but all these quantities are vectors. The instantaneous velocity is the rate of change of the *displacement vector*. Its direction is along the path of the particle at a particular point in time. Acceleration is the rate at which the *velocity vector* is changing. For acceleration to occur, the velocity vector must change in either magnitude or direction—or both.

Projectile Motion An object that is falling freely under gravity and also moving horizontally is called a projectile. Except that they happen at the same time, the two components of the motion are independent of each other: the object falls vertically just as if there were no motion in the horizontal plane, and the horizontal motion is unaccelerated since the acceleration due to gravity is only in the vertical plane. Since each component separately is one that we know how to treat as a motion in one dimension, we have a complete description of what happens.

Centripetal Acceleration Whenever the direction of motion of an object is changing, its acceleration is not zero. For the special case of an object moving in a circle of radius r at speed v, the acceleration has magnitude v^2/r and is directed toward the center of the circle. This is called the *centripetal acceleration*. (If the speed of an object moving along a curved path is not constant, the acceleration vector also has a component *along* the direction of motion.)

Relative Motion The velocity of a moving object is defined relative to some coordinate system. This system may be moving with respect to another coordinate system. In that case, the velocity of the object relative to the second system is the vector sum of the object's velocity relative to the first system and the first system's velocity relative to the second.

II. Numbers and Key Equations

Numbers

There are no conversion factors for this chapter.

Key Equations

Vector components

$$A_x = A \cos \theta \qquad A_y = A \cos \theta$$

Vector magnitude

$$A = \sqrt{A_x^2 + A_y^2}$$

Vector direction

$$\tan \theta = A_y/A_x$$

Projectile motion

$$v_{0x} = v_0 \cos \theta \qquad v_{0y} = v_0 \sin \theta$$

$$v_x = v_{0x} \qquad \Delta x = v_{0x}t$$

$$v_y = v_{0y} - gt \qquad \Delta y = v_{0y}t - \frac{1}{2}gt^2$$

Horizontal range of a projectile

$$R = \frac{v_0^2}{g} \sin 2\theta$$

Vector kinematics

$$\mathbf{v}_{av} = \frac{\Delta \mathbf{r}}{\Delta t} \qquad \mathbf{a}_{av} = \frac{\Delta \mathbf{v}}{\Delta t}$$

Centripetal acceleration

$$a_c = \frac{v^2}{r}$$

III. Possible Pitfalls

Don't forget that when we add (or subtract) vectors, we can't simply add (or subtract) their magnitudes! We must add (or subtract) them graphically, or by adding (or subtracting) corresponding components. The equation $C = A + B$ *does not necessarily* mean that $C = A + B$; this would be true *only* if **A** and **B** were parallel.

The displacement between two points is defined as the straight-line magnitude and direction between them "as the crow flies." Its magnitude is not necessarily equal to the distance actually traveled between the points.

The projectile motion equations apply only when an object is in free-fall, that is, when its only acceleration is that due to gravity. If anything else is affecting the object's motion, you can't use those equations.

The constant-acceleration equations can be applied only separately to one or the other component of the motion, or to the speed of the projectile. For instance, it is only v_y that is zero at the top of the arc; v_x is constant throughout the projectile's motion, as is the acceleration (it is g, downward). The speed of the projectile is minimum (but not zero!) at the top of the arc.

For a motion in one dimension, constant speed meant that the acceleration was zero. This *isn't* true in general! Velocity is a vector quantity; so whenever the direction *or* the speed of an object is changing, its motion is accelerated.

IV.   Questions and Answers

Questions

1. You throw a baseball from the outfield to a friend at home plate. In general, the distance the ball travels is not equal to the magnitude of its displacement vector. Which of the two is larger?

2. Two displacement vectors, S_1 and S_2, add to give a resultant of zero. What can one say about the two displacements?

3. What kind of motion are we looking at when the velocity vector and the acceleration vector of a particle are (instantaneously) perpendicular to one another?

4. Is it possible to drive your car around a curve without accelerating?

5. Suppose you have "sighted in" a rifle so that, on a target 150 m away over level ground, it will hit whatever the sights are aimed directly at. (*a*) To shoot at a target 90 m away, should you aim above, below, or directly at the target? (*b*) To shoot at a target 150 m away but downhill, should you aim above, below, or directly at the target?

6. Describe briefly what kind of motion a particle is undergoing when (*a*) the position vector changes in magnitude but not in direction; (*b*) the velocity vector changes in magnitude but not in direction; (c) the position vector changes in direction but not in magnitude; and (d) the velocity vector changes in direction, but not in magnitude.

7. A ball is (a) thrown straight upward at 7 m/s, (b) thrown horizontally at 7 m/s, and (c) dropped from rest. In which case (or cases) does it strike the ground in the shortest time?

8. A bead slides at constant speed along the flat spiral wire in Figure 3-3. Describe how the acceleration vector of the bead changes with time.

Figure 3-3

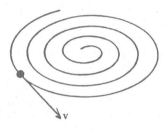

Answers

1. The displacement is the straight-line distance between the two points, so the actual distance cannot be less than the displacement. They would be equal only if the path of the ball were a straight line. Since the ball must rise and fall some in flight, the distance it travels must be greater than the displacement.

2. The equation $S_1 + S_2 = 0$ can only mean that $S_1 = -S_2$; thus, the two displacements are equal and opposite. So if these are the two legs of a trip, then on the second the moving object returned to its starting point.

3. If the acceleration is perpendicular to the velocity, it follows that the change in velocity is at right angles to the velocity vector. The velocity vector thus is *turning*. This means that the moving particle is turning, at a (instantaneously) constant speed.

4. The question points up the difference between everyday usage and the correct meaning of the term "accelerating." Properly speaking, you are accelerating whenever the magnitude or direction of the velocity vector is changing; so since your velocity is at least changing in direction, you cannot drive around a curve without accelerating.

5. When you "sight in" a rifle at a particular distance, you are adjusting the sights to compensate for the distance by which the bullet falls below its initial line of flight. The actual line of the barrel thus is aimed above the sight line. (a) When the target is closer than the distance for which the rifle was sighted, the bullet falls a shorter distance. If you aim at the target, the bullet will be high when it gets there; so you should aim *below* the target to compensate. (b) Surprisingly, you must also aim below the target when you shoot downhill. Because the shooter is aiming down, the bullet follows a straighter trajectory, and so doesn't fall as far below its initial line of flight as you allowed for when you sighted in the rifle. You must aim *below* the target to compensate.

6. (*a*) In this case, the particle is moving in a straight line toward or away from the origin. (*b*) The particle is moving in a straight line (since the direction of *v* isn't changing) with variable speed. (*c*) This would mean that the moving particle stays at the same distance from the origin; so it must be travelling in a circle around the origin. (*d*) This indicates that whatever the moving particle is doing, it's doing it at constant speed.

7. In cases (*b*) and (*c*), the ball has no initial velocity component in the vertical direction, so it hits the ground after the same amount of time. In case (*a*), there is an initial-velocity component in the vertical direction upward; the ball must come to a stop before it starts falling downward, so it takes longer to hit the ground.

8. The bead is moving at constant speed and, at any instant, in a nearly circular path. The acceleration at any instant is of magnitude v^2/r, directed nearly toward the center of the spiral. Since *v* is constant and *r* is constantly decreasing, the magnitude of the acceleration vector is steadily increasing.

V. Problems and Solutions

Problems

1. At 3:00 P.M. you pass mile marker 160 as you are driving due south on Interstate 77. At mile marker 138, you turn off on U.S. 54. At 3:40 P.M. you have gone 14 miles due southwest on U.S. 54. What was your average velocity (magnitude and direction) over this 40-min interval?

How to Solve It
- Where is your car at the end of the 40 min?

- What was the net displacement of your car in this interval?

- Average velocity is this displacement divided by the time interval. Pay attention to units when you do the arithmetic.

2. A car is traveling along a mountain road. At a certain instant it is moving due west across level ground at a speed of 18 m/s; two seconds later it is moving north, down a steep hill at an angle of 15° below the horizontal, at 10 m/s. Calculate the magnitude of the car's average acceleration over this time interval.

How to Solve It
- Calculate the *change* in the car's velocity over the two-second time interval.

- The change in velocity divided by the time interval is the definition of average acceleration.

3.   A pilot starts a trip from St. Louis to Memphis, which is 385 km due south. She flies due south at her maximum airspeed of 90 m/s, but fails to correct for a crosswind of 15 m/s directed due east. (*a*) What is her speed relative to the ground? (*b*) At the time when she expects to reach Memphis, where will she actually be? (*c*) How much longer will it take her to get to Memphis?

How to Solve It
- The airplane's velocity with respect to the air is 90 m/s due south. Add this vectorially to the wind velocity to get the airplane's ground velocity.

- How long did the pilot expect it to take her to get to Memphis?

- At the time when she thought she would get to Memphis, how far away is she in fact?

4.   The current at a certain point in the Wabash River, where it is 80 m wide, flows at 0.4 m/s. A swimmer sets out for a point directly across the river. (*a*) If the swimmer's maximum speed is 0.75 m/s relative to still water, in what direction should he swim to go directly to his goal? (*b*) How long will it take him to get there?

How to Solve It
- The direction of the swimmer's velocity relative to the river bank is the vector sum of the river's current and the swimmer's velocity through the water.

- Use this to find the direction in which he should swim.

- When you know this, you can calculate how far he has to go through the water and therefore how long it will take him.

5.   A ball rolls off a tabletop 0.9 m above the floor and lands on the floor 2.6 m away from a point that is directly under the edge of the table. At what speed did it roll off?

How to Solve It
- Remember that the horizontal and vertical components of the ball's motion can be treated independently.

- How long did it take the ball to fall 0.9 m?

- If it moved horizontally 2.6 m in this time, what was its horizontal speed?

6. A diver leaps from a springboard 6 m above the water with a velocity of 6 m/s at an angle 30° from the vertical. How far from a point directly beneath the board, and at what speed, does she hit the water?

How to Solve It
- Remember that the horizontal and vertical components of her motion can be treated independently.

- What is the vertical component of her velocity? Given this, find how long it takes her to hit the water.

- How far does she move horizontally in this time?

7. A basketball player takes a shot when he is standing 24.0 ft from the 10-ft high basket. If he releases the ball at a point 6 ft from the floor and at an angle of 40° above the horizontal, with what speed must the ball be thrown if it is to go through the hoop?

Figure 3-4

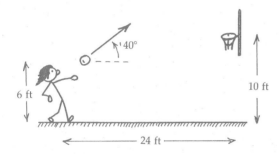

How to Solve It
- In terms of the unknown speed v_0, how much time does it take the ball to reach the basket 24 ft away horizontally?

- When it reaches the basket, you know that the ball is 4 ft above the point from which it was thrown. Write the equation for the vertical component of the ball's motion in terms of this time.

- This equation can be solved for v_0.

8. General Lee's artillery is taking aim on Union cavalry on the cliffs across the river, 300 m away. (See Figure 3-5.) The muzzle velocity of the mortars is 67 m/s. In order to hit the enemy, the mortars have to be fired at an angle of elevation of 31°. How high are the cliffs?

How to Solve It
- This is a straightforward application of the projectile motion equations. But notice that you can't just plug into the range formula,

because the mortar rounds don't land at the same vertical height they were fired from.

Figure 3-5

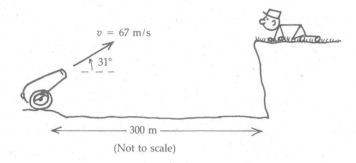

v = 67 m/s

31°

—— 300 m ——

(Not to scale)

- First, calculate the time a cannonball takes to reach the other side of the river, 300 m away.

- At this time, what is the altitude of the projectile above the point from which it was fired?

- Notice that in both Problems 7 and 8, although you aren't asked for the time, it comes in as an intermediate calculated quantity. If you like algebra, you can eliminate the time between the x and y equations. The result is an equation for y in terms of x, which you can apply directly to this sort of problem.

9. A quarterback throws a pass from his 20 yard line with an initial velocity of 22 m/s at an angle of 40° above the horizontal. 1.0 s before the pass is thrown, the receiver starts running downfield from the 30-yard line. With what speed must he run in order to catch the pass? (Assume that the pass is caught at the same height above the ground at which it was thrown. If you need to convert units, 1 m = 1.094 yd.)

How to Solve It
- In this problem you need to know how long the ball is in the air and how long the receiver has to run for the pass.

- Calculate where the pass lands and, from this, how far the receiver has to go.

- The distance he has to run, divided by the time he has to do it in, gives you his velocity. Is the answer reasonable?

10. A batter hits a pop fly into shallow center field. The ball is hit 1.1 m above the ground at a speed of 29.5 m/s in a direction 35° above the horizontal. Unfortunately, the center fielder was playing deep, 116 m from home plate. Assume that the ball was hit straight toward the center fielder and that he starts running at the instant the ball is hit. How fast must the fielder run to catch the ball?

How to Solve It
● Calculate where the ball will hit the ground. How far does the fielder have to run to reach it?

● How long is the ball in the air?

● Now calculate how fast the fielder must run to reach the ball before it reaches the ground.

11. A car rounds a 90° highway curve, turning from east to north, whose radius of curvature is 60 m. The car's constant speed is 25 m/s. Calculate (a) the instantaneous acceleration of the car when it is halfway around the curve and (b) the average acceleration of the car around the 90° curve.

How to Solve It
● Calculate the car's centripetal acceleration. At the point when it is halfway around the curve, what is the direction of its acceleration?

● How long does it take the car to go around the curve?

● Calculate the change in the car's velocity from the beginning to the end of the 90° curve, and, from this, its average acceleration.

12. An earth satellite is in a circular orbit 2700 km above the surface of the earth. (Take the radius of the earth to be 6360 km.) It completes a circuit of the earth every 144 min. What is the acceleration due to gravity at the altitude of this satellite?

How to Solve It
● Use the orbital radius to calculate the orbital speed of the satellite.

● From the speed and the orbital radius, calculate the centripetal acceleration, which is the acceleration due to gravity at this distance from the earth.

● How does this compare to the acceleration due to gravity near the earth?

Figure 3-6

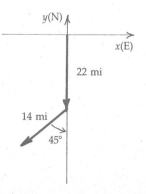

Solutions

1. Before turning, you went 22 miles south (from mile marker 160 to mile marker 138). Call this displacement s_1. After turning, you went 14 miles southwest. Call this displacement s_2. If we call east the x direction and north the y direction, then in components

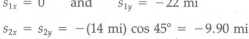

$$s_{1x} = 0 \quad \text{and} \quad s_{1y} = -22 \text{ mi}$$

$$s_{2x} = s_{2y} = -(14 \text{ mi}) \cos 45° = -9.90 \text{ mi}$$

The total displacement is $\mathbf{s} = \mathbf{s}_1 + \mathbf{s}_2$ and its components are

$$s_x = s_{1x} + s_{2x} = 0 + (-9.90 \text{ mi}) = -9.90 \text{ mi}$$

$$s_y = s_{1y} + s_{2y} = (-22 \text{ mi}) + (-9.90 \text{ mi}) = -31.9 \text{ mi}$$

So

$$s = \sqrt{s_x^2 + s_y^2} = \sqrt{9.90^2 + 31.9^2} = 33.4 \text{ miles}$$

Its direction, the angle θ, is given by

$$\tan \theta = \frac{s_x}{s_y} = \frac{-9.90 \text{ mi}}{-31.9 \text{ mi}} = 0.310$$

$$\theta = 17°$$

So your net displacement $\mathbf{s}$ was 33.4 mi directed 17° west of south. Since $\mathbf{s} = \Delta\mathbf{r}$, the average velocity time is

$$v_{av} = \frac{\Delta r}{\Delta t} = \left(\frac{33.4 \text{ mi}}{40 \text{ min}}\right)\left(\frac{60 \text{ min}}{1 \text{ h}}\right)$$

$$= \underline{50.1 \text{ mi/h}} \text{ directed 17° west of south.}$$

2. $a_{av} = \underline{10.3 \text{ m/s}^2}$, directed <u>east-north-east and about 7° below the</u> <u>horizontal.</u>

3. (*a*) Let $\mathbf{V}$ be the wind velocity; that is, the velocity of the air relative to the ground. Here $\mathbf{V}$ is 15 m/s due east. The pilot's velocity relative to the air $\mathbf{v}_a$ is 90 m/s due south. Her velocity relative to the ground $\mathbf{v}_g$ is the vector sum of this and the air's velocity relative to the ground $\mathbf{V}$:

Figure 3-7

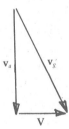

$$v_g = \sqrt{v_a^2 + V^2}$$

$$= \sqrt{15^2 + 90^2} = \underline{91.2 \text{ m/s}}$$

and

$$\tan \theta = \frac{15}{90}$$

so $\theta = 9.5°$ east of due south.

   (*b*) Neglecting the crosswind, she expected the trip to take (385 km)/90 m/s = 4280 s = 1.19 h. Her actual ground velocity, however, is 90 m/s south and 15 m/s east, so in this time she has moved (15 m/s)(4280 s) = 64,200 m = <u>64.2 km eastward</u> of her intended flight path. Thus, she still has to go 64.2 km west to get to Memphis.

(c) To finish her trip she has to fly directly upwind; her maximum velocity is thus 90 m/s − 15 m/s = 75 m/s and it takes

$$\left(\frac{64.2\ \text{km}}{75\ \text{m/s}}\right)\left(\frac{1000\ \text{m}}{1\ \text{km}}\right) = \underline{856\ \text{s}}$$

or a little under 15 minutes to get there.

4. He must swim 32.2° upstream from straight across; it takes him 126 s.

5. Since the ball is falling freely with zero initial (vertical) velocity, we can use the equation $y = y_0 + v_0 t - \frac{1}{2} g t^2$ to find the time it is in the air:

$$0 = (0.9\ \text{m}) + 0 - \frac{1}{2}\,(9.81\ \text{m/s}^2)t^2$$

so $\quad t = 0.429\ \text{s}$

In this time, the ball moved 2.6 m horizontally at constant speed; so

$$v_x = \frac{x}{t} = \frac{2.6\ \text{m}}{0.429\ \text{sec}} = \underline{6.07\ \text{m/s}}$$

6. 5.27 m; 12.4 m/s

7. Let the horizontal component of the ball's initial velocity be

$$v_x = v_{0x} = v_0 \cos \theta = v_0 \cos 40° = .7660\ v_0$$

Then the time the ball is in the air is

$$t = \frac{24\ \text{ft}}{v_x} = \frac{24\ \text{ft}}{.7660\ v_0} = \frac{31.3\ \text{ft}}{v_0}$$

In this time, it has moved vertically

$$y = v_{0y}t - \frac{1}{2} g t^2$$

where

$$v_{0y} = v_0 \sin \theta = v_0 \sin 40° = .6428\ v_0$$

Using g = 32.2 ft/s², we obtain

$$4\ \text{ft} = .6428 v_0\,(31.3/v_0) - 16.1\,(31.3/v_0^2)$$

$$16.12 = 15800/v_0^2$$

$$v_0 = \underline{31.3\ \text{ft/s}}$$

8. 46.4 m

9. If we call the level at which the ball is thrown and caught $y = 0$, then we can use the equation $\Delta y = v_{0y}t - \frac{1}{2}gt^2 = 0$ to find the time it is in the air. Rearranging this equation yields

$$t = \frac{Zv_{0y}}{g} = \frac{Zv_{0y} \sin \theta}{g} = \frac{(2)(22 \text{ m/s})\sin 40°}{(9.81 \text{ m/s}^2)} = 2.88 \text{ s}$$

In this time, the ball goes a horizontal distance

$$x = v_x t = v_0 \cos \theta \, t = 48.6 \text{ m} \left(\frac{1.094 \text{ yd}}{1 \text{ m}}\right) = 53.2 \text{ yd}$$

The receiver must therefore cover 43.2 yd = 39.5 m in (2.89 s + 1 s) = 3.89 s, and so would have to run at

$$v_x = \frac{x}{t} = \frac{39.5 \text{ m}}{3.89 \text{ s}} = \underline{10.2 \text{ m/s}}$$

This may be just possible, but it's awfully good speed!

10. <u>8.82 m/s</u>

11. (a) The car's centripetal acceleration is

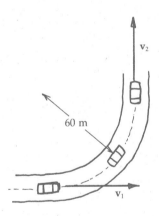

Figure 3-8

$$a_c = \frac{y^2}{r} = \frac{(25 \text{ m/s})^2}{60 \text{ m}} = \underline{10.4 \text{ m/s}^2}$$

(Notice that this is larger than g; the car couldn't really take the turn this fast!) When the car is halfway around the turn, it is heading northeast and its acceleration is therefore directed northwest, toward the center of the curve.

(b) Rounding the quarter-circle curve, the car goes a distance

$$d = \frac{1}{4}(2\pi r) = \frac{1}{4}(2\pi)(60 \text{ m}) = 94.2 \text{ m}$$

at a speed of 25 m/s. The time it takes to round the curve is

$$t = \frac{d}{v} = \frac{94.2 \text{ m}}{25 \text{ m/s}} = 3.77 \text{ s}$$

In this time, the change in the car's velocity has been

$$\mathbf{v} = \mathbf{v}_2 - \mathbf{v}_1 = (25 \text{ m/s, N}) - (25 \text{ m/s, E})$$

This calculates out to 35.4 m/s, NW. (If you have trouble seeing this, draw the vector subtraction diagram.) Its average acceleration is therefore

$$\mathbf{a}_{av} = \frac{\mathbf{v}}{t} = \frac{(35.4 \text{ m/s, NW})}{3.77 \text{ s}} = \underline{9.38 \text{ m/s}^2, \text{ NW}}$$

12. <u>4.79 m/s²</u>. This is a little less than 0.5 g.

CHAPTER 4

Newton's Laws

I. Key Ideas

Newton's Laws of Motion These three laws are the fundamental laws of classical mechanics. One way of stating them is: (1) A body remains at rest or moves with a uniform velocity unless acted upon by external forces. (2) The acceleration of a body is inversely proportional to its *mass* and directly proportional to the *resultant external force* that acts upon it. (3) Interacting bodies always exert *equal and opposite* forces upon one another.

Addition of Forces A corollary of Newton's laws states that forces add as vectors; that is, the *net force*, or resultant, of more than one external force on a body is the *vector sum* of all forces that act upon it.

Law of Inertia The first law says that an isolated object has a natural tendency to remain at rest or in uniform motion. This intrinsic property of all matter is called *inertia*. If we could remove all external forces on an object it would remain at rest or keep moving forever in a straight line with uniform speed. To change the state of motion of an object requires the action of an external agency.

Force and Acceleration An object which is not moving uniformly is accelerating. Newton's second law states that its acceleration is directly proportional to the resultant—the vector sum—of all the forces that act upon it. We define *force*, and measure forces, in terms of the acceleration that is produced on a standard body. *Mass* is an intrinsic property of an object which measures its inertia. Given a known force, the mass of an object is the net force on it divided by the acceleration that results.

Acceleration Due to Gravity The earth's gravitational pull on an object is, empirically, directly proportional to its mass. Thus, the acceleration produced by the earth's gravity acting alone is the same for any object. At any point near the surface of the earth, this acceleration is $g = 9.81 \text{ m/s}^2$.

Weight The gravitational force on an object is called its *weight*. It is not perfectly constant, as there are slight variations in g from place to place. Weight is thus not an intrinsic property of an object. What we actually sense as our weight is not the pull of the earth's gravity. Instead it is the force or forces acting on us (to hold us in place) in opposition to gravity: the upward force of a floor, a chair, etc. This balancing force is what is meant by *apparent weight.*

Units The SI unit of mass is the kilogram, the mass of a standard reference body in Sèvres, France. The unit of force thereafter is the *newton*, where $1 \text{ N} = 1 \text{ kg·m/s}^2$. Nonstandard unit systems include other units for mass (gram, slug) and force (dyne, pound).

Interaction According to Newton's third law, forces always occur in pairs. Thus, objects which interact always exert equal and opposite forces on one another. Note that these two forces never cancel out when the net force on an object is being calculated, because they always act on different bodies.

Contact Forces Except for the force of gravity, most of the forces we encounter in our everyday experience are exerted by objects in contact with one another. When an elastic solid is deformed, stretched, or compressed, it exerts a restoring force on whatever deformed it. If the displacement is small then the restoring force is proportional to displacement. (This is Hooke's law.) Strings, ropes, etc., behave in the same way under tension, but are flexible and so cannot be compressed. Objects in contact with each other exert forces both *normal* to (that is, perpendicular to) and also parallel to the surface of contact. Those perpendicular are called *normal forces* and those parallel are called *frictional forces*. The latter act to prevent or retard the relative motion of objects in contact.

Problem Solving A very useful technique in using Newton's laws to solve problems is to draw a *free-body diagram* for each object involved. Such a diagram shows each object by itself and indicates every force that acts on it. The vector sum of all these forces is equal to the object's mass multiplied by its acceleration. Newton's second law is a vector equation; it is true in any direction. Forces should be resolved into their components along each coordinate axis. Any coordinate system will do, but in many problems a natural choice is one that makes the acceleration and/or some of the forces parallel to one of the axes.

II.   Numbers and Key Equations

Numbers

1 newton (N) = 1 kg·m/s²

1 kg = 1000 g

1 slug = 14.6 kg

1 dyne (dyn) = 1 g·cm/s² = 10^{-5} N

1 pound (lb) = 4.45 N

Key Equations

Acceleration due to gravity

$$g = 9.81 \text{ m/s}^2$$

Newton's second law

$$\mathbf{F} = m\mathbf{a}$$

Weight

$$w = mg$$

III.   Possible Pitfalls

Notice that the action and reaction forces to which Newton's third law refers always act on *different* bodies. Thus, they never cancel when you are calculating the resultant force on a body.

Be sure you put only those forces on a free-body diagram for which you can identify an external physical source. Acceleration is not a force, and the *m*a in Newton's second law is not a separate force. The forces that go on your diagram are gravitational forces, those due to gravity to springs and strings, to contact, and to the other pushes and pulls that are acting on the body.

Be sure you put on a particular free-body diagram only the forces which act on that particular body.

Be sure you know along *what* mutually perpendicular coordinate axes you are resolving forces in a given problem. When you actually write out equations using Newton's second law, each equation should contain only components along one axis direction.

Friction is that component of the contact force between two surfaces which acts *along* the interface in such a direction as to *retard* their motion. In a given problem, if you know that the surfaces are frictionless, then the only contact force between them must be *normal to* the interface.

The force or forces that oppose the weight of an object may or may not be equal to its weight. They are equal only if the object has zero vertical acceleration; clearly they are not equal if the object has a nonzero vertical acceleration.

IV.   Questions and Answers

Questions

1. If a body has zero acceleration, can we conclude that no forces act on it?

2. Suppose that only one force acts on a body. Can you tell in what direction the body is moving?

3. A definition sometimes given of the inertia or mass of an object is that it is a measure of the quantity of matter. How does this definition compare with that discussed in the chapter?

4. Under what circumstances can the apparent weight of an object be less than its true weight?

5. When I jump up in the air, I have (for some time) an acceleration upward. What is it that is exerting an upward force on me? In this case, what is the equal and opposite force that Newton's third law refers to?

6. Is your mass greater in Mexico City or in Los Angeles? In which city is your weight greater?

7. A certain rope will break under any tension greater than 800 N. How can it be used to lower an object weighing 850 N over the edge of a cliff without it breaking?

8. A car is being driven up a hill at a constant velocity. Discuss the forces that are acting on it.

9. One way to define the magnitude of a given force is to measure the distance that it will stretch a spring. We tend to assume that Hooke's law holds, that is, that the extension of the spring is proportional to the force on it. Can you think how you might check this? Suppose a force F stretches the spring a distance Δx and another force F' stretches it by $2 \Delta x$. How can you tell if F' is in fact equal to $2F$?

10. Explain why you are thrown forward when a moving vehicle, in which you are riding or standing, suddenly brakes to a stop.

11. An object of mass m is being weighed in an elevator which is moving upward with an acceleration a. What is the result if the weighing is done using (*a*) a spring balance and (*b*) a pan balance?

Answers

1. No. All this tells us is that the resultant, the vector sum, of whatever forces are acting on the body is zero.

2. Not necessarily. The one force (or, in general, the resultant force) tells you the direction of the body's *acceleration*, but this is not necessarily the direction in which it is *moving*.

3. The official definition of the kilogram is as the mass of some standard object. Another object which has twice the mass would require, by definition, twice the force to cause a given change in its state of motion; so it possesses twice the inertia. Certainly two identical objects possess both twice as much mass and twice as much matter as does one. The question is really one of definition: is there anything more we mean by the "quantity of matter" in an object than its mass?

4. If the object is accelerating downward, then the resultant force on it is downward, and the force "holding it up" (its apparent weight) is clearly less than the force of gravity pulling it downward (its weight).

5. During the jump, while my feet are still in contact with the ground, it exerts an upward force on me. By Newton's third law, this force is equal and opposite to the downward force I exert on the ground. Thus, if I shove down on the ground hard enough, I am pushed upward hard enough to overbalance the downward force of gravity and so move upward.

Figure 4-1

6. Your mass is the same anywhere, of course, because mass is an *intrinsic* property of an object. But since Mexico City is at a much higher altitude than Los Angeles, it is farther from the earth's center, so the force of gravity is slightly less there (by about 0.025%, if altitude is the only consideration). Thus, your weight is very slightly less there.

7. The rope will not break if the object is lowered in such a way that it has a nonzero downward acceleration (that yields a tension of less than 800 N (see Figure 4-1). If the object's downward accel-

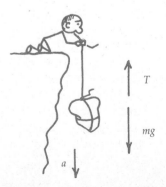

eration is a, then by Newton's second law the tension in the rope is $T = m(g - a)$. Here if the downward acceleration a is at least 0.57 m/s^2, the rope won't break. (Of course, you won't be able to get the thing back up!)

8. The forces acting on the car are shown in Figure 4-2. The frictional force of the ground on the wheels exerts an uphill force f on the car (the wheels are pushing backward on the ground). The ground also exerts a normal force F_n, which supports the car, and the earth's gravity pulls it directly downward with a force of magnitude mg. Since the car is moving at constant velocity, the resultant of these forces must be zero.

9. Recalling Newton's second law, you can check to see whether the acceleration produced by F' is in fact twice that produced by F when the two forces act on the same mass. The definitions of force and mass must rest on Newton's laws, whereas acceleration can be measured directly.

10. Well, really, you aren't thrown forward. Your body's inertia causes it to tend to continue forward at a constant velocity, while the brakes produce a rearward acceleration of the vehicle around you.

11. (*a*) A spring balance measures the apparent weight of the object, by measuring the compression its weight causes in a spring. Thus, its reading will be greater when it must provide extra force to accelerate the object upwards. (*b*) A pan balance compares the weights of unknown and known masses. The apparent weight of both will increase with acceleration, but the comparison will be unaffected.

Figure 4-2

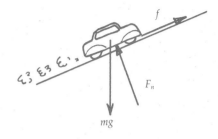

V. Problems and Solutions

Problems

1. A car that weighs 3750 lb is moving at 49.2 ft/s. It is braked uniformly to rest in a time of 5.5 s. What is the value of the braking force acting on the car? What exerts this force?

How to Solve It
- Don't overlook the fact that the problem is stated in mixed units. Be sure to convert everything to one unit system or another before you start calculating!

- From the kinematic data given, calculate the acceleration of the car.

- Mass times acceleration gives the resultant force on the car.

- The car is accelerating in a horizontal direction. What horizontal force is being exerted on the car by something else?

2. In Figure 4-3, Chris is dragging a crate on rollers across the floor of his garage. (Friction in the rollers can be neglected.) He pulls with a constant force of 170 N. The crate, which started at rest, is moving at 1.0 m/s by the time it has traveled 6.5 m across the floor. What is the mass of the crate?

Figure 4-3

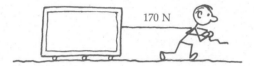

How to Solve It
- From the kinematic data given, calculate the acceleration of the crate.

- Draw a free-body diagram and determine the net force on the crate.

- The net force on the crate is equal to its mass times its acceleration.

3. A balloon with a total mass of 220 kg is descending with a uniform acceleration of 0.2 m/s² downward. If 20 kg of ballast are thrown overboard, what is the balloon's acceleration thereafter?

How to Solve It
- Draw a free-body diagram showing the forces that act on the balloon.

- The resultant force on the balloon is the vector sum of its weight and the upward buoyant force of the air.

- When the ballast is thrown overboard, which of these two forces has changed in value?

- Mass times acceleration gives the resultant force on the balloon.

4. In Figure 4-4, Herman is shoving a couple of cartons across a floor. (Friction with the floor can be neglected.) The carton he is pushing on has a mass of 40 kg, the one in front has a mass of 32 kg. (*a*) If he pushes with a force of 180 N, what is the acceleration of the system? (*b*) Draw diagrams showing the forces that act on each of the two cartons. (*c*) What net force acts on the 32-kg box?

Figure 4-4

How to Solve It

● Draw a free-body diagram for each of the cartons.

● The net force on each carton is equal to its mass times its acceleration, but in this case, the two cartons accelerate together.

● Newton's second law gives you two equations that you can solve for the acceleration and the force on the front carton.

5. A car is proceeding at a speed of 14 m/s when it collides with a stationary car in front. During the collision, the first car moves a distance of 0.3 m as it comes to a stop. The driver is wearing her seat belt, so she remains in her seat during the collision. If the driver's mass is 52 kg, how much force does the belt exert on her during the collision?

How to Solve It

● Use the data given to find the acceleration of the stopping car.

● Since the driver's seat belt holds, she stops at the same rate as does the car.

● The force the seat belt exerts on the driver is her mass times her acceleration.

6. In Figure 4-5, a tractor is pulling a sledge full of stones. The total mass of the sledge and the stones is 6400 kg, and there is a frictional force of 5.1×10^4 N exerted on the sledge by the ground. It takes 25 s to bring the tractor to a speed of 1.2 m/s; thereafter it moves with constant speed. What is the maximum force exerted by the tractor on the cable that pulls the sledge?

Figure 4-5

How to Solve It

- You're asked for the maximum force. Is the force greater while the tractor comes up to speed, or while it is moving with constant speed?

- What total force acts on the sledge? Draw a free-body diagram that shows all the forces that act on it.

- The total force is the sledge's mass times its acceleration.

7. Two ladies are trying to pull their stuck station wagon out of a muddy field by hauling on ropes as shown (from above) in Figure 4-6. The wagon's mass is 600 kg. Alice and Beth, by exerting forces of 580 N and 720 N, respectively, just manage to keep the tractor moving at a constant speed. Find the frictional force (magnitude *and* direction) exerted on the wagon by the ground.

Figure 4-6

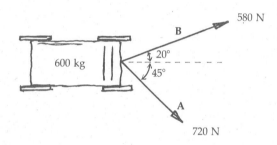

How to Solve It

- Since the wagon moves at a constant speed, the net force on it must be zero.

- In the horizontal plane, the forces with which Alice and Beth pull and the frictional force on the tractor must add up to zero.

- The frictional force is thus the negative of the vector sum of the other two.

8. A canal boat is being pulled along the canal by two mules walking along the banks, as shown in Figure 4-7. Assume that the tow ropes exert the only horizontal forces that act on the boat. If the mass of the boat is 1100 kg and it is accelerating straight up-stream at the rate of 1.1 m/s², find the magnitudes of the forces exerted by the two ropes.

How to Solve It

- Since the boat is accelerating upstream, the net force on it must be in that direction.

Figure 4-7

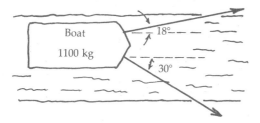

- You can get the magnitude of the net force from Newton's second law.

- The cross-stream components of the two forces must add to zero; and their upstream components, to the net force on the boat. Since you have the directions of the forces, you can work out their components.

9. A 2-kg block rests on top of an 8-kg block, which rests on a rough table top (see Figure 4-8). The 8-kg block is pulled horizontally with a force of 42 N. The force of friction exerted on the 8-kg block by the table top is 32 N, and the surfaces of the blocks are rough enough so that the top block does not slide on the bottom one. Find the acceleration of the blocks *and* the magnitude of the frictional force that is exerted by each block on the other.

Figure 4-8

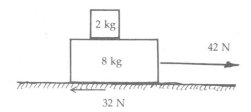

How to Solve It
- Draw a free-body diagram for each block, showing the horizontal forces that act on it.

- The net force on each block is its mass times its acceleration.

- The frictional force exerted by the 8-kg block is the only (horizontal) force on the 2-kg block, and thus is the force that accelerates it.

10. A 2-kg block rests on top of an 8-kg block, which rests on a rough table top. The blocks are connected by a string passing over a frictionless pulley, as shown in Figure 4-9. The force of friction exerted on the 8-kg block by the table top is 32 N. When the 8-kg block is pulled horizontally with a force of 42 N, the two blocks move at constant velocity. Find the magnitude of the frictional force that is exerted by each block on the other.

Figure 4-9

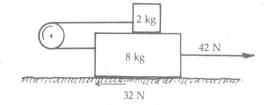

32 N

How to Solve It
• Draw a free-body diagram for each block, showing the horizontal forces that act on it.

• Since the pulley is frictionless, the tension is the same throughout the string (on both sides of the pulley).

• Since the blocks move at constant speed, the resultant force on each is zero.

11. In Figure 4-10, a 5-kg box is on a frictionless 25° incline, held at rest by a spring whose force constant is 1000 N/m. (*a*) How far is the spring stretched beyond its equilibrium length? (*b*) How much force would you have to exert in order to pull the box 5 cm farther down the incline?

Figure 4-10

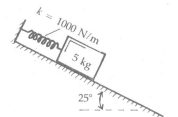

How to Solve It
• The restoring force in a stretched spring is its force constant multiplied by the length it has been stretched.

• What other forces act on the spring in the direction up or down the incline?

• Stretching the spring further in (*b*) requires additional force down the plane. How much additional force?

12. In Figure 4-11, Two mountain climbers are working their way up a glacier when one falls into a crevasse. The icy slope can be considered frictionless. Sue's weight is pulling Paul up the 45° slope. If Sue's mass is 66 kg and if she falls 2 m in 10 s (starting from rest), find (*a*) the tension in the rope joining them and (*b*) Paul's mass.

How to Solve It
• Find the acceleration of the two climbers from the data on Sue's fall.

Figure 4-11

- Draw a free-body diagram for Sue, and calculate the tension in the rope connecting them.

- Draw a free-body diagram for Paul, and calculate from it the component of his weight directed along the slope.

13. In Figure 4-12, the tension in the string is accelerating the 5-kg block at 0.8 m/s². If the table top is frictionless and the frictional force that each block exerts on the other is 1.1 N, (*a*) what is the tension in the string? (*b*) What is the acceleration of the 2-kg block?

Figure 4-12

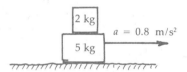

How to Solve It

- Draw free-body diagrams for both blocks, identifying at least all the horizontal forces that act on each.

- You know the force that the 2-kg block exerts on the 5-kg block, and you know the net force on the 5-kg block from its acceleration. Calculate the force exerted on it by the string.

- Remember Newton's third law!

14. An express elevator descends 72 m from the top story of an office building to the ground. As it starts, it accelerates from rest to its top speed of 3.6 m/s² in moving the first 2.5 m downward. It continues at constant speed until it reduces its speed to zero as it moves the final 2.5 m. Riding down in it is a weight-conscious editor whose mass is 59 kg. What is the greatest value of her apparent weight during the elevator ride? When does it have this value?

How to Solve It

- Calculate the acceleration at each end of the ride from the distance that the car goes while speeding up or slowing down.

- Calculate the editor's apparent weight during each of the three legs of the trip.

15. A truck is moving forward. Inside its (closed) trailer, a pendulum hangs from the ceiling at rest with respect to the truck, as shown in Figure 4-13. It makes an angle of 12° with the vertical. Can you calculate the speed of the truck? Its acceleration? Or both?

Figure 4-13

How to Solve It

● Draw a free-body diagram for the mass at the end of the pendulum. Be sure to include only those forces that act on it.

● The pendulum is moving with the truck. If the truck were moving at constant velocity, the net force on the pendulum would be zero. Is this possible?

16. A boy is pulling his sled at constant speed up a 20° incline in the manner shown in Figure 4-14. Friction between the sled runners and the hillside is negligibly small. If the mass of the sled is 6.6 kg, with what force must the boy pull?

Figure 4-14

How to Solve It

● Draw a free-body diagram for the sled. There are forces exerted on it by gravity, the hillside, and the rope that the boy pulls on.

● Resolve these forces into components along and perpendicular to the incline.

• Since the sled is moving at constant speed, the net force on it is zero.

Solutions

1. The car's mass is

$$m = \frac{w}{g} = \frac{3750 \text{ lb}}{32.2 \text{ ft/s}^2}$$

and its acceleration is

$$a = \frac{\Delta v}{\Delta t} = \frac{0 - 49.2 \text{ ft/s}}{5.5 \text{ s}} = -8.95 \text{ ft/s}^2$$

Therefore the force acting on the car is

$$F = ma = \frac{w}{g} \cdot a = \frac{3750 \text{ lb}}{32.2 \text{ ft/s}^2} (-8.95 \text{ ft/s}^2) = -1.04 \times 10^3 \text{ lb}$$

The minus sign means merely that the force is opposite in direction to the car's motion. The only external agency that can be exerting a retarding force on the car is the frictional force of the road surface on the car's tires.

2. 2210 kg. It's a big crate.

3. A free-body diagram of the balloon is shown in Figure 4-15. When the balloon is falling, the net force on it is

$$\mathbf{B} - \mathbf{w} = m\mathbf{a}$$

so

$$B = w + ma = mg + ma = m(g + a)$$
$$= (220 \text{ kg})[(9.81 \text{ m/s}^2) - (0.20 \text{ m/s}^2)] = 2114 \text{ N}$$

is the upward buoyant force on the balloon. (The acceleration is negative because it is directed downward.) This is determined primarily by the size of the balloon, and is essentially unchanged by throwing some small amount of ballast over. After the ballast is discarded, the weight of the balloon is w'. Now

$$B = 2112 \text{ N} = w' + m'a = (200 \text{ kg})(9.81 \text{ m/s}^2 + a)$$

and solving,

$$a = \frac{2114 \text{ N}}{200 \text{ kg}} - 9.81 \text{ m/s}^2 = 0.76 \text{ m/s}^2$$

Figure 4-15

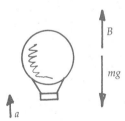

4.   (*a*) 2.5 m/s²   (*b*) The free-body diagrams are shown in Figure 4-16.   (*c*) 80 N

Figure 4-16

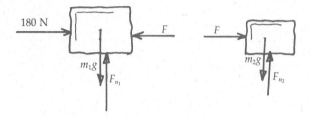

5.   First we need to know the car's acceleration. Its velocity changes from 14 m/s to 0 in the time it takes it to move 0.3 m. Thus,

$$v^2 - v_0^2 = 2ax$$

so

$$a = (v^2 - v_0^2)/(2x)$$

$$= \frac{(0) - (14 \text{ m/s})^2}{(2)(0.3 \text{ m})} = -327 \text{ m/s}^2$$

The minus sign just means that the acceleration is in the $-x$ direction. The driver stops with the car, so the force the seat belt exerts on her is

$$F = ma = (52 \text{ kg})(327 \text{ m/s}^2) = 1.70 \times 10^4 \text{ N}$$

or just a little less than two tons.

6.   5.13 × 10⁴ N, while it is accelerating.

7.   Since the wagon moves at constant velocity, the resultant of all the forces acting on it must be zero. There are three forces in the horizontal plane that must add to zero (see Figure 4-17*a*): **A** + **B** + **F** = 0. We know the components of **A** and **B** are

$$A_x = (580 \text{ N}) \cos 45° = 410 \text{ N}$$

$$A_y = -(580 \text{ N}) \sin 45° = -410 \text{ N}$$

$$B_x = (720 \text{ N}) \cos 20° = 677 \text{ N}$$

$$B_y = (720 \text{ N}) \sin 20° = 246 \text{ N}$$

so

$$F_x = -A_x - B_y = -410 \text{ N} - 677 \text{ N} = -1087 \text{ N}$$

$$F_y = -A_y - B_y = +410 \text{ N} - 246 \text{ N} = 164 \text{ N}$$

which gives

$$F = \sqrt{F_x^2 + F_y^2} = \underline{1099 \text{ N}}$$

The direction of F (see Figure 4-17b) is $\tan^{-1}(164 \text{ N}/1087 \text{ N}) = \underline{8.6°}$
north of west.

8. $\underline{814 \text{ N}}$ and $\underline{503 \text{ N}}$

9. Free-body diagrams for the two blocks are shown in Figure
4-18. (I didn't bother to show vertical forces as we know the blocks
are going to move horizontally.) The (unknown) frictional force f is
the force exerted on each block by the other. Applying Newton's
second law to each block gives

$$42 \text{ N} - f - 32 \text{ N} = (8 \text{ kg})(a) \qquad f = (2 \text{ kg})(a)$$

Figure 4-18

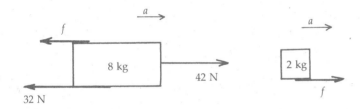

Adding the two equations gives

$$10 \text{ N} = (8 \text{ kg} + 2 \text{ kg})(a)$$

so

$$a = \underline{1.0 \text{ m/s}^2} \quad \text{and} \quad f = (2 \text{ kg})(1.0 \text{ m/s}^2) = \underline{2.0 \text{ N}}$$

10. $\underline{5 \text{ N}}$

11. (a) A free-body diagram for the box is shown in Figure 4-19.
Since it is at rest, the net force on it is zero. The forces along the
incline are due to gravity and the spring. Thus

$$kx = mg \sin \theta = 0$$

$$x = \frac{mg \sin \theta}{k} = \frac{(5 \text{ kg})(9.81 \text{ m/s}) \sin 25°}{1000 \text{ N/m}} = \underline{0.0207 \text{ m}}$$

Figure 4-17

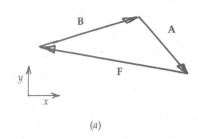

(a)

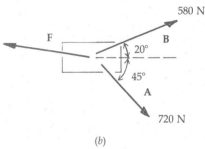

(b)

Figure 4-19

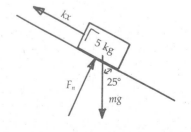

(b) To move the box 5 cm farther down the plane means that the spring must be stretched a total of $x = 0.0707$ m from equilibrium. Thus,

$$kx = mg \sin \theta - F = 0$$

where F is the additional force required down the plane. So

$$F = kx - mg \sin \theta$$

$$= (1000 \text{ N/m})(0.0707 \text{ m}) - (5 \text{ kg})(9.81 \text{ m/s}^2) \sin 25°$$

$$= \underline{50.0 \text{ N}}$$

12.   (a) $\underline{645 \text{ N}}$ (b) $\underline{92.4 \text{ kg}}$

13. Free-body diagrams for the two blocks are shown in Figure 4-20. Since the table top is frictionless, the only horizontal forces on the 5 kg block are the pull, or tension T, of the string and the frictional force f (1.1 N) of the 2-kg block. The net force on the block is

$$F_{net} = (5 \text{ kg})(0.8 \text{ m/s}^2) = 4 \text{ N}$$

We know

$$F_{net} = T - f, \text{ so}$$

$$T = F_{net} + f = 4 \text{ N} + 1.1 \text{ N} = \underline{5.1 \text{ N}}$$

Since only the frictional force f acts on the 2-kg block, its acceleration is

$$a = \frac{f}{m} = \frac{1.1 \text{ N}}{2 \text{ kg}} = \underline{0.55 \text{ m/s}^2}$$

14. $\underline{731 \text{ N}}$ while the elevator is $\underline{\text{coming to rest}}$

15. A free-body diagram for the mass on the end of the pendulum is shown in Figure 4-21. The only two forces acting on it are gravity and the tension of the string; the latter must act along the string. There is no way to know what the truck's speed is from the information given. However, we can calculate the truck's acceleration. Newton's second law must hold in each direction, so the horizontal forces are

$$T_x = T \sin 12° = 0.208 \ T = ma$$

and the vertical forces are

$$T_y - mg = T \cos 12° - mg = 0.978 \ T - mg = 0$$

or

$$0.978 \ T = mg$$

Figure 4-20

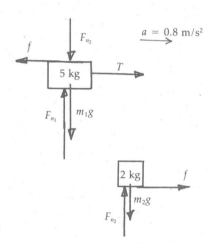

Figure 4-21

Dividing the two component equations gives

$$\frac{ma}{mg} = \frac{0.208\ T}{0.978\ T} = 0.213$$

so

$$a = 0.213g = \underline{2.09\ \text{m/s}^2}$$

16. $\underline{23.6\ \text{N}}$

CHAPTER 5

Applications of Newton's Laws

I. Key Ideas

Static Friction The force between two objects in contact ordinarily has a component parallel to the surface of contact. When the objects are at rest relative to one another, this force—called *static friction*—will oppose any attempt to slide them against each other. This force is "self-adjusting"; that is, it has whatever value is necessary to oppose the applied force up to a maximum value that depends on the surfaces and the normal force of contact. In order to cause the objects to move relative to one another, you must overcome this force.

Kinetic Friction When objects in contact are sliding against one another, they exert a force of *sliding* or *kinetic friction* on one another. This force is also parallel to the surface of contact and acts to oppose the relative motion of the surfaces.

Coefficients of Friction The *maximum force* that can be exerted by static friction between two surfaces is *independent* of the area in contact and *proportional* to the normal force of contact. The ratio of this force to the normal force is called the *coefficient of static friction* μ_s. It is a constant determined by the nature of the two surfaces. Similarly, the force of kinetic friction is proportional to the normal force, and the ratio of the two is the *coefficient of kinetic friction* μ_k. At low speeds, μ_k is approxi-

mately independent of the speed of the relative motion. Empirically, μ_s is always less than μ_k for given surfaces.

Torque The *torque* of a force upon an object expresses the tendency the force has to cause the object to *rotate* about any given point. It is defined as the force multiplied by its lever arm. The *lever arm* is the perpendicular distance from the line of action of the force to the point about which the object may rotate.

Static Equilibrium An extended object is in static equilibrium when it is at rest and remains so. In order for static equilibrium to exist, both the resultant force on the object *and* the resultant of the torques of all the forces about any point on the object must be zero. Notice that the total torque about *any* point must be zero, so in a given problem, we can choose to take torques about whatever point is most convenient.

Center of Gravity The force of gravity acts at every point of an extended object. For any body, there is a point with the property that the sum of the torques about that point exerted by gravity is zero; this point is the body's *center of gravity*. For many purposes, we can consider the center of gravity to be the point at which gravity acts on the whole object. We can locate the center of gravity of an irregularly shaped object by suspending it from various points; when it is in equilibrium, the center of gravity will hang directly below any suspension point. The center of gravity of a symmetric object will lie on the plane or line of symmetry.

Stability The equilibrium of a body can be stable, unstable, or neutral. Equilibrium is *stable* if the forces or torques that result from a small displacement of the body from equilibrium tend to move the body back toward its equilibrium position. It is *unstable* if they tend to move it farther away from its equilibrium position. A body resting on a surface is in equilibrium if its center of gravity lies over its base of support. The stability or balance of a body can be improved by lowering its center of gravity or by widening its base of support.

Centripetal Force When any object moves in an arc of a circle, it is accelerating toward the center of the circle at a rate given by v^2/r. Whatever net force acts on it in the radial direction to cause this acceleration is called the *centripetal force*. Its magnitude is mv^2/r. If the speed of the object is changing as it moves around the arc, there is also a tangential component to its acceleration, so there must also be a component of force on it parallel to its direction of motion.

Fluid Drag There is a *drag force* on an object moving through a fluid that tends to retard its motion. Unlike friction between solid surfaces, fluid drag forces are strongly velocity-dependent; they also depend on the shape and size of the moving object and on the properties of the fluid. Characteristically, they increase with increasing speed. An object driven by a constant force, therefore, such as a falling body, eventually reaches a *terminal speed*, where the drag force equals the driving force and the acceleration of the object is zero.

II. Key Equations

Key Equations

Frictional forces

$$f_s \leq \mu_s F_n \qquad f_k = \mu_k F_n$$

Torque

$$\tau = F_\perp r = Fr \sin \theta = FL$$

Conditions for static equilibrium

$$\Sigma F = 0 \qquad \Sigma \tau = 0$$

Centripetal force

$$F = \frac{mv^2}{r}$$

Velocity of circular motion

$$v = \frac{2\pi r}{T}$$

Fluid drag force

$$F = bv^n$$

Terminal speed

$$v = (mg/b)^{1/n}$$

III. Possible Pitfalls

When you isolate a body to draw a free-body diagram, make sure that you don't include any forces that act on something else. If I pull on a rope and the rope raises a scaffold and there's a brick on the scaffold, it's only the scaffold (not I or the rope) that exerts a force on the brick.

The tension in a string or rope and the force it exerts are always along the direction of the rope. Such things are flexible and cannot sustain any force perpendicular to their length nor any compressive force. Of course, the same is *not* true of a solid object.

You cannot always assume that the normal force on an object is equal to its weight, even if the object is at rest. If I sit on a sled that rests on the ground, for example, the normal force of the ground on the sled is not equal to the sled's weight. Your free-body diagram should make clear what the normal force is in a given problem.

Devices such as massless and frictionless pulleys or pegs, massless strings, and the like can be considered to change the direction of a force without changing its magnitude. If a string is effectively massless and no forces act along its length, the tension in it must be the same throughout.

The value of the force of static friction in general is not μ_s multiplied by the normal force of contact of the surfaces; this is its maximum value. Static friction is a self-adjusting force that has whatever value is required to keep the object at rest, up to this maximum.

Remember that static friction always *opposes the intended motion*. Thus, to determine in what direction static friction acts, you must judge how the system would move *in the absence* of friction.

Centripetal force is not a separate force; don't make the mistake of adding it to your free-body diagram as if it were. Whatever forces cause the circular motion are already there—friction, the tension in a string, the earth's gravity, or whatever. Centripetal force is just a name for whatever resultant force is causing the object to move in a circle.

In circular motion, don't imagine that there is some outward force that balances the centripetal force so that the circulating object is in equilibrium. The object is not in equilibrium; it's accelerating toward the center of the circle.

An unbalanced centripetal force is present in rotational motion even if the speed of the motion is constant. The centripetal acceleration is the component of acceleration that is due to the changing *direction* of the velocity.

IV. Questions and Answers

Questions

1. Why can't the coefficient of static friction for some surfaces be less than that of kinetic friction?

2. Beyond a certain point, further polishing of two surfaces may actually increase the coefficient of friction between them. Why is this?

3. Why does a pregnant woman's posture change as she comes near to delivering?

4. If the vector sum of all the forces that act on an object is zero, is it necessarily in static equilibrium?

5. Explain why curves in roads, running tracks, rail beds, and the like are banked.

6. When you drive a car, the frictional force of the road on your tires is what accelerates the car. Is it static or kinetic friction?

7. A car is moving along a curving highway in such a way that it has a centripetal acceleration but no tangential acceleration. How can this be?

8. Can a car move along a curving highway in such a way that it has a tangential acceleration but no centripetal acceleration? Explain.

9. Why does mud or snow fly off a rapidly spinning tire?

10. You are swinging some object in a vertical circle on the end of a string. For some point that is neither the exact top nor the exact bottom of its circular path, draw a diagram showing the forces that act on the object and its resultant acceleration.

11. The object in the previous question is not moving at constant speed. How can you be sure of this? As you drew the diagram, is its speed increasing or decreasing?

12. At low speeds, the drag force on an object moving through a fluid is proportional to its velocity. According to Newton's second law, force is proportional to acceleration. As acceleration and velocity aren't the same quantity, is there a contradiction here?

13. James Bond is thrown out of an airplane at 15,000 ft without his parachute. Ten seconds later his assistant, who was following behind in another plane (and who remembered her parachute, as well as his) leaps after him. If they are both falling "out of control" through the air, how is it possible for her to catch up with him and save him?

Answers

1. The coefficient of static friction determines how much force is required just barely to start the surfaces moving against each other, but if the coefficient of kinetic friction were to be larger, this force would not be enough to keep the objects moving. This would mean that the force required to start an object moving isn't enough to keep it moving.

2. As the irregularities in the surfaces are made very small, more of the surface molecules of the two objects come close enough to one another for short-range intermolecular forces to bond them to each other, thus increasing the force required to make them slide against each other.

3. For stability, we stand in such a way as to keep our center of gravity over our base of support—our feet. As a woman comes nearer to term and the fetus, which is carried more or less in front, becomes larger, her center of gravity moves forward, perhaps by as much as 4 or 5 cm. To compensate for the forward shift of her center of gravity, she tends to stand, and walk, leaning backward somewhat.

4. No. For static equilibrium to exist, the total torque on the object must also be zero.

5. Banking the road, for example, causes the normal force it exerts on the vehicle to have a component directed radially inward (see Figure 5-1). This radial force provides some or all of the centripetal force on the car. As a result, friction between the tires and the road is not the sole source of the centripetal force needed to keep the car moving around the curve. Thus, a car can go around a curve faster, without slipping, if the curve is banked.

Figure 5-1

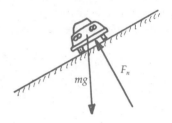

6. It is static friction that accelerates the car. The parts of your tires that are (instantaneously) in contact with the road surface are not sliding against it—unless you're in a skid.

7. This will be the case whenever the car's speed is constant.

8. This cannot be, unless the piece of the road on which the car is moving (at least at this particular moment) is straight. Whenever the direction of motion is changing, there is a centripetal acceleration.

9. It flies off because the forces holding it onto the tire are not strong enough to provide the centripetal force required to keep it in circular motion on the spinning tire.

10. Your sketch should look like Figure 5-2. **T** is the tension in the string and m is the mass of the object. The acceleration is in the direction of the resultant force $\mathbf{F}_{net}$.

Figure 5-2

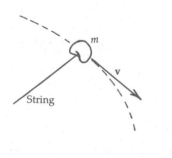

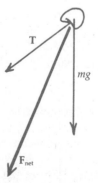

11. Because the net force and therefore the acceleration **a** are obviously not directed toward the center of the circle in Figure 5-2, the acceleration has a component along the direction of motion. Thus, the speed is not constant.

12. Newton's second law says that the acceleration of a body is proportional to the resultant force on it, regardless of what exerts that force, what its detailed behavior is, or what its magnitude depends on. Saying that there is a force proportional to the velocity of a particle is no more unreasonable or contradictory than saying that there is one proportional to the distance of the particle from some equilibrium position, as in Hooke's law.

13. Neither is completely out of control. By changing body position and orientation, each can change his or her terminal velocity by changing the value of the constant b in the force equation. If Bond spreads out in a horizontal position as he falls, while his assistant dives, she may fall up to twice as fast as he. As it takes 5 or 6 s to fall each 1000 ft, she may well be able to reach him in time.

V.   Problems and Solutions

Problems

1. A flatbed truck moving at 29 m/s is carrying an unsecured load of steel I-beams. The beams are at rest on the truck bed. The coefficient of static friction between the beams and the truck bed is 0.42. The driver sees a road accident ahead and brakes suddenly to avoid it. What is the least distance in which he can stop without the beams sliding forward at him from behind?

How to Solve It
- Find the maximum force that static friction can exert on one of the beams.

- From this, you can determine the maximum (magnitude of) acceleration that static friction can cause the beams to have.

- If the truck stops with a constant acceleration of this magnitude, it will cover the shortest stopping distance that is possible without the beams slipping.

Figure 5-3

2. In Figure 5-3, a 2-kg block of wood rests on a 5-kg block, which in turn rests on a tabletop. The blocks are connected by a light string that passes over a frictionless pulley. The coefficients of fric-

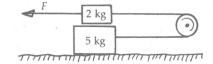

tion at both surfaces are $\mu_s = 0.47$ and $\mu_k = 0.33$. The 2-kg block is being pulled to the left by a force F, and the lower block is pulled to the right by the string that passes around the pulley. The blocks are moving at constant speed. What is the value of F?

How to Solve It
- Draw free-body diagrams showing all the forces that act on each block.

- Don't forget that the normal force of the tabletop on the bottom block will *not* be equal to the block's weight. Why not?

- You can now get the tension in the connecting string from the diagram for the bottom block.

3.   A car is moving at 33 m/s down a 19° grade. If the coefficient of static friction between the tires and the road is 0.58, what is the shortest time in which the car can stop?

How to Solve It
- In order to stop the car in the shortest time, we must have the largest possible net force directed uphill. The limitation is the force of static friction between the tires and the road.

- What is the maximum possible force of static friction? The maximum possible net force uphill?

- Use this force to calculate the acceleration and the stopping time.

4.   In Figure 5-4, a block of mass 3 kg is connected to a fixed point at the top of the 37° incline by a spring. The force constant of the spring is 200 N/m, and the coefficient of static friction between the block and the incline is 0.24. Starting with the block in the position such that the spring is unstretched, how much farther down the plane can you put it and have it still remain at rest?

Figure 5-4

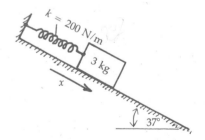

How to Solve It
- Draw a free-body diagram for the block showing all the forces that act on it.

- Calculate the maximum force that can be exerted by static friction.

- The farthest downhill point you can put the block is the point at which static friction has this maximum value.

- How much must the spring be stretched, in that case, to make the net force along the incline equal to zero?

Figure 5-5

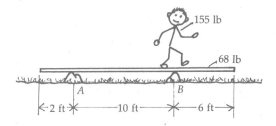

5. In Figure 5-5, a 155-lb man is standing on a uniform plank 18 ft long that is supported on two rocks, 2 ft and 6 ft from the ends. The plank weighs 68 lb. How far to the right can the man walk before the plank tips out from under him?

How to Solve It
- Draw a free-body diagram for the plank showing all the forces that act on it.

- The man is stable as long as the left-hand rock is exerting some upward force; since it can't exert a downward force, the plank will tip up if he goes beyond the point where its upward force becomes zero.

- So assume the support force of the left-hand rock is zero, and calculate where the man must be.

6. A painter's scaffold consists of a uniform horizontal plank 6 m long, of mass 13 kg, which is supported by a vertical rope at each end. The painter is standing 2.2 m from one end, and her mass is 58 kg. Calculate the tension in each of the two ropes.

How to Solve It
- Draw a free-body diagram for the plank showing all the forces that act on it.

- Use the condition that the vertical forces must equal zero in static equilibrium to find the sum of the two tensions.

- Calculate the torques about one of the support points to find the tension in one of the ropes. Use the condition that the torques about any point must equal zero for a body in static equilibrium.

- Subtract this from the sum of the tensions in the two ropes to get the other tension.

Figure 5-6

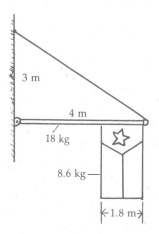

7. The flagpole in Figure 5-6 is a uniform aluminum pipe 4 m long, of mass 18 kg, fastened at one end by a hinge to the wall of a building and supported at the other by a steel wire. The wire is attached to the building at a point 3 m above the end of the flag-pole. A flag 1.8 m wide, with a mass of 8.6 kg, hangs from the end of the pole. What are (*a*) the tension in the wire and (*b*) the force exerted on the end of the pole by the hinge at the wall?

How to Solve It
- Draw a free-body diagram of the flagpole showing all the forces that act on it.

- The unknown force exerted on the pole by the hinge has both horizontal and vertical components.

- Take torques about the point where the pole is fastened to the wall since two of the three unknown force components act there.

- Remember that the weight of the flag acts at its center of mass.

- Once you've found the tension in the wire, calculate the components of the force at the hinge from the fact that the forces must add up to zero.

8. In Figure 5-7, Murray the mountain climber is hauling his buddy Basil over the edge of a cliff. Basil's mass is 83 kg, and his feet contact the cliff face at a point 1.2 m below their clasped hands. Murray is pulling in a direction 30° above the horizontal with a force of 158 N. Where (as far as you can find) is Basil's center of gravity?

How to Solve It
- There are three unknown quantities that act on Basil: two components of the force on his feet and the torque due to his weight.

- If you calculate torques about the point where Basil's feet contact the cliff face, you eliminate two of three unknowns.

Figure 5-7

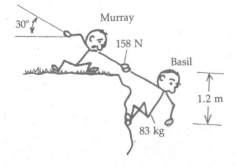

● This will give you the torque of Basil's weight about his feet. Calculate what you can about the location of his center of mass from this. You will *not* be able to calculate completely where his center of mass is.

9. A coin of mass 25 g rests on a phonograph turntable that rotates at 78 rev/min. The coin is 13 cm from the turntable axis. If the coin does not slip, what is the minimum value of the coefficient of static friction between the coin and the turntable surface?

How to Solve It

● If the coin doesn't slip, it's moving in a circle. Calculate the centripetal force that acts on it.

● Static friction must supply this force.

10. A pilot of mass 56 kg pulls her stunt airplane out of a vertical dive at an air speed of 180 m/s. At the bottom of the arc, her apparent weight is 7.5 times larger than her true weight. What is the radius of curvature of her path?

How to Solve It

● Calculate the pilot's centripetal acceleration at the bottom of the arc from her apparent weight. Be careful! What does apparent weight mean?

● Calculate the radius of her path from the known speed and the centripetal acceleration.

11. A model airplane with a mass of 1.8 kg is tethered to a string 40 m long and is flying at a speed of 20 m/s in a horizontal circle of radius 18 m (see Figure 5-8). Assuming the lift force of the air on the plane to be directed straight up, calculate the tension in the string and the lift force.

How to Solve It

● Draw a free-body diagram showing all the forces that act on the plane.

● Calculate the centripetal force on the airplane from the known radius and speed of its path.

● This centripetal force is provided by the only horizontal applied force in the problem. What is that?

● Calculate the tension force from its known component and direction and the lift force from the tension.

Figure 5-8

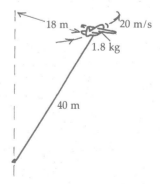

12. In a carnival ride, passenger gondolas are attached by cables 10 m long to a circular horizontal structure of radius 2 m on top of a tall pole (see Figure 5-9). The pole is rotated rapidly, swinging the gondolas around in a circle. How fast are the gondolas moving when the cables make a 55° angle with the vertical?

Figure 5-9

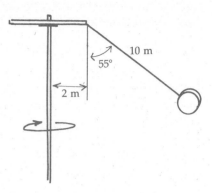

How to Solve It
- Draw a free-body diagram showing all the forces that act on a gondola.

- Calculate the horizontal component of the tension.

- The motion is in a horizontal circle, so the horizontal component of the tension in the cable provides the centripetal force.

- The vertical component of the tension balances the weight of the object. Find the tension from this information.

- Calculate the centripetal acceleration and, from it, the speed of the circling gondolas.

13. Assume that the drag force of the air on an object moving through it is proportional to v^2. A skydiver named Helen, whose mass is 63 kg, jumps from a plane and falls freely for a while, reaching a speed of 70 m/s; then her parachute opens. Her terminal speed when falling with the parachute open is 6 m/s. How much force does the parachute exert on Helen immediately after it opens?

How to Solve It
- From the terminal speed of her fall with the parachute open, find the value of the drag constant b that applies in this case.

- If she is (instantaneously) falling at 70 m/s when the parachute opens, find the upward force it exerts on her.

14. Assume that the drag force of the air on an object moving through it is proportional to v^2. My 55-kg cousin drops from an airplane (with a parachute!) and reaches a terminal speed of 5.0 m/s. When I jump with the identical parachute, I reach a terminal speed of 7.3 m/s, which makes for a somewhat less comfortable landing. What is my mass?

How to Solve It
- From the terminal speed of my cousin's fall with the parachute, find the drag constant b produced by this parachute.

- For the same value of b, what mass produces a terminal speed of 7.3 m/s?

Solutions

1.   The normal force of the truck bed on a beam of mass m is mg. The maximum force that static friction can exert on it to keep it at rest with respect to the truck bed is therefore

$$f_{s(max)} \le \mu_s F_n = \mu_s mg = (0.42)(m)(9.81 \text{ m/s}^2)$$

$$= (4.12 \text{ m/s}^2)m = ma$$

Static friction therefore cannot provide an acceleration of magnitude greater than 4.12 m/s². If the truck stops with a (negative) acceleration of this magnitude, it covers a distance Δx given by the equation

$$v^2 - v_0^2 = 2a \, \Delta x$$

$$\Delta x = \frac{v^2 - v_0^2}{2a} = \frac{0 - (29 \text{ m/s})^2}{(2)(-4.12 \text{ m/s}^2)} = \underline{102 \text{ m}}$$

2.   <u>35.6 N</u>

3.   The free-body diagram for the car is shown in Figure 5-10. The force of gravity on it is resolved into components along and perpendicular to the incline. The normal force is

$$F_n = mg \cos 19° = 0.9455mg$$

Thus, the frictional force is

$$f \le \mu_s F_n = (0.58)(0.9455mg) = (5.38 \text{ m/s}^2)m$$

From the diagram, the net force along the slope is therefore

$$f - mg \sin 19° \le (5.38 \text{ m/s}^2)m - m(9.81 \text{ m/s}^2)(\sin 19°)$$

$$= (2.19 \text{ m/s}^2)m = ma$$

Figure 5-10

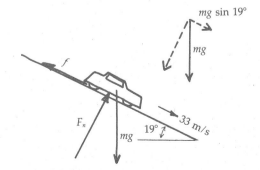

So, the car can't accelerate at a greater rate than 2.19 m/s². It is moving downhill, so we take this as the positive direction. Then as it stops

$$v = v_0 + at_{min}$$

$$0 = 33 \text{ m/s} - (2.19 \text{ m/s}^2)t_{min}$$

so

$$t_{min} = \frac{33 \text{ m/s}}{2.19 \text{ m/s}^2} = \underline{15.1 \text{ s}}$$

4. <u>11.7 cm</u>

5. The free-body diagram for the plank is shown in Figure 5-11. What we want to know is the force exerted on the plank by the rock at A. When this is zero, our man is as far to the right as he can go without tipping the board since the rock at A clearly can't provide a downward force. Assume that he is at that point when he is a distance x to the right of the right-hand rock. Then if we calculate torques around point B,

$$(68 \text{ lb})(3 \text{ ft}) = (155 \text{ lb})(x)$$

so

$$x = \underline{1.32 \text{ ft}}$$

If his center of gravity moves more than 1.32 ft to the right of support point B, the plank will tip up and he will fall.

6. <u>424 N</u> and <u>272 N</u>

Figure 5-11

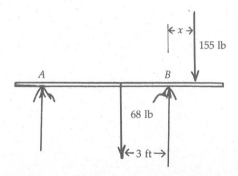

7. Figure 5-12 shows the forces that act on the pole. The weight of the flag is treated as if it were located at the center of the flag. Since we don't know the force F that acts at B, it is drawn in whatever direction seems reasonable. (If we guess wrong, we will know because one or more of our answers will come out negative.)

(*a*) Consider first, for convenience, the torques around point B. Taking counterclockwise torques as positive, we have

$$(T)(4 \text{ m})(\sin 37°) - (18 \text{ kg})(g)(2 \text{ m}) - (8.6 \text{ kg})(g)(3.1 \text{ m}) = 0$$

$$T = \underline{255 \text{ N}}$$

Figure 5-12

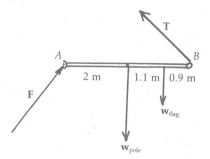

(*b*) The components of F are F_x and F_y. The vertical and horizontal forces must each add to zero, so

$$F_x - T \cos 37° = 0$$

$$F_x = (255 \text{ N})(0.7986) = \underline{204 \text{ N}}$$

and

$$F_y + T \sin 37° - (18 \text{ kg})(g) - (8.6 \text{ kg})(g) = 0$$

$$F_y = \underline{107 \text{ N}}$$

Then

$$F = \sqrt{F_x^2 + F_y^2} = \sqrt{204^2 + 107^2} = \underline{230 \text{ N}}$$

It is directed

$$\tan \theta = \frac{F_y}{F_x} = \frac{107 \text{ N}}{204 \text{ N}} = 0.5245$$

$$\theta = \underline{27.7° \text{ above the horizontal}}$$

8. Horizontally, Basil's center of gravity is $\underline{20.2 \text{ cm to the right}}$ of the point at which his feet contact the cliff face. There's no way to tell from the information given where it is vertically.

9. If the coin does not slip, it is moving in a circle of radius 0.13 m at a speed of

$$v = \frac{2\pi r}{T} = \frac{(2\pi)(0.13 \text{ m})}{(1/78 \text{ min})(60 \text{ s/min})} = 1.062 \text{ m/s}$$

Its centripetal acceleration is therefore

$$a_c = v^2/r = \frac{(1.062 \text{ m/s})^2}{0.13 \text{ m}} = 8.67 \text{ m/s}^2$$

This acceleration is caused by the force of static friction that holds the coin in place on the turntable. Thus, $f_s = ma_c$, but $f_s \leq \mu_s mg$, so

$$a_c \leq \mu_s g$$

$$8.67 \text{ m/s}^2 \leq (\mu_s)(9.81 \text{ m/s}^2)$$

and

$$\mu_s \leq \underline{0.884}$$

This is a pretty large value, but not an unthinkable one.

10. <u>508 m</u>

11. The angle α in the free-body diagram in Figure 5-13 is

$$\tan \alpha = \frac{18 \text{ m}}{40 \text{ m}} = 0.4500$$

$$\alpha = 26.7°$$

The centripetal force on the model airplane is horizontal and of magnitude

$$F = \frac{mv^2}{r} = \frac{(1.8 \text{ kg})(20 \text{ m/s})^2}{18 \text{ m}} = 40 \text{ N}$$

Since the lift force is vertical, the horizontal component of the tension in the string must be providing the centripetal force. Thus

$$T \sin 26.7° = 40 \text{ N}$$

so

$$T = \underline{89.0 \text{ N}}$$

Since the airplane is moving horizontally, the vertical forces must add up to zero. Thus,

$$F_{\text{lift}} = T \cos 26.7° + mg$$

$$= (89.0 \text{ N})(0.8934) + (1.8 \text{ kg})(9.81 \text{ m/s}^2) = \underline{97.2 \text{ N}}$$

Figure 5-13

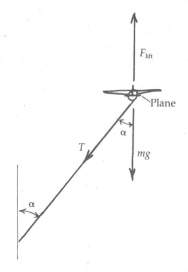

12. <u>11.9 m/s</u>

13. The drag force is $F_{\text{drag}} = bv^2$. When Helen reaches her terminal speed,

$$bv_t^2 = mg$$

Her terminal speed is therefore

$$v_t = \sqrt{mg/b}$$

When she is falling with the parachute open, $v_t = 6$ m/s. Thus,

$$6 \text{ m/s} = \sqrt{(63 \text{ kg})(9.81 \text{ m/s}^2)/b}$$

which gives

$$b = 17.2 \text{ kg/m}$$

This value of b applies to falling with the parachute open. If when it opens her velocity is 70 m/s, then the air drag force is

$$F_{\text{drag}} = bv^2 = (17.2 \text{ kg/m})(70 \text{ m/s})^2 = \underline{8.41 \times 10^4 \text{ N}}$$

This result is unrealistic, as it assumes that the parachute opens instantaneously. Actually, it would take a few seconds to open. That's just as well, as this is 93 tons of force, well over 1000 times her weight. If it opened instantaneously, either the parachute straps or Helen would break.

14. <u>117 kg</u>

CHAPTER 6

Work and Energy

I. Key Ideas

Work When a force acts on something that moves through some distance and the force has a nonzero component along the direction of motion, we say that the force has done work. The work done is defined as the component of the force along the displacement times the displacement. When there has been no motion, no work has been done; when the force is directed perpendicular to the displacement, it does no work. The work done is positive if the force component is in the same direction as the displacement and negative if it is in the opposite direction. Work is a scalar quantity and has no direction.

Work and Kinetic Energy The usual result of doing work on something is a change in its motion; that is, an increase or decrease in its speed. We introduce the idea of energy as a quantity that is *conserved* in physical processes, and we define the *kinetic energy*—the energy of motion—of an object as half its mass times the square of its speed. Energy of one kind or another results from doing work; to do work, energy must be supplied. It follows from Newton's second law that the net work done on an object by all the forces that act on it is equal to the change in its kinetic energy. This result is known as the *work-energy theorem*.

Work Done by a Variable Force The definition of work as force times displacement applies only if the value of the force remains constant during the displacement. If the force is not constant, a more general (and more complex) definition must be used. If we know how the force varies with *position*, the work it does in causing a particular displacement can be defined as the area in that interval under the graph of force versus displacement. (This is equivalent to calculating the work done in intervals small enough for the force during each interval to be considered constant and then adding up the work done in all the intervals.) The work-energy theorem holds for a variable as well as a constant force.

Conservative and Nonconservative Forces We can distinguish types of forces according to how the work that they do is calculated. When an object moves, for example, the work that gravity does on it is simply the weight of the object multiplied by its net vertical displacement; the path it takes does not matter. A force of this kind is called a *conservative force*. Friction, on the other hand, is an example of a *nonconservative force*. Since it always opposes an object's motion, the work that it does depends on just what path the object takes from one point to another.

Potential Energy An object raised vertically against gravity has potential energy equal to the work required to raise it because, if we let the object fall freely, the work done on it by gravity will deliver that much kinetic energy to it. *Poten-*

tial energy, then, is energy that is potentially available because of an object's position. Notice that only differences in potential energy are defined; the zero point of potential energy can be anything we like. In general, this holds for a conservative force, such as gravity or the restoring force of a stretched spring. For conservative forces, we can define the change in an object's potential energy as the negative of the work that the force has done. When I lift something against the force of gravity, I do work that does not appear immediately as kinetic energy; it has been "stored" as gravitational potential energy, which will appear as kinetic energy later if the object is allowed to fall back. Most generally, the concept of potential energy applies not to a single particle but to an entire system of particles.

Conservation of Mechanical Energy If all the forces that act on a particle or a system are conservative forces, then whatever increases in kinetic energy occur as the system moves are compensated for by decreases in potential energy, or vice versa. Thus, the sum of the potential and kinetic energies—the total mechanical energy of the system—remains constant. This is the *conservation of mechanical energy*.

Nonconservative Work In real macroscopic systems, nonconservative forces are always present; the conservation of mechanical energy is thus a limiting case, an abstraction. The work done to overcome nonconservative forces can't be recovered mechanically; this work is energy that is dissipated from the system, usually as heat. If we define the potential energy for all the conservative forces that act on a given system, we can state the work-energy theorem in a more general form: the net work done on a system by nonconservative forces over any interval is equal to the change in its total mechanical energy.

Machines Simple machines are devices in which small forces are converted into larger forces—they effectively multiply forces. Examples are the lever, the pulley, the wheel and axle, and the inclined plane. The ratio of the force exerted at the output of the machine to that exerted at the input is called the *mechanical advantage* of the machine. In an ideal machine, there are no energy losses, so the work output is the same as the work input. In this case, the ideal mechanical advantage is also the ratio of the distance moved at the input to that moved at the output.

Power Power is the rate at which energy is transferred or the rate at which work is done. If a force F acts on an object that is moving at velocity v in the direction of the force, the power that it delivers is given by Fv.

Thermal Energy The conservation of energy is a much broader idea—that is, it applies much more generally—than the conservation of mechanical energy. The energy that is dissipated by nonconservative forces appears as other forms of energy, usually thermal energy or heat. The source of energy for human (and animal) actions is derived from chemical energy that is developed by "burning" foodstuffs; burning enough to consume one litre of oxygen delivers the energy equivalent of about 2×10^4 J. The rate at which the body converts energy is called the *metabolic rate*. A typical metabolic rate for a human body at rest is around 120 W.

II.   Numbers and Key Equations

Numbers

1 joule (J) = 1 N·m

1 watt (W) = 1 J/s

1 kilowatt-hour (kW·h) = 3.60×10^6 J

1 J = 0.738 foot-pound (ft·lb)

1 horsepower (hp) = 746 W

1 calorie (cal) = 4.184 J

1 "food calorie" = 1 kcal = 10^3 cal

1 Btu = 1054 J

Key Equations

Definition of work

$$W = F_x \, \Delta x = F \, \Delta x \cos \theta$$

Kinetic energy

$$E_k = \tfrac{1}{2}mv^2$$

Work-energy theorem

$$W_{\text{net}} = \Delta E_k \quad \text{or} \quad W_{nc} = \Delta E_{\text{total}}$$

Potential energy

$$\Delta U = -W$$

Gravitational potential energy

$$U_{\text{grav}} = mgh$$

Potential energy of a spring

$$U_{\text{spr}} = \tfrac{1}{2}kx^2$$

Conservation of mechanical energy

$$E_k + U = \text{constant}$$

Ideal mechanical advantage

$$MA_{\text{ideal}} = \frac{F_{\text{out}}}{F_{\text{in}}} = \frac{S_{\text{in}}}{S_{\text{out}}}$$

Power

$$P = \frac{W}{t} = \frac{\Delta E}{\Delta t}$$

$$P = F_s v$$

III. Possible Pitfalls

This chapter is full of words—*work, energy, power, machine*, and so forth—with meanings in everyday usage that are different, sometimes very different, from their definitions in mechanics. Make sure you understand what their precise definitions are, and don't let yourself be confused by other usages.

Notice that work can be calculated simply as a force times a displacement *only if* the force is constant throughout the displacement. If the force is not constant, the calculation of the work becomes more complicated; it can be calculated, for example, from a graph of force versus displacement.

Notice that a force does work only when it has a component along the direction of the displacement. Forces that act to constrain the motion of an object to a particular path, like the force exerted by the string on a pendulum bob, act perpendicular to the path and do no work.

Work and energy are scalar quantities; they have no direction. The signs of these quantities express the sense of an energy change—that is, whether the energy is increased or decreased.

It's very important to understand that only *changes* in potential energy are defined. The choice of the point at which the potential energy or the total mechanical energy is zero is always arbitrary; you choose the zero point that is convenient in a particular situation. In a particular situation, however, there may be one convenient choice that is usually used.

The formula for the potential energy of a spring—$U_{\text{spr}} = \tfrac{1}{2}kx^2$—holds only if x is the distance by which the spring is stretched or compressed *from its equilibrium length*.

Be careful of the sign in the definition of potential energy. The change in potential energy corresponding to a force is the *negative* of the work that force does in causing a given displacement. Thus, gravitational potential energy increases as an object is raised against gravity, but gravity does negative work on the object as it is raised.

IV. Questions and Answers

Questions

1. I pick up my typewriter from my desk top, carry it to the other end of the house, and set it down on a kitchen table that is the same height as the desk. Have I done work on the typewriter?

2. A bicyclist is pedaling downhill into a strong headwind at a constant speed. Discuss what is happening to energy of various kinds in this situation.

3. The moon's motion around the earth is essentially a uniform circular motion, so its kinetic energy is not changing. But there is certainly an unbalanced force—the gravitational attraction of the earth—acting on the moon. Does this contradict the work-energy theorem?

4. When you climb a mountain, is the work done on you by gravity different if you take a short steep trail or a long gentle one? If not, why would one climb be easier than the other?

5. The resultant force on a particle is not zero, but the net work done on the particle over some interval is zero. Is it possible that the particle has moved in a straight line? Explain your answer.

6. You get into your car and drive away, accelerating from rest to some speed v. The kinetic energy of the car has increased. Where did the energy come from? An increase in kinetic energy always means that some work has been done. What exactly is it that does work on your car?

7. What does it mean for an object to have a total mechanical energy that is less than zero?

8. An object slides down a rough incline and hits a spring at the bottom. It bounces back and comes to rest (instantaneously) part of the way up the slope. At this point, what has happened to the potential energy that the object has lost?

9. How can the work-energy theorem explain why the force of sliding friction always reduces the kinetic energy of an object?

10. Is it *always* true that frictional forces (static as well as kinetic) reduce a body's kinetic energy?

11. Guess reasonable numbers for some familiar situation, and estimate the maximum power output of a human body in horsepower.

Answers

1. Zero net work has been done on the typewriter since it starts and finishes with the same kinetic energy. Gravity has done zero work on it since it ends at the same vertical height. Thus, the work I have done on the typewriter is also zero.

2. The bicyclist is moving at constant speed; therefore, she has constant kinetic energy. She is doing work pedaling and so is putting mechanical energy into the system. (Ultimately, this energy comes from her body's conversion of chemical energy.) As she moves downhill, her gravitational potential energy is decreasing. Since her kinetic energy is not increasing, the work that she is doing and that gravity is doing is being dissipated by friction in the bicycle mechanism and by the drag force of the air.

3. There is a force on the moon, but it is directed toward the center of the moon's orbit, that is, toward the earth. Thus, at every point the force is perpendicular to the moon's direction of motion, which means that it does zero work on the moon. It is work, not force, that changes kinetic energy. There is no contradiction here.

4. No, the work done by gravity is the same regardless of which trail you take. You will find the gentle trail less tiring because you are doing the work over a longer time and so are demanding a smaller power output from your body.

5. Yes, it can happen in a number of ways. For example, the force may be perpendicular to the path of the particle. If there is a force along the particle's direction of motion, its direction must reverse (at least once) during the interval. In this case, the total work would be zero, with the negative work done in some parts of the interval canceling the positive work done in other parts.

6. The energy was supplied by oxidizing (burning) gasoline in the engine, which turned a crankshaft, and so on. But what actually does work on the car? This is a little tricky. It is tempting to say "the frictional force between the tires and the road surface," and indirectly, perhaps, this is true; but there is no motion at that point, so this force is doing no work. What actually does work on the car body is the rolling wheel shoving forward on the axle (see Figure 6-1).

Figure 6-1

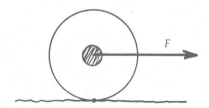

7. It means that the total mechanical energy is less than it would be for an object at rest at whatever point you chose to define potential energy to be zero, that is, the object does not have enough energy to make it possible for it to reach that point. For instance, my total mechanical energy is negative when I am standing in a deep ditch, if we have defined potential energy to be zero at ground level.

8. It has been dissipated by nonconservative forces—in this case, the friction of the rough incline.

9. Sliding friction always exerts a force that is directed against the motion, so it always does negative work. Negative work means a decrease in the object's kinetic energy.

10. Not always. If you sit on a flatbed truck or a flat wagon as it starts up, it is static friction between the seat of your pants and the truck bed that sets you moving forward.

11. A typical flight of stairs is around 3 m high; suppose a large man (100 kg) runs up it in 5 s. The power he is putting out would be about 600 W—about 0.8 hp! He would be capable of this only for very short bursts of power, however. For sustained effort over several hours, it's doubtful that anyone can exceed about 0.1 hp.

V.   Problems and Solutions

Problems

1. A girl is pulling her 7.2-kg sled across a snow-covered road, as in Figure 6-2. You can assume that the force of friction is negligibly small. Starting from rest, she pulls with a constant force of 20 N on a rope that makes an angle of 40° with the horizontal. When she reaches the other side of the road, she is moving at a speed of 5.6 m/s. (*a*) How much work did she do on the sled? (*b*) How wide is the road?

Figure 6-2

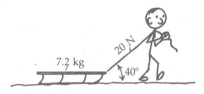

How to Solve It
- Calculate how much the kinetic energy of the sled has changed.

- Assume that the width of the street is x. How much work has the girl done on the sled?

- Calculate the distance the sled goes from the work done and the known force. Don't forget that only the component of a force along the direction of motion counts in calculating the work it does!

2. Your car has stalled out and you have to push it to the side of the road. The car's mass is 1300 kg, and you have to exert a horizontal force of 480 N to overcome rolling friction and just barely keep it rolling at constant speed. When you have pushed the car (from a standing start) a distance of 60 m down a level road, it is moving at 1.4 m/s. At this point, (*a*) how much net work has been done on the car? (*b*) How much work has friction done on it? (*c*) How much work have you done? (*d*) Assuming that you exert a constant force on the car along the direction of its motion, find this force.

How to Solve It

- Calculate how much the kinetic energy of the car has changed. This is equal to the net work done on the car.

- The data in the problem give you the magnitude of the frictional force.

- The forces that act on the car are friction, the force with which you are pushing it, gravity, and the normal force of the road. The last two do no work.

3.   A 50-kg sled slides 5.5 m down a straight slope that is inclined 20° to the horizontal. The coefficient of kinetic friction between the sled and the hillside is 0.18. Find (a) the work done by gravity on the sled, (b) the work done by friction, and (c) the change in the sled's kinetic energy.

How to Solve It

- The work done by gravity is the component of the sled's weight directed down the plane, multiplied by the distance the sled has moved.

- The frictional force acts along (but opposite to) the direction of the sled's motion.

- The change in the sled's kinetic energy is equal to the net work done by all the forces that act on it.

4.   A 35-kg boy slides down a snowbank that is 2.6 m high and has a slope of 20° and out onto the frozen surface of a pond (see Figure 6-3). The coefficient of kinetic friction between the boy's behind and snow or ice is 0.15. If he starts from rest at the top of the bank, how far out on the ice does he slide?

Figure 6-3

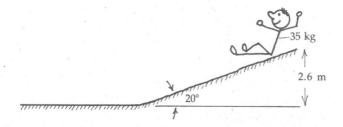

How to Solve It
- When the boy is on the slope, friction and gravity are doing work on him.

- When he is on the level ice, only friction is doing work on him.

- Since he starts and ends at rest, the net work that has been done on him must be zero.

5. A boulder of mass 305 kg falls from a height of 1.2 m onto a platform that is mounted on a vertical spring (see Figure 6-4). The force constant of the spring is 2.0×10^6 N/m. What is the maximum distance to which the boulder compresses the spring when it lands?

Figure 6-4

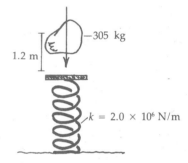

How to Solve It
- Gravity does work on the boulder throughout its motion.

- The spring, however, does work on the boulder only while it is in contact with it.

- The boulder starts from rest and finishes at rest (instantaneously) when the spring is at its maximum compression. Thus, zero net work has been done on it between these two points.

6. At the end of a railway track, you will often find a spring mounted so as to bring the train to rest if the driver overshoots his stopping point a little. Suppose a train of mass 6×10^5 kg hits such a stopping spring at a speed of 0.50 m/s. If it travels 12 cm further while being brought to rest, find the force constant k of the spring.

How to Solve It
- Calculate the initial kinetic energy of the train.

- At the point where the train stops, all of this kinetic energy has been turned into potential energy of the compressed spring.

7. A mass of 1 kg is hanging from a spring. When it hangs at rest, the spring is stretched 10 cm beyond its equilibrium length. The mass is pulled down 6 cm below this point, further stretching the spring, and released. How high does it rise?

How to Solve It
- The equilibrium data let you calculate the force constant of the spring; a force equal to the weight of the 1-kg mass stretches it 0.1 m.

- Calculate the potential energy of the stretched spring just before it is released. Don't forget that it's the total distance the spring is stretched that determines this.

- The total potential energy must be the same at the beginning and end of the motion, since at both these points the mass is at rest. Thus, the increase in the gravitational potential energy of the mass as it rises must be equal to the decrease in the potential energy of the spring.

8. Two masses are hung over a pulley by a string, as shown in Figure 6-5. Friction in the pulley and the masses of the pulley and the string can be neglected. The system is released from rest in the position shown, the 6-kg mass falling 1.5 m to the ground and pulling the 5-kg mass up. To what maximum height does the 5-kg mass rise above its position?

Figure 6-5

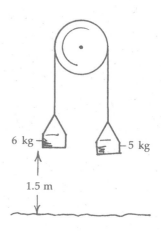

How to Solve It
- While the two masses are moving freely, their potential energy is just that due to gravity.

- But be careful; mechanical energy is lost when the heavier block hits the ground.

- Use the kinetic energy of the lighter mass alone at that point to calculate how high it rises.

9. In Figure 6-6, a roller-coaster car of mass 5000 kg starts from rest at point A and rolls along the track until it comes to rest at point C. (*a*) Find the force of friction acting on the car, assuming it is constant. (*b*) How fast is the car moving at point B?

Figure 6-6

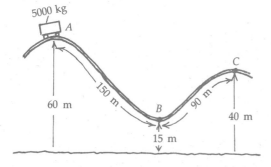

How to Solve It
- The kinetic energy of the car is zero at A and again at C, so the work done by friction over that distance must be equal to the change in the car's potential energy.

- Knowing the work friction does and the distance the car goes, you can calculate the force of friction.

- Calculate the work friction does from A to B; this must be equal to the change in the car's total mechanical energy.

10. A driver finds that, if he switches off his engine and puts the car in neutral at 30 mi/h on a level road, friction brings the car to rest in a distance of 650 ft. Assuming the frictional force remains the same, how far would the car go in coming to rest from 30 mi/h if the road were not level but had a 4° upward grade?

How to Solve It
- Be careful to keep track of units; some conversion is necessary.

- The work done by friction is the change in the car's kinetic energy, as the motion is horizontal and there is no change in the car's potential energy. From this, you can calculate the force of friction in terms of the unknown mass of the car.

- When the car is going uphill, some of its kinetic energy is dissipated doing work against friction, and some is transformed into potential energy.

11. Figure 6-7 shows the approximate dimensions of my elbow and forearm. What must be the tension in the biceps tendon in order for me to hold a 15-kg mass in my hand? Neglect the weight of the arm itself. Considering my forearm to be a lever, what is its mechanical advantage?

How to Solve It
- My forearm is in static equilibrium, so both the forces on it and the torques on it must add to zero.

Figure 6-7

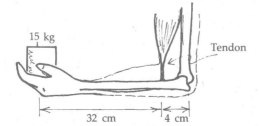

15 kg   Tendon

32 cm   4 cm

- To calculate the tension in the tendon most easily, take torques about the elbow.

- The ratio of the weight of the 15-kg mass to the tension is the mechanical advantage of this "lever."

12. You are delivering a refrigerator to the loading dock of a hardware store, which is 1.3 m above street level. Suppose you use a plank 3.5 m long to make an inclined ramp to roll the refrigerator up to the loading dock on a handcart. The refrigerator and cart together have a total mass of 240 kg. Neglecting any friction involved in rolling the cart, how much force is needed to push the refrigerator up the ramp?

How to Solve It
- Find the angle of the incline that you've made with the plank and, from it, the ideal mechanical advantage.

- If friction is negligible, the force you must use is the weight of the refrigerator divided by its ideal mechanical advantage.

13. The drag force on objects moving through the air is approximately proportional to the square of the speed. Suppose that a certain car needs power of 7.5 hp from its engine in order to move at 30 mi/h. If the engine's maximum power capability is 140 hp, what is the car's top speed?

How to Solve It
- The power that the engine delivers to the moving car is the force on the car multiplied by the car's velocity.

- If the force increases proportionally to the square of the speed, the power that the engine delivers is proportional to the speed cubed.

- Therefore, v^3 increases by the same factor as the power; solve for v.

14. An elevator car and its occupants together have a mass of 1400 kg. The car is raised by a motor with a maximum power output of 10 kW. When the car is moving, there is a constant frictional force of 1350 N impeding its motion. What is the maximum speed at which the motor can raise the elevator?

How to Solve It
- The motor has to do work against two external forces: friction and gravity. The total rate at which it can do work is 10 kW.

- Calculate the total force the motor must exert on the car in order to raise it at constant velocity.

- This force times the velocity of the car is equal to the power.

15. Suppose that a certain bicyclist's normal metabolic rate at rest is 150 W and that he can increase it (for short periods) to 900 W when riding at a constant 7.5 m/s on level ground. Assume that this is his maximum output. The mass of the bicycle and rider together is 86 kg. (*a*) What is the frictional force acting on him? (*b*) How fast can he ride up a 3° grade, assuming that the same frictional force acts?

How to Solve It
- Assume that the extra 750 W is going into moving the bicyclist and bicycle upward.

- Force times velocity gives the rate at which work is being done.

- When he is moving uphill, some of his power output is going into increasing his gravitational potential energy, so less is available for moving him against the frictional force.

16. A certain hiker's mass is 52 kg, and her normal metabolic rate at rest is 110 W. Suppose that she hikes up a hill 380 m high in 20 min. (*a*) What total power is her body delivering while she is climbing, if she converts the energy content of food into mechanical energy with an efficiency of 19 percent? (*b*) How many kilocalories has she burned during the climb?

How to Solve It
- Calculate the additional power the hiker uses to increase her gravitational potential energy.

- This plus the 110 W needed to supply basic biological activity is her total power output.

- Convert the total energy she expends over the 20-min interval to kilocalories.

Solutions

1.   (*a*) When the sled gets to the other side of the road, its kinetic energy is

$$E_k = \tfrac{1}{2}mv^2 = \tfrac{1}{2}(7.2 \text{ kg})(5.6 \text{ m/s})^2 = 113 \text{ J}$$

Since it started from rest, this is the change in the kinetic energy of the sled, which is equal to the net work that has been done on it, so the work the girl does is 113 J.

(b) The work is done by the horizontal component of the force exerted by the rope on the sled. If the distance the sled moves is x, the work done is

$$W = (20 \text{ N})(\cos 40°)(x) = (15.3 \text{ N})(x)$$

Since friction is negligible, this is the only force that does work on the sled; therefore,

$$113 \text{ J} = (15.3 \text{ N})(x)$$

or

$$x = \frac{113 \text{ J}}{15.3 \text{ m}} = 7.37 \text{ m}$$

2.   (a) 1274 J   (b) 2.88×10^4 J   (c) 3.01×10^4 J
(d) 501 N

3.   (a) The vertical distance h that the sled has moved is

$$h = (-5.5 \text{ m})(\sin 20°) = -1.88 \text{ m}$$

The minus sign means that the displacement was downward. Thus, the work done by gravity is

$$W_{\text{grav}} = -mgh = -(50 \text{ kg})(9.81 \text{ m/s}^2)(-1.88 \text{ m}) = 923 \text{ J}$$

(b) The force of kinetic friction on the sled is

$$f = \mu_k F_n = \mu_k mg \cos \theta$$

$$= (0.18)(50 \text{ kg})(9.81 \text{ m/s}^2)(\cos 20°) = 83.0 \text{ N}$$

The work it does is thus

$$W_f = -f \, \Delta x = -(83.0 \text{ N})(5.5 \text{ m}) = -456 \text{ J}$$

The negative sign, of course, is because the frictional force is in the opposite direction from the displacement.

(c) The only other force that acts on the sled is the normal force of the hillside on it. This does no work because it is perpendicular to the displacement. Thus,

$$\Delta E_k = W_{\text{net}} = W_{\text{grav}} + W_f$$

$$= 923 \text{ J} + (-456 \text{ J}) = 467 \text{ J}$$

4.   10.2 m

5.   The total energy of the system remains constant because the only two forces that act—gravity and the restoring force of the spring—are both conservative. Call $y = 0$ at the initial level of the platform. Then initially the spring is not compressed, and we have

$$U = U_{grav} = mgy = (305 \text{ kg})(9.81 \text{ m/s}^2)(1.2 \text{ m}) = 3590 \text{ J}$$

Suppose that when the boulder hits, it compresses the spring to a final height $y = -y_f$. At this point, its potential energy is

$$U = U_{grav} + U_{spr} = mg(-y_f) + \tfrac{1}{2}k(-y_f)^2$$

$$= -(305 \text{ kg})(9.81 \text{ m/s}^2)y_f + (0.5)(2.0 \times 10^6 \text{ N/m})y_f^2$$

$$= (10^6 \text{ N/m})y_f^2 - (2992 \text{ N})y_f$$

The kinetic energy is the same (zero) at both the starting point and the point of maximum compression. Since the total energy is constant, the potential energy must be the same at these two points:

$$(10^6 \text{ N/m})y_f^2 - (2992 \text{ N})y_f = 3590 \text{ J}$$

This simplifies to

$$y_f^2 - (0.002992 \text{ m})y_f - 0.00359 \text{ m}^2 = 0$$

The positive solution of this quadratic equation is $y_f = \underline{0.0614 \text{ m}}$ or 6.14 cm.

6. $\underline{1.04 \times 10^7 \text{ N/m}}$

7.   In equilibrium, the restoring force of the spring when it is stretched 10 cm $= 0.10$ m must be equal to the downward pull of gravity on the 1-kg mass. Thus,

$$k \, \Delta y = mg$$

$$(k)(0.10 \text{ m}) = (1 \text{ kg})(9.81 \text{ m/s}^2)$$

so

$$k = 98.1 \text{ N/m}$$

The spring is stretched $\Delta y = 10$ cm when the mass hangs in equilibrium and a further 6 cm before it is let go. Thus, it is stretched a total of 0.10 m + 0.06 m $= 0.16$ m beyond its unstretched length. The potential energy of the stretched spring at this point is

$$U_{spr} = \tfrac{1}{2}ky^2 = \tfrac{1}{2}(98.1 \text{ N/m})(0.16 \text{ m})^2 = 1.256 \text{ J}$$

If we measure height y upward from the position at which the mass and spring are in equilibrium, then at this point

$$U_{grav} = mgy = (1 \text{ kg})(9.81 \text{ m/s}^2)(-0.06 \text{ m}) = -0.589 \text{ J}$$

At this point, the kinetic energy is zero, so the total mechanical energy is

$$E = U = U_{spr} + U_{grav} = 1.256 - 0.589 = 0.667 \text{ J}$$

This total energy remains constant as the mass moves upward because only conservative forces are acting. When the mass has reached its maximum height y' above the equilibrium point, the spring is stretched by a distance 10 cm $- y'$. (Remember that it is stretched 10 cm when the mass is at its equilibrium position.) At this point, the kinetic energy is again (instantaneously) zero, so

$$E = U = U_{spr} + U_{grav} = \tfrac{1}{2}k(0.1 \text{ m} - y')^2 + mgy'$$

$$0.667 \text{ J} = \tfrac{1}{2}(98.1 \text{ N/m})(0.1 \text{ m} - y')^2 + (1 \text{ kg})(9.81 \text{ m/s}^2)(y')$$

With a little simplification, this becomes

$$0.490 \text{ J} - (9.81 \text{ N})(y') + (49.0 \text{ N/m})(y')^2$$
$$+ (9.81 \text{ N})(y') = 0.667 \text{ J}$$

$$y'^2 - 0.00360 \text{ m}^2 = 0$$

$$y' = \pm 0.060 \text{ m}$$

The negative root is the release point—the kinetic energy is zero there too!—and the positive root is the point we are looking for. Thus, the motion stops when the mass is 6.0 cm above the point at which the mass hangs at rest.

8.   It will rise 1.636 m above its starting point.

9.   What we use here is the fact that the change in the total mechanical energy in a nonconservative system is equal to the work done by nonconservative forces. Here, that is the work done by friction.

     (a) The roller-coaster car has traveled a distance of 240 m from A to C, with a constant frictional force f on it directed opposite to its motion. Thus, the (nonconservative) frictional force on it does work

$$W_{fr} = -fx = -(f)(240 \text{ m})$$

The decrease in its gravitational potential energy in descending a net distance of 20 m is

$$\Delta U = mg \, \Delta y$$
$$= (5000 \text{ kg})(9.81 \text{ m/s}^2)(-20 \text{ m}) = -9.81 \times 10^5 \text{ J}$$

There is no change in the car's kinetic energy from A to C, so

$$\Delta E = \Delta U = W_{fr}$$

$$-9.81 \times 10^5 \text{ J} = -(f)(240 \text{ m})$$

so

$$f = \underline{4.09 \times 10^3 \text{ N}}$$

(b) From point A to B,

$$\Delta E = \Delta E_k + \Delta U = W_{fr}$$

$$\Delta E_k + (5000 \text{ kg})(9.81 \text{ m/s}^2)(-45 \text{ m})$$

$$= -(4.09 \times 10^3 \text{ N})(150 \text{ m})$$

$$= 1.59 \times 10^6 \text{ J}$$

Thus, at point B

$$E_k = \tfrac{1}{2}mv^2 = \tfrac{1}{2}(5000 \text{ kg})(v^2) = 1.59 \times 10^6 \text{ J}$$

so

$$v = \underline{25.2 \text{ m/s}}$$

10. No, you don't need the mass of the car. It goes <u>259 ft</u>.

11. The weight of the 15-kg mass is

$$W = mg = (15 \text{ kg})(9.81 \text{ m/s}^2) = 147.2 \text{ N}$$

The torques about the elbow are $(T)(0.04 \text{ m})$ and (147.2 N) $(0.36 \text{ m}) = 53.0 \text{ N·m}$, so

$$T = \frac{53.0 \text{ N·m}}{0.04 \text{ m}} = \underline{1324 \text{ N}}$$

This large result (about 300 lb) is typical of the skeletal "levers" in large mammals because they typically have mechanical advantages much less than 1. Here,

$$MA = \frac{F_{out}}{F_{in}} = \frac{147.2 \text{ N}}{1324 \text{ N}} = \underline{0.111}$$

12. <u>874 N</u>

13. The drag force acting on the car is $F = bv^2$. The power needed to overcome this force is therefore

$$P = Fv = (bv^2)(v) = bv^3$$

We can determine the constant b from the 30-mi/h data:

$$7.5 \text{ hp} = (b)(30 \text{ mi/h})^3$$

$$b = \frac{7.5 \text{ hp}}{(30 \text{ mi/h})^3} = 2.78 \times 10^{-4} \text{ hp·h}^3/\text{mi}^3$$

(Yes, the units are weird, but they won't cause any trouble as long as we keep using the same units for P and v.) Thus, at top speed

$$140 \text{ hp} = (2.78 \times 10^{-4} \text{ hp·h}^3/\text{mi}^3)(v^3)$$

$$v^3 = \frac{140 \text{ hp}}{2.78 \times 10^{-4} \text{ hp·h}^3/\text{mi}^3}$$

giving $v = \underline{79.6 \text{ mi/h}}$.

14. $\underline{0.664 \text{ m/s}}$

15.   (*a*) When the bicyclist is riding at constant speed on level ground, the 750 W is all going into work done against frictional forces—friction in the bike's innards, air drag, or whatever. Thus,

$$P = fv$$

$$750 \text{ W} = (f)(7.5 \text{ m/s})$$

so

$$f = \underline{100 \text{ N}}$$

(*b*) When he is riding uphill, some of his power output is going into increasing his gravitational potential energy, so less is available to counteract friction. Riding up a 3° grade,

$$P = (mg)(\sin \theta)(v) + (f)(v)$$

$$= [(86 \text{ kg})(9.81 \text{ m/s}^2)(\sin 3°) + (100 \text{ N})](v)$$

$$= (944 \text{ N})(v) = 750 \text{ W}$$

so

$$v = \underline{5.20 \text{ m/s}}$$

16.   (*a*) $\underline{963 \text{ W}}$ (*b*) $\underline{275 \text{ kcal}}$

CHAPTER 7

Impulse, Momentum, and Center of Mass

I. Key Ideas

Impulse and Momentum The forces that act between objects that collide are typically very large and act only in a very short time interval. We define the *impulse* of a force as the average value of the force times the time interval. It follows from Newton's second law that the impulse delivered to a body in any interval is equal to the change in the body's momentum. *Momentum* is a vector quantity defined as a body's mass multiplied by its velocity; it can be thought of as measuring the difficulty of bringing the body to rest. If we can estimate the time in which a collision happens, we can use the impulse-momentum equation to estimate the average force that acts.

Conservation of Momentum The acceleration and the change in the total momentum of a system of particles are determined by the net external force on the system. If the net external force on a system is zero, its momentum does not change; that is, the total momentum of an isolated system is constant. This is the *law of conservation of momentum*. It is a more generally applicable principle than the conservation of mechanical energy because it is true regardless of the nature of the forces between the objects in a system as long as the net external force is zero. It even holds (approximately) during collisions that occur in the presence of external

forces for the very short time of the actual collision.

Center of Mass There is a point in any system of particles that moves just as if it were a particle having the entire mass of the system and acted on only by the net external force on the system. This point is the center of mass of the system. The *center of mass* is defined as the average location of all the mass of the system. (The center of gravity and the center of mass of a system are the same unless the system is so large that the gravitational field is not the same at every point in it.) An equivalent definition is to say that the total momentum of a system is its total mass times the velocity of the center of mass. It is in this sense that the center of mass "moves as a particle" having the total mass of the system.

Collisions In a collision, the total momentum of the two colliding objects is conserved. If their total mechanical energy is also conserved, the collision is said to be *elastic*. The collision of billiard balls, for example, is nearly elastic. In an elastic collision, the *relative velocity* of the two objects is simply reversed. The other extreme, a collision in which the objects stick together and move together afterward (with the velocity of the center of mass), is a *perfectly inelastic* collision. In a perfectly inelastic collision, the maximum possible loss of energy that is consistent with the conservation of momentum occurs. Real collisions lie somewhere between these two extremes.

Analysis in the Center-of-Mass Reference Frame Collisions in one dimension are particularly simple to analyze in a frame of reference in which the center of mass of the two colliding bodies is at rest. (In this frame, the total momentum of the two bodies is always zero, and so it is sometimes called the "zero-momentum" reference frame.) In the center-of-mass frame, in a perfectly inelastic collision the two bodies are at rest at the center of mass after the collision. In an elastic collision, the velocities of the two objects simply reverse directions.

General Case More generally, in an arbitrary frame of reference, the two bodies in a perfectly inelastic collision move together with the center of mass after the collision. In this case, the loss of kinetic energy in the collision is maximum, as the velocity of the center of mass is unaffected by the collision. Most real collisions are intermediate cases. The fraction of the relative velocity that disappears in the collision due to nonconservative forces is called the *coefficient of restitution*. Collisions in more than one dimension (except for the perfectly inelastic case) are more complicated to analyze.

Jet Propulsion The propulsion of jets and rockets is a very direct application of the momentum conservation principle. The center of mass of the rocket and the exhausted fuel moves in response to external forces; in isolation, the center of mass moves at constant speed. As fuel is ejected, momentum is transferred from it to the "payload," accelerating both.

II. Key Equations

Key Equations

Momentum of a particle

$$\mathbf{p} = m\mathbf{v}$$

Impulse and momentum

$$\text{Impulse} = \mathbf{F}_{av}\,\Delta t = m\,\Delta\mathbf{v} = \Delta(m\mathbf{v}) = \Delta\mathbf{p}$$

Conservation of momentum

$$\mathbf{p} = \mathbf{p}_1 + \mathbf{p}_2 = \text{constant}$$

Center of mass

$$M\mathbf{R}_{cm} = m_1\mathbf{r}_1 + m_2\mathbf{r}_2 + \cdots$$

$$M\mathbf{V}_{cm} = m_1\mathbf{v}_1 + m_2\mathbf{v}_2 + \cdots$$

$$\mathbf{F}_{net,\ ext} = M\mathbf{A}_{cm}$$

Elastic collision

$$v_{2f} - v_{1f} = -(v_{2i} - v_{1i})$$

General collision

$$v_{2f} - v_{1f} = -e(v_{2i} - v_{1i})$$

Rocket equation

$$v_r = v_e\,\ln(M_i/M_f)$$

III. Possible Pitfalls

The law of conservation of momentum is a vector statement. If the net external force on a system is zero, the component of the system's momentum in any direction is constant.

On the other hand, if the net external force is not zero but its component in some particular direction is zero, then the component of the system's momentum in that direction is constant.

Momentum is conserved in a system if and only if the net external force is zero. We ignore external forces and apply the conservation of momentum in collision problems, but this is an approximation that is valid only in this special case. In this case, we can apply momentum conservation only from *just before* to *just after* the collision.

The conservation of momentum applies only to the system *as a whole*; one part of the system may transfer momentum to another. In fact, this is what usually occurs.

Momentum is conserved in collisions, but energy may or may not be.

The center of mass of a system does not necessarily coincide with the position of any of the particles or objects that make up the system, and it is not necessarily fixed relative to any particle or object in the system.

The kinetic energy of center-of-mass motion cannot change in a collision because, as a consequence of momentum conservation, the velocity of the center of mass is unaffected by the collision. The maximum possible energy loss in a collision occurs in a perfectly inelastic collision, in which, after the collision, the two bodies stick together and move with the center of mass.

Beware of the old trap of thinking that a rocket can't move without something external to "push against." It accelerates part of itself in a vacuum by exhausting the rest of itself in the opposite direction.

The "ln" in the rocket equation is the so-called *natural logarithm*. Don't confuse it with the common logarithm to base 10.

IV. Questions and Answers

Questions

1. Explain why we usually ignore external forces such as friction and gravity when doing collision problems.

2. In what way might an air bag be a safer form of restraint than a seat belt or a chest harness?

3. Imagine that you are stranded in the middle of a frozen pond on which the ice is perfectly smooth. How can you get to shore?

4. A man jumps from a boat onto a dock (see Figure 7-1). Explain why he has to jump "harder"—that is, with greater force—than he would if he were jumping from the dock into the boat? Neglect the likely difference in vertical height; the vertical motion is not the point here.

5. Years ago when I was small, my father built us a playhouse in the back yard. The edge of its roof was 5 or 6 ft high. Of course we used to climb all over it, jump off the roof, and all. I can remember that landing, even from this very short distance, was painfully jarring until someone showed me how to flex my knees when I landed. Why did this make such a big difference?

6. Can you tell in what direction the center of mass of the system (the pulley and blocks) in Figure 7-2 will accelerate? Explain.

7. Can a system have zero kinetic energy and nonzero momentum? Can it have zero momentum and nonzero kinetic energy?

Figure 7-1

Figure 7-2

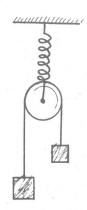

8. You are becalmed on your sailboat out in the middle of the lake; the air is perfectly still. Can you get yourself moving by blowing air on the sails with an electric fan?

9. A bowling ball hits a pin that is initially at rest. The collision is partially inelastic. Is it possible for both the ball and the pin to be at rest just after the collision?

10. You set an open bucket sliding across perfectly smooth ice—no friction—in a rainstorm. How does the velocity of the bucket change?

11. Is it possible for all the kinetic energy of a system of two colliding objects to be dissipated in a collision?

12. A woman is walking around inside a house trailer on wheels. The trailer is free to move—that is, friction is negligible. The woman and the trailer are initially at rest. Can she move the trailer? Can she set the trailer moving?

Answers

1. Because it is usually the case that, during the very short time in which the collision occurs, the internal collision forces are very much stronger than such external forces as gravity or friction. Thus, during the very short time in which the collision occurs, the collision forces can be assumed to cause all the transfer of momentum.

2. When your car collides with something and is brought to rest, something is going to bring you to rest too. Your momentum will be changed by a given amount, and so something must deliver an impulse to you. If the impulse is delivered in a very short time, as by a tight seat belt or chest harness, the force exerted on you must be correspondingly large—large enough to crack ribs, disrupt internal organs, and the like. The air bag lets you travel a foot or so in the process of coming to rest and so exerts a much smaller force for a much longer time.

3. If the ice is perfectly frictionless, your momentum is conserved, and your center of mass is stuck there in the middle of the frozen pond. If one part of this "system" moves in one direction, however, then the rest of it must move in the opposite direction in order for the center of mass to stay at rest. So you might take off your boots and throw them in a direction opposite to the one in which you want to move, or you might even just spit in that direction, in principle.

4. When the man jumps from the boat onto the dock, the center of mass of the system consisting of him and the boat stays put, so the boat moves in the opposite direction. He must therefore exert enough force and do enough work to set both himself and the boat moving. When he jumps from the fixed dock, he only has himself to accelerate.

5. This question is very like the one about the air bag. Letting my knees flex as the ground exerted its stopping impulse on my feet allowed the momentum transfer to occur over a longer time interval, so the force exerted on my feet was much smaller than when I landed stiff-legged. Of course, no one explained it to me this way when I was five.

6. No. The acceleration will, of course, be in the direction of the net external force, but this might be either upward or downward depending on how much the spring is stretched.

7. No and yes, respectively. If there is momentum, then something is moving and there must be kinetic energy. But, the total momentum will be zero if the center of mass is at rest, even if parts of the system are moving.

8. Sure, if you happen to have taken an electric fan and a motor-generator with you when you set out in the sailboat. This is like being stuck out on the frozen pond. If you use the fan to move air in one direction, the rest of the system—the system here consisting of you, the boat, and the air that you have moved—must move in the opposite direction. In fact, you don't even need to blow on the sails; just use the fan to move air! (This is the principle of the jet engine.)

9. No. The center of mass of the ball and pin is moving before the collision, and it will be moving with the same velocity after.

10. The bucket slows down. As it collects rainwater, the bucket imparts a forward momentum to the water that has fallen into it. This momentum transfer must be balanced by a decrease in the forward momentum of the bucket.

11. Yes, but only if the center of mass of the colliding objects was at rest before the collision.

12. The woman can move the trailer, in the sense of displacing it, by walking from one end of the trailer to the other. As she does so, the center of mass of herself and the trailer does not move; therefore, if one part of this system goes east, the other has to go

Figure 7-3

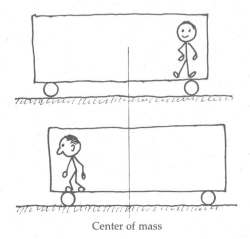

Center of mass

west, as shown in Figure 7-3. But she can't "set the trailer moving" in the sense of imparting a velocity to it permanently, again because the center of mass remains at rest.

V. Problems and Solutions

Problems

1.   A rifle fires a bullet of mass 10 g at a muzzle speed of 650 m/s. The rifle's mass is 3 kg. (*a*) If the rifle were free to move, at what speed would it recoil? (*b*) Instead, you fire from the shoulder, and the rifle's recoil shoves your shoulder back a distance of 3 cm. What average force does the rifle exert on your shoulder while it (the rifle) is being brought to rest?

How to Solve It
- Momentum is conserved in firing the bullet.

- The momentum of bullet and rifle, together, is zero before the rifle is fired; thus, their total momentum is zero after.

- In part (*b*), find the time in which the rifle's recoil momentum is transferred to your shoulder. From the time and the impulse, you can calculate the force.

2.   A baseball of mass 140 g reaches the batter at a speed of 35 m/s and is hit directly back at the pitcher at a speed of 50 m/s. (*a*) What impulse is delivered to the ball by the bat? (*b*) High-speed photographs reveal that the ball is in contact with the bat for 0.6 ms. What average force does the bat exert on the ball?

How to Solve It
- Calculate by how much the momentum of the ball has changed.

- The impulse is equal to the change in the ball's momentum.

- This impulse is the average force exerted on the ball by the bat multiplied by the time during which they are in contact.

3.   Fast Freddie is rolling along in his soapbox racer at 4.4 m/s when his little sister Nellie, who is initially at rest, hops aboard. After she jumps on, the racer is rolling at 2.6 m/s. If Freddie and the racer together have a mass of 41 kg, what is the mass of little Nellie?

How to Solve It
- The momentum of the system—consisting of Freddie, Nellie, and the racer—is conserved.

- Calculate the initial momentum of the system.

- The total momentum of the system after Nellie hops on is the same as it was before; use this to calculate her mass, which is the only unknown quantity.

4.   An 18-g bullet traveling at 500 m/s strikes and imbeds itself in a 1.2-kg wooden block that is initially at rest. The bullet penetrates 5 cm into the wood before it stops. (*a*) How fast is the block moving just after the bullet strikes? (*b*) What average force is exerted on the block by the bullet?

How to Solve It
- The total momentum of bullet and block after the collision is the same as it was before.

- Everything moves together after the collision, so the speed of the system is just the momentum divided by the total mass.

- Calculate the time it takes the bullet to stop; use this and the impulse delivered to the block to calculate the average force.

5.   When we do problems in which things bounce up from the ground, we never worry about the effect on the earth. If conservation of momentum is true, surely the earth must recoil from the collision. We know that the reason we neglect the change in earth's motion is that its mass is so large. In this problem, assume that a meteorite of mass 10^5 kg strikes the earth at a speed of 10^3 m/s. The mass of the earth is 5.98×10^{24} kg. At what speed does the earth recoil from the collision?

How to Solve It

- Take the earth to be "at rest" before the meteorite hits. The initial momentum of the system is that of the incoming meteorite.

- After the collision, this momentum is that of the earth and the meteorite moving together.

6. An astronaut whose mass (including his space suit) is 105 kg has been working outside his spaceship, using a small, hand-held rocket "gun" to change his velocity in order to move around. After a while he finds that he has been careless: his propulsion gun is empty, and he is out of reach of his spaceship and is drifting away from it at 0.7 m/s. The empty gun has a mass of 2.6 kg. How can he get back to his ship?

How to Solve It

- The astronaut, his propulsion gun, and the spaceship are an isolated system. Calculate its momentum.

- He wants to get himself moving toward the ship. How can he do this, given that the total momentum of the system is constant?

- He should throw the empty gun directly away from the ship. How fast should he throw it?

7. Four children, each having a mass of 35 kg, are skating on a pond (see Figure 7-4). Bob is at rest, and Carol is skating toward him from the west at a speed of 9 m/s. Ted is skating toward Bob from the east at 8 m/s, and Alice is skating toward Bob from the south at 5 m/s. They collide all at once, grabbing and holding on to one another. (*a*) After the collision, how fast is the group of chil-

Figure 7-4

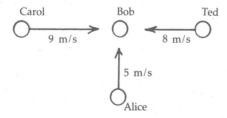

dren moving? (*b*) What fraction of the group's initial kinetic energy has been dissipated in the collision?

How to Solve It
- This is a simple conservation of momentum problem, but remember that momentum (mass times velocity) is a vector quantity.

- Calculate the initial momenta of Carol, Ted, and Alice individually.

- The vector sum of these momenta is equal to the momentum of the quartet moving together after they collide.

8. A car, initially going east at 80 km/h, and a pickup truck, initially going north at 60 km/h, collide at an intersection. The masses of car and truck are 1500 kg and 2900 kg, respectively. After the collision, the car and truck are mangled together into a single object called a wreck. What is the velocity (magnitude and direction) of the wreck just after the collision?

How to Solve It
- This is a simple conservation of momentum problem, but remember that momentum (mass times velocity) is a vector quantity.

- Calculate the initial momenta of the car and the truck individually.

- The vector sum of these momenta is equal to the momentum of the wreck after the collision.

9. Four hoboes, each weighing 150 lb, sit on an empty railway flatcar 55 ft long and weighing 4200 lb. Three of them sit on one end and the fourth on the other end of the flatcar. (*a*) Where is the center of mass of the system comprising the four hoboes and the flatcar? (*b*) Assume the track on which the flatcar rests is frictionless. One of the three hoboes gets up and walks to the other end and sits down. How far does the car move?

How to Solve It
- The system consists of five "particles" including the flatcar. Assume that the mass of the flatcar acts at its center.

- Measuring x from one end of the flatcar, calculate x_{cm}.

- How much does the distance of the center of mass from the end of the flatcar change when there are two hoboes at each end?

10. A truck of mass 8500 kg, traveling west at 20 m/s, overtakes a Honda of mass 700 kg, which is traveling at 12 m/s in the same direction, and hits it. (*a*) What is the velocity of the center of mass of the truck-car system before the collision? (*b*) After the collision, the Honda is moving west at 25 m/s. Now what is the velocity of the center of mass of the system? (*c*) At what velocity is the truck moving now?

How to Solve It

- The velocity of the center of mass of a system is the total momentum of the system divided by its total mass.

- The total momentum of the system and the velocity of its center of mass are unaffected by the collision.

- Calculate the final momentum of the car. The total momentum of the system is unchanged, so you can then calculate that of the truck.

11. In Figure 7-5, a birdhouse of mass 230 g hangs from a tree branch by a string 40 cm long. A 50 g wren lands on the birdhouse and sets it swinging back and forth on its string. If it swings to an angle of 16° away from the vertical, at what speed did the bird land?

Figure 7-5

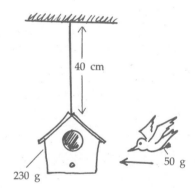

How to Solve It

- This problem is much the same as that dealing with the ballistic pendulum in the text.

- The "collision" of the wren with the birdhouse is inelastic. Using the conservation of momentum, calculate how fast the bird and the birdhouse together are moving just after she lands.

- Since the bird and the birdhouse swing out like a pendulum, use the conservation of energy to relate how far out they swing to their kinetic energy just after the bird lands.

12. An 85-kg running back moving at 7 m/s has a head-on collision with a 105-kg linebacker who is initially at rest. Just after the collision, the running back is still moving forward, but at a speed of 0.7 m/s.(*a*) What is the linebacker's velocity just after the collision? (*b*) How much kinetic energy was lost in the collision?

How to Solve It

- This collision isn't elastic, as you will see if you try to use the elastic-collision formulas here.

- You can still use conservation of momentum, however.

- The momentum of the linebacker just after the collision is equal to the momentum lost by the running back.

13. A woman is driving her car along at 15 m/s when a small boy in the road dead ahead throws a rubber ball at her car at a speed of 30 m/s. The ball bounces straight back toward the boy from the front of the car. If the collision is elastic, at what speed does the ball bounce back?

How to Solve It
- The simplest way to do this is to use the fact that the relative velocity of ball and car is simply reversed in the elastic collision.

- The car is so much heavier than the ball that you can assume that the velocity of the car isn't changed by the collision.

- You can also solve the problem directly from the equations for the conservation of momentum and kinetic energy.

14. In Figure 7-6, two pendulums hang side by side, in contact. Their masses are 1.5 kg and 2.5 kg. The heavier one is pulled aside until it makes a 30° angle with the vertical and is then released to collide with the other. Assuming they collide elastically, to what angle does each pendulum swing out after the collision?

Figure 7-6

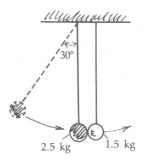

2.5 kg   1.5 kg

How to Solve It
- Use the conservation of mechanical energy to find the speed at which the 2.5-kg pendulum strikes the other.

- Just during the collision, the strings can be ignored and you can treat the problem as a collision between free bodies.

- Knowing that the collision is elastic tells you that the relative velocity of the two objects simply reverses in the collision. From this you can get the velocities at which the two pendulum bobs move just after they collide.

- Finally, use the conservation of energy again to determine the maximum angle to which each pendulum swings after they collide.

15. A bowling ball of mass 7.25 kg strikes a man of mass 72.5 kg in the chest and bounces directly back the way it came with a speed of 2 m/s. If the initial speed of the ball was 6.5 m/s, what is the coefficient of restitution of the collision?

How to Solve It
- This collision isn't elastic, as you will see if you try to use the elastic-collision formulas here.

- Use the conservation of momentum to find the speed imparted to the man.

- The coefficient of restitution is the fraction of the initial relative speed of the two objects that remains after the collision.

Solutions

1. (a) Let v be the speed at which the rifle would recoil if it were free and m be its mass; let m_b and v_b be the mass and speed of the bullet. Then, since the total momentum of rifle and bullet is zero,

$$p_{after} = mv + m_b v_b = 0$$

$$(0.010 \text{ kg})(650 \text{ m/s}) + (3.0 \text{ kg})(v) = 0$$

from which

$$v = -\frac{6.50 \text{ kg·m/s}}{3 \text{ kg}} = \underline{-2.17 \text{ m/s}}$$

(b) The impulse that must be exerted on the rifle to stop it is equal to the change in its momentum. This is

$$\Delta(mv) = m \,\Delta v = (3 \text{ kg})(2.17 \text{ m/s}) = 6.50 \text{ kg·m/s}$$

The rifle moves a distance of 3 cm as it is brought to rest from a speed of 2.17 m/s. We don't know any details of how it slows down; if we assume it decelerates uniformly to rest, that is,

$$v_{av} = \tfrac{1}{2}v = \frac{2.17 \text{ m/s}}{2} = 1.08 \text{ m/s}$$

then the time it takes is

$$\Delta t = \frac{0.03 \text{ m}}{1.08 \text{ m/s}} = 0.0278 \text{ s}$$

Impulse is average force multiplied by time, so

$$F_{av} = \frac{6.50 \text{ kg·m/s}}{0.0278 \text{ s}} = \underline{234 \text{ N}}$$

This is a substantial force—around 50 pounds. The kick of a high-powered rifle is quite noticeable!

2. (a) $\underline{11.9 \text{ N·s}}$ (b) $\underline{1.98 \times 10^4 \text{ N}}$

3. The motion of the system is horizontal, so we won't worry about vertical forces. We can neglect the external force of friction for the very short time it takes for Nellie to jump into the racer.

The total momentum before she jumps in is

$$p = (41 \text{ kg})(4.4 \text{ m/s}) + (m)(0) = 180.4 \text{ kg·m/s}$$

where m is Nellie's mass. Since the total momentum of the system must be the same after the "collision,"

$$p = 180.4 \text{ kg·m/s} = (41 \text{ kg} + m)(2.6 \text{ m/s})$$

so

$$m = \frac{180.4 \text{ kg·m/s}}{2.6 \text{ m/s}} - 41 \text{ kg} = \underline{28.4 \text{ kg}}$$

4.   (a) $\underline{7.39 \text{ m/s}}$ (b) $\underline{4.43 \times 10^4 \text{ N}}$

5.   This is a simple collision between two objects (the earth and the meteorite) in which momentum is conserved. In a frame of reference in which the earth is at rest before the collision, the total momentum before and after the collision must be the same:

$$m_1 v_1 + m_2 v_2 = (m_1 + m_2)(v')$$

$$(10^5 \text{ kg})(10^3 \text{ m/s}) + (5.98 \times 10^{24} \text{ kg})(0) = (5.98 \times 10^{24} \text{ kg})(v')$$

giving

$$v' = \underline{1.67 \times 10^{-17} \text{ m/s}}$$

This is half a millimetre per million years, which should make it clear why, in practice, we neglect the earth's recoil.

6.   The astronaut should throw the empty gun directly away from the ship at a speed greater than 29.0 m/s, the speed that would leave him at rest relative to the ship. This is about 65 mi/h—he needs a good arm!

7.   (a) The individual momenta of the three skaters are

$$\mathbf{P}_C = (35 \text{ kg})(9 \text{ m/s, E}) = 315 \text{ kg·m/s, E}$$

$$\mathbf{P}_T = (35 \text{ kg})(8 \text{ m/s, W}) = 280 \text{ kg·m/s, W}$$

$$= -280 \text{ kg·m/s, E}$$

$$\mathbf{P}_A = (35 \text{ kg})(5 \text{ m/s, N}) = 175 \text{ kg·m/s, N}$$

Their total momentum is thus the vector sum of 315 kg·m/s, E − 280 kg·m/s, E = 35 kg·m/s, E and 175 kg·m/s, N. (See the vector triangle in Figure 7-7.) This is

$$p = \sqrt{p_x^2 + p_y^2} = \sqrt{(35 \text{ kg·m/s})^2 + (175 \text{ kg·m/s})^2}$$

$$= 178.5 \text{ kg·m/s}$$

$$\tan \theta = \frac{35 \text{ kg·m/s}}{175 \text{ kg·m/s}} = 0.20$$

$$\theta = 11°$$

Their velocity after the collision is $(178.5 \text{ kg·m/s})/(4)(35 \text{ kg}) = 1.28$ m/s directed 11° east of north.

(b) The kinetic energy of the four children after the collision is

$$E_k = \tfrac{1}{2}mv^2 = \tfrac{1}{2}(140 \text{ kg})(1.28 \text{ m/s})^2 = 113.8 \text{ J}$$

Their total energy before the collision was

$$\tfrac{1}{2}(35 \text{ kg})[(9 \text{ m/s})^2 + (8 \text{ m/s})^2 + (5 \text{ m/s})^2] = 2975 \text{ J}$$

The fraction of the initial kinetic energy dissipated in the collision is

$$\frac{2975 \text{ J} - 114 \text{ J}}{2975 \text{ J}} = \underline{0.962} \text{ or } \underline{96.2\%}$$

Figure 7-7

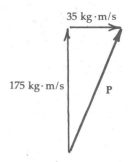

8. <u>48.0 km/h, 35° E of N</u>

9. (a) Take $x = 0$ at the end where three hoboes are sitting. If the flatcar is uniform, its mass acts effectively at its center, $x = 27.5$ ft. Approximate the hoboes as point masses. The position of the center of mass is given by

$$x_{cm} = \frac{m_1 x_1 + m_2 x_2 + \cdots}{m_1 + m_2 + \cdots}$$

Here the weights rather than masses are given, but this makes no difference as weight and mass are proportional. Thus,

$$x_{cm} = \frac{(3)(150 \text{ lb})(0) + (4200 \text{ lb})(27.5 \text{ ft}) + (150 \text{ lb})(55 \text{ ft})}{(4)(150 \text{ lb}) + 4200 \text{ lb}}$$

$$= \underline{25.8 \text{ ft}}$$

(b) When the hoboes are divided two and two, the position of the center of mass becomes

$$x_{cm} = \frac{(2)(150 \text{ lb})(0) + (4200 \text{ lb})(27.5 \text{ ft}) + (2)(150 \text{ lb})(55 \text{ ft})}{(4)(150 \text{ lb}) + 4200 \text{ lb}}$$

$$= \underline{27.5 \text{ ft}}$$

We could have seen this directly from symmetry. However, if the track is frictionless, the center of mass, not the flatcar, remains at rest. Thus, the car has moved 27.5 ft − 25.8 ft = 1.7 ft.

10. (a) 19.4 m/s   (b) 19.4 m/s   (c) 18.9 m/s

11. The bird "collides inelastically" with the birdhouse at a speed v_1. By the conservation of momentum

$$m_1v_1 + m_2v_2 = (m_1 + m_2)v'$$

$$(50 \text{ g})(v_1) + (230 \text{ g})(0) = (50 \text{ g} + 230 \text{ g})(v')$$

so

$$v' = \frac{50}{280} v_1 = 0.1786v_1$$

The kinetic energy of the system just after the wren lands is

$$E_k = \tfrac{1}{2}(m_1 + m_2)(v')^2$$

$$= \tfrac{1}{2}(0.280 \text{ kg})(0.1786v_1)^2 = (0.00446 \text{ kg})v_1^2$$

As the birdhouse with the wren swings up after the collision, this kinetic energy is turned into gravitational potential energy. It moves upward against gravity a distance (see Figure 7-8)

$$y = (L)(1 - \cos \theta)$$

The potential energy increase is thus

$$U = mgy = mgL(1 - \cos \theta)$$

$$= (0.280 \text{ kg})(9.81 \text{ m/s})(0.40 \text{ m})(1 - \cos 16°) = 0.0426 \text{ J}$$

Since the potential energy at the end of the swing is equal to the kinetic energy at the beginning,

$$(0.00446 \text{ kg})v_1^2 = 0.0426 \text{ J}$$

$$v_1^2 = 9.54 \text{ (m/s)}^2$$

$$v_1 = 3.08 \text{ m/s}$$

Figure 7-8

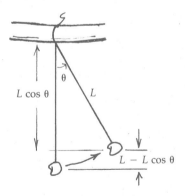

12. (a) 5.1 m/s   (b) 696 J, or about one third of the running back's initial kinetic energy

13. Before the collision,

$$v_r = v_b - v_c = 30 \text{ m/s} - (-15 \text{ m/s}) = 45 \text{ m/s}$$

taking the ball's initial direction as positive. Since this collision is elastic, the relative velocity of the ball and the car simply reverses in the collision. After the collision, therefore,

$$v'_r = v'_b - v'_c = -45 \text{ m/s}$$

The car's mass is very large compared to that of the ball, so its velocity is essentially unchanged by the collision. Thus, $v'_r \approx -45$ m/s, so

$$v'_b = v'_r + v'_c = (-45 \text{ m/s}) + (-15 \text{ m/s}) = -60 \text{ m/s}$$

The ball bounces back toward the kid at a speed of 60 m/s.

14. After the collision, both pendulums swing in the same direction, the heavier to an angle of 7.4° and the lighter to 37.8 °.

15. The momentum of the system (the man and the bowling ball) is unchanged in the collision. Thus if v is the man's speed just after the ball strikes him,

$$(7.25 \text{ kg})(6.5 \text{ m/s}) + 0 = (7.25 \text{ kg})(-2 \text{ m/s}) + (72.5 \text{ kg})(v)$$

(The minus sign is used because the ball's velocity has changed direction.) So

$$v = \frac{61.6 \text{ kg·m/s}}{72.5 \text{ kg}} = 0.850 \text{ m/s}$$

The relative speed of ball and man just before the ball hits is 6.5 m/s. After the collision it is

$$0.850 \text{ m/s} - (-2 \text{ m/s}) = 2.850 \text{ m/s}$$

The coefficient of restitution is the ratio of the relative speed after the collision to the relative speed before,

$$e = \frac{2.85 \text{ m/s}}{6.5 \text{ m/s}} = 0.438$$

CHAPTER 8

Rotation

I. Key Ideas

Rotation of a Rigid Body about a Fixed Axis This is the fundamental type of rotational motion problem that we will deal with. If the rotating body is rigid, then although different points in the body move at different speeds, *every point turns through the same angle* in a given time as it rotates. Thus, we can describe rotational motion in terms of angular displacements.

Angular Velocity and Acceleration We define the kinematic quantities in rotational motion very much as we did those in linear motion. Thus, the angle, in radian measure, through which the rotating body turns per unit time is its *angular velocity*. The angular velocity unit is the radian per second; this is the same as s^{-1} since the radian is a dimensionless quantity. Likewise, the rate at which the angular velocity changes in time is the *angular acceleration*.

Relation to Linear Quantities Any point in a rotating rigid body is moving in a circle around the axis of rotation. The *linear speed* of this circular motion is the distance of the point from the rotation axis multiplied by the angular velocity. Likewise, multiplying the angular acceleration by the distance of a point from the axis gives the *tangential component*—that is, the component along the direction of motion—of the point's linear acceleration. The tangential component is never the whole acceleration of the point, however; since the point moves in a circle, there is always a centripetal acceleration directed radially inward toward the axis.

Analogies to Straight-Line Motion The relationships among the definitions of angular displacement, angular velocity, and angular acceleration are exactly the same as those for the corresponding linear quantities. Therefore, many of the equations for linear motion are analogous to equations for rotational motion. The set of equations for uniformly accelerated motion, for example, correspond to equations for rotation with uniform angular acceleration.

Rotational Dynamics Newton's second law gives the fundamental equation of dynamics. However, for the rotation of a rigid body about a fixed axis, it is not force but *torque* that determines the change in the body's rotational motion (that is, in its angular acceleration). The torque of a force is the force multiplied by the perpendicular distance from its line of action to the rotation axis. The inertial quantity for rotation, which is a measure of a body's resistance to an angular acceleration, is its *moment of inertia*. The moment of inertia is a property determined by the distribution of the body's mass about the rotation axis. Formulas for the moments of inertia of various simple shapes are found in Table 8-1 in the text.

Energy and Angular Momentum Rotational kinetic energy and power are defined in terms of the moment of inertia and the kinematic quantities, just as for linear motion. For a rotating rigid body, the *angular momentum* is the moment of inertia times the angular velocity. The rotational version of Newton's second law states that the net torque is equal to the rate at which the angular momentum is changing.

Conservation of Angular Momentum It follows from Newton's second law that, if there is zero net external torque on a system, then its angular momentum is constant. This *law of conservation of angular momentum*, like the other conservation laws, is more general than the systems that we use to define it. It is valid even on the subatomic scale, where newtonian mechanics is not.

Rolling Rolling is a combination of rotational and translational motions. If a body of circular outline rolls without slipping along a surface, the contact point on the surface and the center of the body move the same distance. If the rolling is without slipping, the rolling body is instantaneously at rest at the point of contact, usually due to static friction.

Vector Nature of Rotation In general, all the rotational quantities we have discussed are vector quantities. The direction of the angular velocity, for example, is along the rotation axis in a "right-handed" sense. Rotational motion in three dimensions is very much more difficult to analyze; in fact, many of the equations we have derived do not apply. The *gyroscope* illustrates some vector properties. A spinning gyroscope does not move in the direction that the external force on it would seem to indicate; instead, it moves in the direction such that its change in angular momentum is parallel to the torque.

II. Key Equations

Key Equations

Rotational kinematic quantities

$$\Delta\theta = \frac{\Delta s}{r} \qquad \omega = \frac{\Delta\theta}{\Delta t} \qquad \alpha = \frac{\Delta\omega}{\Delta t}$$

Relation of rotational to linear quantities

$$v = r\omega \qquad a_t = r\alpha \qquad a_c = v^2/r = r\omega^2$$

Constant angular acceleration

$$\omega = \omega_0 + \alpha t$$

$$\theta = \theta_0 + \omega_0 t + \tfrac{1}{2}\alpha t^2$$

$$\omega^2 = \omega_0^2 + 2\alpha(\theta - \theta_0)$$

Moment of inertia

$$I = \Sigma m_i r_i^2$$

Newton's second law (rotational version)

$$\tau_{net} = I\alpha$$

Energy and power

$$E = \tfrac{1}{2}I\omega^2 \qquad P = \tau\omega$$

Angular momentum

$$L = I\omega \qquad \tau_{net} = \frac{\Delta L}{\Delta t}$$

Particle in circular motion

$$L = mvr$$

Conservation of angular momentum

$$L = \text{constant}$$

Rolling without slipping

$$v_{cm} = R\omega \qquad a_{cm} = R\alpha$$

III. Possible Pitfalls

Remember that angular quantities must be in terms of *radian* measure. A number of the equations derived in this chapter—specifically those that relate angular to linear quantities—are valid *only* if the angles are measured in radians. It is a good habit always to convert degrees, revolutions per minute, and the like into radian units before you work a problem.

Remember also that quantities such as torque, angular momentum, and angular acceleration are defined in terms of a specific axis of rotation. The rotational motion equations relate only quantities that are defined about the same axis (or center) of rotation.

Some of the equations we have seen in this chapter are valid *only* for the special case of rotation about a fixed axis. This is the "one-dimensional case" of rotational motion. The equation $L = I\omega$, for example, is not true in general for a rotation in three dimensions.

For a rotating rigid body, $r\omega$ is the linear velocity of a point at distance r from the axis. Don't fall into the trap of thinking that αr is the acceleration of the point. This is only one component (the tangential component) of the acceleration vector; there is also always a centripetal component, directed radially inward.

A centripetal force is always present on something that is rotating; it is the force required to turn its velocity vector. But the centripetal force is equal to the net force only if the rotation is at constant speed. Otherwise, the force also has a tangential component (that is, a component along the direction of motion).

What determines the torque is not how far it is from the point at which the force acts to the rotation axis. It is, rather, how far the axis is from the *line* along which the force acts. If the line of action of the force goes through the axis, then no matter where the force is applied, its torque is zero.

IV.   Questions and Answers

Questions

1.   Consider two points on a disk rotating (not necessarily at constant anything) about its axis, one at the rim and the other halfway from the rim to the center. Which point has the greater (*a*) angular acceleration, (*b*) tangential acceleration, (*c*) radial acceleration, and (*d*) centripetal acceleration?

2.   Can there be more than one value for the moment of inertia for a given rigid body?

3.   A figure skater starts a spin with her arms outstretched. As she draws them in toward her body, her rate of rotation increases. Why? What force is it that accelerates her?

4.   A cat's ability to land on its feet is legendary. But how does the cat do it? Suppose you hold a cat feet upward (wearing gloves is highly recommended) and let her fall. If you dropped her from rest, she has no rotation about her axis, and while she falls there is no external torque on her. How can she turn over?

5.   In Figure 8-1, a man is hanging from one side of a Ferris wheel, which swings him down toward the ground. Can you use constant-angular-acceleration formulas to calculate the time it takes him to reach the bottom?

6.   Does applying a positive net torque to a body necessarily increase the body's kinetic energy?

Figure 8-1

7. If the main rotor blade of the helicopter sketched in Figure 8-2 is rotating counterclockwise as seen from above, which way should the smaller rotor be pushing the tail of the helicopter? Why?

8. A ball, a disk, and a hoop, all with the same mass and the same radius, are set rolling up an incline, all with the same initial speed. Which goes farthest? If they are all set rolling with the same initial kinetic energy instead, which goes farthest?

9. Figure 8-3 represents a yo-yo sitting on edge on a rough surface. If its string is pulled as indicated, does the yo-yo roll to the left or to the right? Does your answer depend on the direction in which the string is pulled?

10. If polar ice caps melted tomorrow, what would be the effect on the earth's rotation?

11. The propeller of a light airplane is rotating clockwise, as seen from the rear. If the pilot pulls up out of a dive without making any attempt to control his left-right heading, in which direction will the plane tend to turn.

Figure 8-2

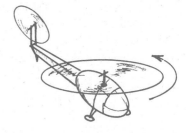

Figure 8-3

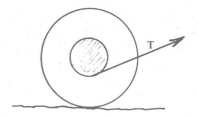

Answers

1. (*a*) Both points have the same angular acceleration. This is the point of describing rotational motion in terms of angular quantities. (*b*) The tangential acceleration of the point at the rim is larger because it is farther from the axis. (*c*)(*d*) The radial acceleration and the centripetal acceleration (they are the same thing) of the point at the rim are larger for the same reason.

2. In general, the moment of inertia is different about every different axis of rotation.

3. The easiest way to explain why the skater speeds up is by the conservation of angular momentum ($L = I\omega$). As she draws in her arms, she is decreasing her I so her ω must be increasing if L is to stay the same. What it is that accelerates her body is less obvious. As she draws in her arms, their velocity is decreased ($v = \omega r$); thus, her body is exerting a force on her arms that tends to produce a negative tangential acceleration of them. By Newton's third law, her arms must then be exerting a force on her body such as to produce a positive tangential acceleration, that is, to increase its speed.

4. What the cat does is twist the front and rear halves of her body in opposite directions. Her angular momentum remains zero, but if she stretches out her hind legs and draws in her front, then her front half turns farther one way than her rear does in the other. She can then reverse the procedure to bring her rear around. Believe it or not, that's more or less what cats do—although where they learn the physics I really can't say.

5. No, because the gravitational torque on the man—and so his angular acceleration also—decreases as he swings down toward the bottom.

6. Not necessarily. It will increase the kinetic energy—and therefore the angular speed—if the work it does is positive; that is, if the torque is in the same sense as the body's rotation. A torque in the opposite sense reduces the body's kinetic energy.

7. The angular momentum vector of the main rotor is directed upward in the figure. The drag force of the air on the main rotor blade produces an external torque directed downward, since its tendency would be to slow the rotor down. If not counteracted by the tail rotor, this torque would cause the helicopter body to rotate clockwise (as seen from above). Thus, the small tail rotor has to push the tail of the helicopter to the left, or counterclockwise.

8. The moments of inertia are

   ball: $\frac{2}{5}MR^2$ disk: $\frac{1}{2}MR^2$ hoop: MR^2

From an example on page 188 of the text, we have for this sort of situation

$$v^2 = \frac{2mgh}{m + I/R^2}$$

where h is the vertical height to which the object will roll. Thus,

$$h = \frac{(m + I/R^2)(v^2)}{2mg} = \frac{E_k}{mg}$$

From this, clearly the one with the largest moment of inertia (here, the hoop) will roll to the greatest height if they all have the same initial velocity v. Of course, if all three start with the same kinetic energy, they will all roll to the same height.

Figure 8-4

9. This is a confusing question only if you let your eye be drawn to the center because the relative sizes of the string force and the frictional force aren't obvious. Instead, consider $\tau = I\alpha$ about point P in Figure 8-4. Plainly, the tension in the string is the only force that exerts a torque about P, so the yo-yo will start rolling in the

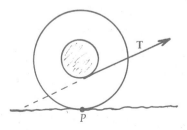

Figure 8-5

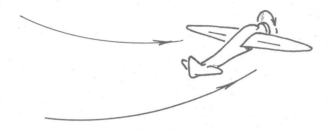

direction of that torque. In the diagram shown here, it will roll to the right, in the direction of the pull. But if you pull at a steep enough angle, the torque of the string will be in the other direction, and the yo-yo will roll backward.

10. If all the water now frozen at the earth's poles distributed itself over the earth's oceans, the earth's moment of inertia would increase, and by conservation of angular momentum, its rotation rate would have to decrease.

11. What this amounts to is a gyroscope question. To pull upward, the pilot changes the configuration of the wing and tail flaps in such a way as to make the air exert an external torque to turn his flight path upward—that is, a torque directed to his right. The angular momentum vector of the airplane is mostly that of the propeller; it is directed forward and thus is changed ($\tau = \Delta L / \Delta t$)—that is, turned—to the right, so the plane will tend to yaw to the right (see Figure 8-5).

V. Problems and Solutions

Problems

In several of these problems, you will need the formulas for the moments of inertia of simple shapes that appear in Table 8-1 on page 179 in the text. Consider a disk to be a solid cylinder.

1.   A bicyclist starts from rest and pedals in such a way that the wheels have constant angular acceleration of 2.8 rad/s². The wheels are 0.75 m in diameter. (*a*) How many revolutions have the wheels made after 15 s? (*b*) What is their angular velocity at that point? (*c*) How far has the cyclist traveled at that point?

How to Solve It
- Use the equation for the angular displacement versus time for a body with uniform angular acceleration.

- Another of the constant-angular-acceleration formulas can be used to find the angular velocity of the wheels.

- If the wheels are not slipping, the distance the cyclist has gone is the same as the distance through which a point on the rim of the wheel has turned.

2. A phonograph turntable accelerates from rest to 33.3 rev/min in exactly 0.5 rev. (*a*) What is its angular acceleration? (*b*) How much time did it take to come up to speed?

How to Solve It
- The data given do not include the time, so for part (*a*) use the uniform-acceleration equation that does not involve the time.

- Once you have the angular acceleration, you can calculate how much time the turntable took to come up to speed.

- Angular acceleration is the change in angular velocity per unit time.

3. A phonograph turntable of mass 1.8 kg and radius 15 cm, rotating initially at 33.3 rev/min, is being braked to a stop. After 5 s, it has lost half of its initial angular velocity. Assuming constant angular acceleration, (*a*) how long altogether does it take to come to rest? (*b*) Through how many revolutions does it turn while coming to rest? (*c*) How much work has to be done on the turntable to stop it? Approximate the turntable as a uniform disk.

How to Solve It
- From the time it takes the turntable to lose half its initial velocity, you can find its angular acceleration. From this, in turn, find the total time it takes to stop.

- Another formula gives you the angle through which it has turned.

- Since you can calculate the initial kinetic energy of the turntable, you can use the work-energy theorem for part (*c*).

4. A girl of mass 55 kg is riding a Ferris wheel that is turning at constant speed. At the topmost point, her apparent weight is 400 N. (*a*) What is her apparent weight when the wheel is at the bottom? (*b*) If the wheel's diameter is 9 m, at what rate is the wheel turning?

How to Solve It

- The girl's apparent weight is the upward force of the seat that supports her. Show all the forces that act on her on a free-body diagram.

- Use Newton's second law to calculate her centripetal acceleration. How does the free-body diagram change when she is at the bottom of the wheel?

- Use the radius of the wheel and the girl's known centripetal acceleration to find the wheel's angular velocity.

5. A penny rests on a turntable as in Problem 3 above, at a point 10 cm from the axis. The coefficient of friction between the penny and the turntable is 0.35. If the turntable starts from rest with a uniform angular acceleration of 1.0 rad/s², how much time passes before the penny starts to slip?

How to Solve It

- There is a limit to the force that can be provided by static friction, but this is the force that accelerates the coin.

- Calculate the total acceleration of the coin—it has two components—as a function of time.

- When this acceleration requires a force that exceeds the maximum that can be supplied by static friction, the coin will start to slip.

6. In Figure 8-6, consider the pulley to be a uniform disk of mass 1.4 kg and radius 6 cm. The two blocks have masses of 1.6 and 2.4 kg. Assuming that there is no friction in the pulley axle and that the string does not slip or stretch, find the acceleration of either block.

How to Solve It

- Draw three free-body diagrams: one for each of the two blocks and one for the pulley. Note that the tensions in the two strings *are not* the same.

- Now it may seem that you don't have enough equations to solve for all the quantities that you don't know. But how is the angular acceleration of the pulley related to the linear acceleration of the blocks?

- Eliminating the tensions in the string gives you the acceleration of the blocks.

Figure 8-6

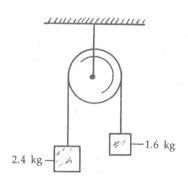

2.4 kg

1.6 kg

7.   Tarzan swings down from a tree branch to kick a bad guy in the head. For the purpose of calculating Tarzan's moment of inertia, consider him to be a uniform thin rod 1.90 m long with a mass of 91 kg. When he starts swinging, his body makes an angle of 40° with the vertical (see Figure 8-7). At what speed are Tarzan's feet moving when they encounter the baddie's head?

Figure 8-7

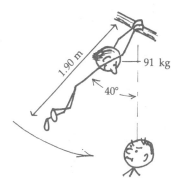

How to Solve It
- You can't apply the conservation of angular momentum here! Gravity exerts an external torque on Tarzan.

- It's not a case of constant angular acceleration, either; the torque decreases as he swings down.

- Apply the conservation of mechanical energy. Tarzan's center of gravity moves downward, so his potential energy turns into kinetic energy.

8.   A rotating grindstone (take it to be a uniform disk) of mass 50 kg and radius 22 cm is brought to rest by the friction of a knife pressed against its edge. If the knife exerts a tangential force of 25 N on the grindstone, and the grindstone makes one and a half complete revolutions in coming to rest, how fast was it turning initially?

How to Solve It
- The most direct way to solve this is to use the work-energy theorem. How much work does the force of the knife do on the grindstone?

- This work is equal to the change in the grindstone's kinetic energy. The grindstone in this case loses all its original kinetic energy.

9.   Consider the sun as a uniform solid sphere having a radius of 1.39×10^9 km. Its mass is 1.99×10^{30} kg, and it rotates on its axis with a period of 26 days. If it continues to rotate at this rate to the end of its natural life and then collapses into a neutron star with a radius of 16 km, what will be its period of rotation then?

How to Solve It
- Calculate the sun's moment of inertia.

- The sun is pretty much an isolated system, so its angular momentum will not change when it collapses.

- Calculate the sun's angular momentum, and then calculate how fast a 16-km sphere with this much angular momentum must be rotating.

10. Two disks are mounted side by side on a common axis but are free to move individually. Initially, one of them (moment of inertia 1.2 kg·m^2) is rotating at 900 rev/min and the other (moment of inertia 2.8 kg·m^2) is at rest. The two disks are pressed together, and they eventually come to rotate at the same angular velocity. Find the final angular velocity of the two disks.

How to Solve It
- If the disks rotate freely on the axis, presumably any torque exerted on them at the axis is negligibly small.

- Thus, the angular momentum of the two-disk system is conserved.

- You know the initial angular momentum; the two disks rotating together afterward have this same angular momentum.

11. A merry-go-round consists of a circular platform mounted on a frictionless vertical axle. Its radius is 2.4 m and its moment of inertia is 210 kg·m^2. Initially, a boy whose mass is 38 kg is standing at the edge of the merry-go-round, which is rotating at an angular velocity of 2.40 rad/s. When the boy begins to run around the edge of the platform at constant speed, the angular velocity of the platform decreases to 1.91 rad/s. How fast is the boy moving relative to the platform?

How to Solve It
- Calculate the total angular momentum of the system (the boy and the platform) initially.

- Apply the conservation of angular momentum; the total angular momentum remains at this value. Calculate the angular momentum of the platform at reduced speed.

- Whatever is left must be the angular momentum of the boy's motion about the axis; from this, you can calculate his speed relative to the ground.

- Calculate his speed relative to the platform from this.

12. A child playing with a 10-kg tire gives it a stiff push to set it rolling. The tire rolls without slipping up a hill 7 m high, just barely reaching the top. Approximate the tire as a hollow cylindrical shell 87 cm in diameter. How fast did the child set it rolling?

How to Solve It
- Calculate the moment of inertia of the tire.

- The tire rolling up the hill gains gravitational potential energy.

Since it rolls without slipping, mechanical energy is conserved; the potential energy at the top must equal its initial kinetic energy.

• Knowing its kinetic energy, you can calculate its speed, but remember that it has rotational as well as translational kinetic energy.

13. A bowling ball (mass 7.5 kg, diameter 20 cm) is set spinning about a horizontal axis at 120 rev/min. In this state, the ball is set down on the floor. It is not given any initial velocity across the floor. The coefficients of static and kinetic friction between the ball and the floor are, respectively, 0.60 and 0.42. How much time passes before the ball begins rolling without slipping?

How to Solve It
• It takes a little thought to see what will happen here. The ball will start sliding in the direction of the spin because of the (external) force of kinetic friction on it.

• Find the linear acceleration of the ball due to the force of kinetic friction.

• At the same time, its spin is decreasing because of the frictional torque on it. Find the angular acceleration of the ball due to the frictional torque on it.

• When the ball's surface comes instantaneously to rest with respect to the floor, it will stop sliding and start rolling.

Solutions

1. (a) Since the cyclist starts from rest, the angle through which a wheel has turned is given by

$$\theta = \theta_0 + \omega_0 t + \tfrac{1}{2}\alpha t^2 = \tfrac{1}{2}\alpha t^2$$

$$= \tfrac{1}{2}(2.8 \text{ rad/s}^2)(15 \text{ s})^2 = \underline{315 \text{ rad}}$$

This is $315/2\pi = 50.1$ complete revolutions.
    (b) Since she starts from rest, her angular velocity after 15 s is

$$\omega = \omega_0 + \alpha t = 0 + (2.8 \text{ rad/s}^2)(15 \text{ s}) = \underline{42 \text{ rad/s}}$$

    (c) The radius of the bicycle wheel is 0.375 m, so after this time she has gone a distance

$$\Delta x = r\theta = (0.375 \text{ m})(315 \text{ rad}) = \underline{118 \text{ m}}$$

Notice that the radians have disappeared from the calculation. A radian is a dimensionless quantity, not really a unit.

2.   (a) 1.94 rad/s   (b) 1.80 s

3.   (a) The initial angular velocity is

$$(33.3 \text{ rev/min}) \frac{2\pi \text{ rad/rev}}{60 \text{ s/min}} = 3.49 \text{ rad/s}$$

Half of this is lost in a time of $t_1 = 5$ s. Thus,

$$\omega_1 = \omega_0 + \alpha t_1$$

so

$$\alpha = \frac{\omega_1 - \omega_0}{t_1} = \frac{1.74 \text{ rad/s} - 3.49 \text{ rad/s}}{5 \text{ s}} = -0.349 \text{ rad/s}^2$$

At this rate, to come all the way to rest will take time

$$t = \frac{\omega - \omega_0}{\alpha} = \frac{0 - 3.49 \text{ rad/s}}{-0.349 \text{ rad/s}^2} = \underline{10 \text{ s}}$$

(b) The angle through which the turntable has turned is

$$\theta = \omega_0 t + \tfrac{1}{2}\alpha t^2 = (3.49 \text{ rad/s})(10 \text{ s}) + \tfrac{1}{2}(-0.349 \text{ rad/s}^2)(10 \text{ s})^2$$

$$= \underline{17.5 \text{ rad}}$$

(c) If we approximate the turntable as a disk, its moment of inertia is

$$I = \tfrac{1}{2}MR^2 = \tfrac{1}{2}(1.8 \text{ kg})(0.15 \text{ m})^2 = 0.0202 \text{ kg·m}^2$$

Its initial kinetic energy was therefore

$$E_k = \tfrac{1}{2}I\omega_0^2 = \tfrac{1}{2}(0.0202 \text{ kg·m}^2)(3.49 \text{ rad/s})^2 = 0.123 \text{ J}$$

In coming to rest, it lost all this kinetic energy. Thus, by the work-energy theorem, the work done was equal to the change in kinetic energy, or -0.123 J.

4.   (a) 679 N   (b) 0.75 rad/s

5.   The angular acceleration α of the turntable is given. At time t, the angular velocity is

$$\omega = \alpha t$$

The *maximum* force that can be exerted by static friction is

$$f_{max} = \mu_s mg$$

The maximum (total) acceleration it can provide is thus

$$a = \mu_s g = (0.35)(9.81 \text{ m/s}^2) = 3.43 \text{ m/s}^2$$

This is the vector sum of two components, the tangential and centripetal accelerations. The tangential acceleration is

$$a_t = \alpha r = (1.0 \text{ rad/s})(0.1 \text{ m}) = 0.10 \text{ m/s}^2$$

Since

$$a = \sqrt{a_t^2 + a_c^2}$$

the centripetal acceleration is

$$a_c = \sqrt{a^2 - a_t^2} = \sqrt{3.43^2 - 0.1^2} = 3.43 \text{ m/s}^2$$

But $a_c = r\omega^2$, so

$$\omega^2 = \frac{a_c}{r} = \frac{3.43 \text{ m/s}^2}{0.1 \text{ m}} = 34.3 \text{ rad/s}^2$$

and

$$\omega = \underline{5.86 \text{ rad/s}}$$

6. $\underline{1.67 \text{ m/s}^2}$

7. Replacing the ape man with a uniform thin rod may seem like a lousy approximation, but we have to make some such approximation here or we can't do the problem. Always look for simplifications that give you an approximate idea of what happens; real problems are almost always too hard to solve! Tarzan's center of gravity is about 0.95 m from the pivot (see Figure 8-8), so it moves

Figure 8-8

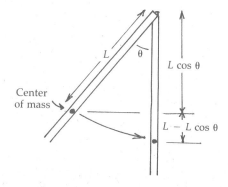

downward a distance

$$\Delta y = L(1 - \cos \theta) = (0.95 \text{ m})(1 - 0.7660) = 0.222 \text{ m}$$

The change in his potential energy is

$$\Delta U = Mg \, \Delta y = (91 \text{ kg})(9.81 \text{ m/s}^2)(-0.222 \text{ m}) = -198 \text{ J}$$

His kinetic energy has increased by the same amount when he reaches the bottom of his swing. The moment of inertia of a uniform thin rod about one end is

$$I = \tfrac{1}{3}ML^2 = \tfrac{1}{3}(91 \text{ kg})(1.90 \text{ m})^2 = 110 \text{ kg·m}^2$$

Thus at the bottom of his swing,

$$E_k = \tfrac{1}{2}I\omega^2$$

$$198 \text{ J} = \tfrac{1}{2}I\omega^2 = (55 \text{ kg·m}^2)(\omega^2)$$

so

$$\omega^2 = \frac{198 \text{ J}}{55 \text{ kg·m}^2} = 3.60 \text{ rad}^2/\text{s}^2$$

and

$$\omega = 1.90 \text{ rad/s}$$

His feet are therefore moving at $v = \omega r = (1.90 \text{ rad/s})(0.9 \text{ m}) = $ 1.71 m/s.

8. 9.26 rad/s

9. The moment of inertia of a uniform solid sphere is

$$I = \tfrac{2}{5}MR^2 = \tfrac{2}{5}(1.99 \times 10^{30} \text{ kg})(1.39 \times 10^9 \text{ m})^2$$

$$= 1.54 \times 10^{48} \text{ kg·m}^2$$

The angular velocity of rotation of the sun currently is

$$\omega = \frac{2\pi}{T} = \frac{2\pi}{(26 \text{ d})(24 \text{ h/d})(60 \text{ min/h})(60 \text{ s/min})}$$

$$= 2.80 \times 10^{-6} \text{ rad/s}$$

so

$$L = I\omega = (1.54 \times 10^{48} \text{ kg·m}^2)(2.80 \times 10^{-6} \text{ rad/s})$$

$$= 4.31 \times 10^{42} \text{ kg·m}^2/\text{s}$$

The moment of inertia of a sphere with the mass of the sun and a radius of 16 km would be 2.04×10^{38} kg·m², so

$$\omega_{final} = \frac{4.31 \times 10^{42} \text{ kg·m}^2/\text{s}}{2.04 \times 10^{38} \text{ kg·m}^2} = 2.12 \times 10^4 \text{ rad/s}$$

Its period would be

$$T = \frac{2\pi}{\omega} = \frac{2\pi}{2.12 \times 10^4 \text{ rad/s}} = \underline{0.000296 \text{ s}}$$

10. $\underline{28.3 \text{ rad/s}}$ or $\underline{270 \text{ rev/min}}$

11. The total angular momentum of the system is that of the platform plus that of the boy. Initially, this is

$$L = (I_{platform} + I_{boy})\omega$$

where

$$I_{boy} = mr^2 = (38 \text{ kg})(2.4 \text{ m})^2 = 219 \text{ kg·m}^2$$

so

$$L = (210 \text{ kg·m}^2 + 219 \text{ kg·m}^2)(2.4 \text{ rad/s}) = 1030 \text{ kg·m}^2/\text{s}$$

When it rotates at the lower speed, the angular momentum of the platform is

$$L' = (210 \text{ kg·m}^2)(1.91 \text{ m/s}) = 401 \text{ kg·m}^2/\text{s}$$

The difference of 629 kg·m²/s is the boy's angular momentum. Thus, his angular velocity is

$$\frac{629 \text{ kg·m}^2/\text{s}}{219 \text{ kg·m}^2} = 2.87 \text{ rad/s}$$

The velocity, relative to the ground, at which he is moving around in a circle is

$$v_b = r\omega = (2.4 \text{ m})(2.87 \text{ rad/s}) = 6.89 \text{ m/s}$$

To get his velocity relative to the platform, we must subtract its velocity, which is

$$v_p = (2.4 \text{ m})(1.91 \text{ rad/s}) = 4.58 \text{ m/s}$$

Thus, relative to the platform the boy is running at 6.89 m/s − 4.58 m/s = $\underline{2.31 \text{ m/s}}$.

12. $\underline{8.3 \text{ m/s}}$

13. The only external (horizontal) force that acts on the ball is that of friction with the floor. This force is due to kinetic friction:

$$f = \mu_k M g = M a_{cm}$$

so

$$a_{cm} = \mu_k g = (0.42)(9.81 \text{ m/s}^2) = 4.12 \text{ m/s}^2$$

Since the ball starts from rest, $v_{cm} = \mu_k g t$. The frictional torque about the center of the ball is equal to the rate at which its angular momentum about the center changes:

$$\tau = -fR = -\mu_k M g R$$

$$= (0.42)(7.5 \text{ kg})(9.81 \text{ m/s}^2)(0.1 \text{ m}) = -3.09 \text{ N·m}$$

$$= \frac{\Delta L}{\Delta t} = I \frac{\Delta \omega}{\Delta t} = I \alpha$$

The rotational inertia of the ball is

$$I = \tfrac{2}{5} M R^2 = (0.4)(7.5 \text{ kg})(0.1 \text{ m})^2 = 0.0300 \text{ kg·m}^2$$

so

$$\alpha = \frac{\tau}{I} = \frac{-3.09 \text{ N·m}^2}{0.0300 \text{ kg·m}^2} = -103 \text{ rad/s}^2$$

The *initial* angular velocity of the ball is

$$\omega_0 = 120 \text{ rev/min} = 12.57 \text{ rad/s}$$

and

$$\omega = \omega_0 + \alpha t = 12.57 \text{ rad/s} + (-103 \text{ rad/s}^2)(t)$$

Rolling without slipping requires that

$$v_{cm} = a_{cm} t = \omega R$$

$$= (4.12 \text{ m/s}^2)(t) = [12.57 \text{ rad/s} - (103 \text{ rad/s}^2)(t)][(0.1 \text{ m})]$$

Solving this for t gives $t = \underline{0.087 \text{ s}}$.

CHAPTER 9

Gravity

I. Key Ideas

Kepler's Laws Observations of the positions of the planets and the sun have been made for thousands of years. Around 1600, Johannes Kepler derived, from many observations, three general, empirical statements about planetary motion: (1) Planetary orbits are ellipses with the sun at one focus. (2) The planets move such that an imaginary line joining a planet and the sun sweeps through *equal areas* in *equal times*. (3) The *square* of the period of each planet is proportional to the *cube* of its mean distance from the sun.

Ellipse The *ellipse* is a mathematical figure defined such that the sum of the distances from each point on the ellipse to two fixed points is the same. The two fixed points are called the *focal points* of the ellipse. If the two focal points coincide, the figure is a circle; the planetary orbits are nearly circular ellipses.

Newton's Law of Gravitation Isaac Newton was the first to propose a force law from which Kepler's empirical rules could be derived. Newton's *law of universal gravitation* states that every body in the universe attracts every other body with a force that is directly proportional to the two masses and inversely proportional to the square

of the distance between them. Newton showed that such a force predicted elliptical orbits and Kepler's third law; Kepler's second law, the law of equal areas, is a direct consequence of the conservation of angular momentum and holds for any central force.

The Earth's Gravity Newton's law describes the gravitational force between particles. Newton was able to prove (although he had to invent integral calculus to do it!) that the force of gravity exerted by any spherically symmetric body on or outside its surface is the same as if all its mass were concentrated at its center. His first check on his theory was to show that the earth's gravitational force on an object at the surface of the earth and that on the moon in its orbit are consistent with the inverse-square law.

Cavendish's Experiment The constant of proportionality in Newton's law of gravitation is called the *gravitational constant G*. It can be measured by measuring directly the force of attraction between two bodies of known mass, although experimentally this is difficult because the force between any two bodies of laboratory size is extremely small. The measurement of G was first accomplished by Cavendish in 1798.

Gravitational and Inertial Mass Newton's law makes it clear that the *mass* of a body measures two quite different properties. It measures the strength of its gravitational interaction with another particle (gravitational mass), and it also measures the particle's resistance to being accelerated (inertial mass). The equivalence of these two properties is by no means obvious, but it has been confirmed to very high precision experimentally. This *principle of equivalence* is very important in the theory of general relativity.

Gravitational Potential Energy Gravity is a conservative force, so we can describe it in terms of potential energy. The *gravitational potential energy* at a certain point is defined as the negative of the work done by the force of gravity on a particle that moves to that point from a great distance away. It follows from this that the gravitational potential energy of the particle is inversely proportional to its distance from the other attracting body. (It is only when a particle is near the surface of the earth that its distance from the earth's center is nearly constant, and so the force of gravity on it is constant.)

Escape Speed If an object leaves the earth's surface at a great enough speed, the earth's gravity will never be able to stop it and pull it back to earth. The object must have a kinetic energy greater than the magnitude of its potential energy at the surface of the earth so that it will still have some kinetic energy when it is an arbitrarily great distance away. The necessary speed is called the *escape speed* of the earth.

Energy and Orbits Potential energy can be measured from any convenient point. Except for problems involving objects very near the earth's surface, it is (conventionally) taken to be zero when the separation of the two attracting bodies is infinite. With this choice, the potential energy is negative for any finite separation. Thus, the *total* energy of a body moving around the sun is negative if the body is in a bound orbit (an ellipse) and positive if it is in an unbound orbit (a parabola or hyperbola).

II. Numbers and Key Equations

Numbers

Gravitational constant

$$G = 6.67 \times 10^{-11} \text{ N·m}^2/\text{kg}^2$$

Key Equations

Newton's law of gravitation

$$F = G \frac{m_1 m_2}{r^2} \qquad g = \frac{GM_E}{R_E^2}$$

Kepler's third law

$$T^2 = \left(\frac{4\pi^2}{GM_s}\right) r^3$$

Gravitational potential energy

$$U = -G \frac{Mm}{r}$$

Earth's escape speed

$$v_e = \sqrt{\frac{2GM_E}{R_E}} = \sqrt{2gR_E}$$

III. Possible Pitfalls

Every formula for gravitational potential energy (or for any potential energy, for that matter) *assumes* a particular zero point. Keep clearly in mind what this point is when working a particular problem. Gravitational potential energy $U = mgy$ if we call the potential energy zero at $y = 0$. It is $U = -GMm/r$ if we call the potential energy zero at a point infinitely far away.

The force of gravity of a spherically symmetric mass on another object is the same as if the spherical mass were a point particle located at its center. This is true *only* for a spherically symmetric object, however. No such simple statement can be made about a mass of any other shape.

Don't forget that, in problems where we need the distances between planets and moons and such, the distances are measured from their *centers*.

A body moving freely in the earth's gravity is not necessarily moving toward the earth! Orbiting bodies, even those in circular orbits, are falling freely toward the earth. Their acceleration is toward the earth no matter in what direction they are moving.

The earth's escape speed is a fairly well-known figure: about 11.2 km/s or 7.0 mi/s. If you remember this number, remember too that it refers only to escape from the earth's *surface*. A body that starts from a higher altitude requires a lower initial velocity to escape.

Every planet—or moon, or star, or whatever—has its own escape speed, determined by its mass and size.

Be very careful of signs, especially when working problems that deal with the energy of orbital motion. According to the conventional definition, the potential energy of an object due to the earth's gravity is zero at an infinite distance from the earth; it is therefore negative at any finite distance. The total energy of a body in a bound orbit is likewise negative. This can throw you off—for instance, as the total energy of an orbiting body decreases, its numerical value *increases*.

For the sake of illustration, we often do problems involving the earth's gravity that neglect other forces such as the air drag. Don't worry if the results are wildly different from reality. An object falling to earth from a great distance would actually impact at a speed an order of magnitude smaller than the escape speed, thanks to air drag.

IV.   Questions and Answers

Questions

1.   Some communications satellites hang stationary over one point on the earth. How is this accomplished?

2.   Must the satellite of Question 1 be in a circular orbit?

3.   The two satellites of the planet Krypton have near-circular orbits whose diameters are in the ratio 1.7 to 1. What is the ratio of their periods?

4.   Estimate the order of magnitude of the force of gravity exerted upon you by a mountain 2 km high. (You'll have to make some simplifying assumptions about the shape of the mountain, its density, and so forth.)

5.   How does the earth's atmosphere affect the escape speed?

6.   The earth's orbit isn't a perfect circle; the earth is a little closer to the sun in the winter than it is in the summer. How can you tell this from the apparent motion of objects in the heavens?

7. As a satellite orbits the earth, the drag of the slight traces of the atmosphere that exist at its orbital altitude dissipate its energy. As a result, the satellite speeds up. How does this come about? Where does the extra kinetic energy come from?

8. The drag force of the atmosphere on an orbiting satellite has a tendency to make the satellite's orbit more nearly circular. Why?

9. What does it mean to say that an astronaut in a satellite orbiting the earth is "weightless"?

Answers

1. If a satellite is in a circular equatorial orbit whose period is equal to that of the earth's rotation, it will appear to be stationary relative to a single spot on the ground. (This is spoken of as a "geosynchronous" orbit.) The altitude of the required orbit works out to be about 35,800 km or 22,400 mi.

2. Yes, the orbit must be circular if the satellite is to appear truly stationary. If the satellite were in an elliptical orbit in which it moves faster than the earth turns part of the time and slower part of the time, relative to an observer on the ground, the satellite would seem to oscillate back and forth about a stationary point.

3. By Kepler's third law, the period T of an orbit is proportional to $R^{3/2}$, where R is the radius of the orbit. Thus, the ratio of the periods here is $(1.7)^{3/2}$ to 1 or 2.22 to 1.

4. The model I used for the mountain had the mass of a cone 2 km high and 10 km in diameter at the base. I assumed that I was standing at a point on the base of the cone, and that the gravity of the mountain acted as from a point one third of the way up its axis. These thoroughly arbitrary assumptions gave me a force of 2×10^{-4} N, which is about 2×10^{-7} times my weight. You may get something quite different, depending on the assumptions you make, but it will be small. Gravity is a weak force!

5. The object will lose some energy to the dissipative force of air drag as it climbs out of the atmosphere, so it must start with some extra energy (and thus some extra initial speed at the ground) if it is to have enough energy to escape the earth.

6. As the earth orbits the sun, an observer on the earth sees the sun's position in the heavens move against the background of the "fixed" stars. By Kepler's second law, we see it move a little faster in the winter than in the summer.

7. As the satellite loses energy, its orbital radius must decrease. But friction with the atmosphere tends to keep the orbit circular (see the next question) and the speed of an object in circular orbit is greater at lower altitude. There's no disagreement with the conservation of energy; the satellite has lost more potential energy than it has gained in kinetic energy. To say it another way, part of its decreased potential energy goes to increased kinetic energy, and part of it is dissipated by air friction.

8. Kepler's law of equal areas tells us that a satellite moves faster at a lower altitude; there is also more residual atmosphere at lower altitudes. Thus, most of the effect of atmospheric drag on the satellite occurs when it is near its closest approach to the earth (perigee). Every time it comes around to perigee, it loses a little speed, and as a result, it doesn't climb as far away from the earth during the next orbit; its path becomes a little more nearly circular each time.

9. Properly speaking, the astronaut isn't weightless since his weight is the force that gravity exerts on him. The earth's gravity hasn't gone away. He is in free fall, however, so his apparent weight is zero.

V. Problems and Solutions

Problems

In several of these problems you will need the mass and radius of the earth: $M_E = 5.98 \times 10^{24}$ kg and $R_E = 6.37 \times 10^6$ m. You may also need to know the mass of the sun (1.99×10^{30} kg) and the radius of the earth's orbit (1.50×10^{11} m).

1. An astronaut is exploring a planet in another solar system. The planet's diameter is 7940 km. He drops a stone from the port of his landing craft and observes that it takes 1.78 s to fall to the surface, a distance of 4.5 m. From these data, what value does he calculate for the mass of the planet?

How to Solve It
• Calculate the local acceleration due to gravity from the time it takes the stone to fall.

- Knowing the planet's radius and the local acceleration of gravity, you can use Newton's law of gravitation to calculate the mass of the planet.

2. The moons of Mars are Deimos and Phobos ("Terror" and "Fear"). Deimos orbits at 14,600 mi from the center of Mars with a period of 30 h. The orbital period of Phobos is 7.7 h. (*a*) What is the radius of the orbit of Phobos? (*b*) What is the mass of Mars?

How to Solve It
- Since you know the orbital periods, you can get the radius of Phobos' orbit from Kepler's third law.

- Use Kepler's third law to calculate the mass of Mars.

- Don't forget that you have to convert units before calculating the mass of Mars.

3. The average distance of Saturn from the sun is 1.40×10^9 km. Assuming its orbit to be circular, find the speed of Saturn in its orbit around the sun in metres per second.

How to Solve It
- If Saturn's orbit is circular, then its orbital speed is constant.

- From the orbital radius of Saturn and the radius and period of the earth's orbit, calculate Saturn's orbital period using Kepler's third law.

- Now you know how far Saturn goes in one orbit and how long this takes, so you can calculate its average speed.

4. The planet Jupiter takes 11.9 years to orbit the sun. What is the radius of Jupiter's orbit?

How to Solve It
- The earth takes one year to orbit the sun.

- Use Kepler's third law to relate the orbital radii and periods of the earth and Jupiter.

5. Suppose a space probe is launched from the earth at the earth's escape speed, directly away from the sun. Does it escape from the solar system as well?

How to Solve It

- The formula for escape speed can be used not only at the surface of the earth or sun but also at any distance from the center of the body.

- Calculate the speed of escape from the sun's gravity for an object at the position of the earth's orbit.

- If this is greater than the speed at which the probe is launched, then even though the probe may escape the earth, it is still bound by the sun's gravity.

6. Take the mass of the moon to be 7.4×10^{22} kg and its radius to be 1.74×10^6 m. What is the escape speed from the surface of the moon?

How to Solve It

- When an object is at escape speed, it has enough energy to get very far away from the attracting body (here, the moon).

- The escape speed is the speed that makes the object's initial total energy nonnegative.

7. Neglecting the resistance of the atmosphere, (*a*) at what speed would an object have to be projected directly upward in order to reach an altitude of 200 km? (*b*) The object's mass is 800 kg. Once it reaches 200 km, how much more energy would it have to be given in order to put it into a circular orbit at this altitude?

How to Solve It

- If air resistance is negligible, the satellite's total energy is conserved.

- In part (*a*), the object must start upward from the ground with enough kinetic energy to supply the additional potential energy it has 200 km up.

- In part (*b*), you must give it additional kinetic energy in order for it to have orbital speed at an altitude of 200 km.

8. An earth satellite of mass 1800 kg is to be placed in a circular orbit at an altitude of 150 km above the earth. (*a*) How much work must be done on the satellite to accomplish this? Neglect the resistance of the atmosphere. (*b*) How much of the work goes into increasing the potential energy of the satellite and how much into increasing its kinetic energy?

How to Solve It
- The change in the satellite's potential energy is due to its increased height, just as calculated in the previous problem.

- The change in its kinetic energy corresponds to its orbital speed.

- The work that a booster rocket must do on the satellite is equal to the change in its total mechanical energy, if air resistance can be neglected.

9.   A research satellite is in an orbit around the earth. Its closest approach to the earth's surface (perigee) is 100 km, whereas its greatest distance (apogee) is 1600 km. Find its speed at each of these two points.

How to Solve It
- This is a hard problem. The reason it's hard is that you have to apply two different ideas to it and then do a bit of algebra. Don't skip it, though; it's good for you to deal with a hard one every now and then.

- The energy of the orbiting satellite is constant, and so is its angular momentum (this is Kepler's law of equal areas).

- From the law of equal areas, we know that the speeds of the satellite at the two extreme points of its orbit will be in an inverse ratio to their distances from the center of the earth.

- The fact that the satellite's total mechanical energy remains constant provides another equation involving the speeds.

10. A satellite is placed in an elliptical orbit around the earth. At its closest approach to the earth (perigee), it is 112 km above the earth's surface and is moving at a speed of 8032 m/s. How high is it at its greatest distance from the earth (apogee)?

How to Solve It
- This problem is hard in the same way that the last one was: you have to use both the conservation of angular momentum and the conservation of energy to solve for two variables.

- From the data given, you can calculate both the angular momentum and the energy at perigee. Both quantities must have the same values at apogee.

- This gives you two equations. Use one to eliminate v, and the other one becomes a quadratic equation that you can solve for r.

Solutions

1. The position as a function of time for something falling freely under gravity is given by

$$y = \tfrac{1}{2}gt^2$$

so the local acceleration of gravity is

$$g = \frac{2y}{t^2} = \frac{(2)(4.5 \text{ m})}{(1.78 \text{ s})^2} = 2.84 \text{ m/s}^2$$

But the acceleration of gravity at the planet's surface is

$$g = GM/R^2$$

so

$$M = \frac{gR^2}{G} = \frac{(2.84 \text{ m/s}^2)(3.97 \times 10^6 \text{ m})^2}{6.67 \times 10^{-11} \text{ N·m}^2/\text{kg}^2}$$

$$= 6.71 \times 10^{23} \text{ kg}$$

2. (a) 5900 mi   (b) 6.5×10^{23} kg

3. Using Kepler's third law, we can compare the orbits of Saturn and the earth:

$$\left(\frac{T_E}{T_S}\right)^2 = \left(\frac{R_E}{R_S}\right)^3$$

$$\left(\frac{1 \text{ year}}{T_S}\right)^2 = \left(\frac{1.50 \times 10^8 \text{ km}}{1.40 \times 10^9 \text{ km}}\right)^3$$

so

$$T_S = 28.5 \text{ years}$$

The circumference of Saturn's orbit is

$$2\pi R_S = (2\pi)(1.40 \times 10^{12} \text{ m}) = 8.80 \times 10^{12} \text{ m}$$

Saturn covers this distance in a time of

$$t = (28.5 \text{ y})(365.25 \text{ d/y})(24 \text{ h/d})(60 \text{ min/h})(60 \text{ s/min})$$

$$= 8.99 \times 10^8 \text{ s}$$

so its speed is

$$v = \frac{d}{t} = \frac{8.80 \times 10^{12} \text{ m}}{8.99 \times 10^8 \text{ s}} = \underline{9.79 \times 10^3 \text{ m/s}}$$

4. $\underline{7.8 \times 10^8 \text{ km}}$

5. The escape speed at distance r from a body of mass M is given by

$$\tfrac{1}{2}mv_e^2 = G\frac{Mm}{r}$$

$$v_e = \sqrt{2GM/r}$$

In this, r is the distance from the center of the earth or sun and M is the mass of the earth or sun. So for the earth

$$v_{eE} = \left(\frac{(2)(6.67 \times 10^{-11} \text{ N·m}^2/\text{kg}^2)(5.98 \times 10^{24} \text{ kg})}{6.37 \times 10^6 \text{ m}}\right)^{1/2}$$

$$= 1.12 \times 10^4 \text{ m/s}$$

At the distance of the earth from the sun, the speed required to escape the sun's gravitational pull is

$$v_{eS} = \left(\frac{(2)(6.67 \times 10^{-11} \text{ N·m}^2/\text{kg}^2)(1.99 \times 10^{30} \text{ kg})}{1.50 \times 10^{11} \text{ m}}\right)^{1/2}$$

$$= 4.21 \times 10^4 \text{ m/s}$$

Thus, the probe <u>cannot</u> escape from the solar system.

6. $\underline{2.37 \times 10^3 \text{ m/s}}$

7. (*a*) The gravitational potential energy of an object at a distance r from the center of the earth is

$$U = -G\frac{Mm}{r}$$

where M is the mass of the earth. Thus, the increase in its potential energy as it moves from ground level ($r = R_E$) to an altitude $y = 200$ km above the ground is

$$\Delta U = GMm\left(\frac{1}{R_E} - \frac{1}{R_E + y}\right) = GMm\left(\frac{y}{R_E(R_E + y)}\right)$$

This must come from its initial kinetic energy, so

$$\tfrac{1}{2}mv^2 = \Delta U = \frac{GMmy}{R_E(R_E + y)}$$

Thus,

$$v^2 = \frac{2GMy}{R_E(R_E + y)}$$

$$= \frac{(2)(6.67 \times 10^{-11}\ \text{N·m}^2/\text{kg}^2)(5.98 \times 10^{24}\ \text{kg})(2 \times 10^5\ \text{m})}{(6.37 \times 10^6\ \text{m})(6.37 \times 10^6\ \text{m} + 2 \times 10^5\ \text{m})}$$

$$= 3.81 \times 10^6\ \text{m}^2/\text{s}^2$$

so

$$v = \underline{1.95\ \text{km/s}}$$

(b) The extra energy that must be supplied to put the object in orbit is just the kinetic energy corresponding to its orbital speed. The orbital speed is given by

$$v_o = \left(\frac{GM}{r}\right)^{1/2}$$

so the extra energy required is

$$E_k = \tfrac{1}{2}mv_o^2 = \tfrac{1}{2}m\left(\frac{GM}{r}\right) = \frac{GMm}{2r}$$

$$= \frac{(6.67 \times 10^{-11}\ \text{N·m}^2/\text{kg}^2)(5.98 \times 10^{24}\ \text{kg})(800\ \text{kg})}{(2)(6.37 \times 10^6\ \text{m} + 2.0 \times 10^5\ \text{m})}$$

$$= \underline{2.43 \times 10^{10}\ \text{J}}$$

8.   (a) $\underline{5.77 \times 10^{10}\ \text{J}}$ (b) The orbital kinetic energy is $\underline{5.51 \times 10^{10}\ \text{J}}$.

9.   At the two "endpoints" of the elliptical orbit, the velocity of the satellite is perpendicular to the radius vector (see Figure 9-1). Thus, at these points the angular momentum is just mvr. Since the angular momentum is a constant of the motion,

$$v_A r_A = v_P r_P$$

$$v_A = \frac{v_P r_P}{r_A} = \left(\frac{6370\ \text{km} + 100\ \text{km}}{6370\ \text{km} + 1600\ \text{km}}\right)v_P = 0.812 v_P$$

Figure 9-1

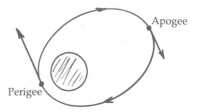

Apogee

Perigee

Likewise, $r_A = r_P/0.812 = 1.232r_P$. The total mechanical energy of the satellite is also constant since the only force on it (gravity) is conservative. It is given by

$$E = \tfrac{1}{2}mv^2 - G\frac{Mm}{r}$$

so

$$E = \tfrac{1}{2}mv_P^2 - G\frac{Mm}{r_P} = \tfrac{1}{2}m(0.812v_P)^2 - G\frac{Mm}{1.232r_P}$$

Eliminating m and rearranging gives

$$\tfrac{1}{2}v_P^2(1 - 0.812^2) = \frac{GM}{r_P}(1 - 0.812)$$

so

$$v_P^2 = 1.104\frac{GM}{r_P}$$

$$= \frac{(1.104)(6.67 \times 10^{-11}\ \text{N·m}^2/\text{kg}^2)(5.98 \times 10^{24}\ \text{kg})}{6.37 \times 10^6\ \text{m} + 1.00 \times 10^5\ \text{m}}$$

$$= 6.80 \times 10^7\ \text{m}^2/\text{s}^2$$

Thus,

$$v_P = \underline{8.25 \times 10^3\ \text{m/s}}$$

and

$$v_A = 0.812v_P = \underline{6.70 \times 10^3\ \text{m/s}}$$

10. $\underline{771\ \text{km}}$

CHAPTER 10

Solids and Fluids

I. Key Ideas

States of Matter Matter in bulk can be either solid or fluid. Solids are more or less rigid, tending to maintain a fixed shape; fluids (liquids and gases) flow more or less freely. This distinction between states of matter is not perfectly sharp, however.

Density The *density* of a substance is its mass per unit volume. One cubic centimetre of water has a mass of 1 gram (this was the original definition of the gram), so the density of water is 1 g/cm³. Something with a lower density than water floats in water; something of greater density sinks. The ratio of the density of a substance to that of water is the *specific gravity* of the substance. Solids and liquids have densities that are roughly independent of external conditions, but the density of a gas is strongly affected by such conditions as pressure and temperature. Thus, T and P must be specified when stating the density of a gas. (*Standard conditions* are 1 atm and 0°C.)

Weight Density *Weight density* is mass density times the acceleration of gravity, that is, weight per unit volume. It is most often used in the U.S. customary unit system.

Elasticity Most solids behave *elastically*; that is, if stretched or deformed by external forces, they tend to return to their original size and shape when the deforming stress is removed. For a

given material, this is true up to some *elastic limit*; exceeding this limit will cause permanent deformation.

Stress and Strain If an applied force tends to stretch an object in one direction, the force per unit area (perpendicular to the force) is called the *tensile stress*. The resulting fractional change in the object's length is the *tensile strain*. Up to the elastic limit, stress is proportional to strain; the constant of proportionality is *Young's modulus*. Under a sufficiently large stress (called the *tensile strength* of the material), the material will break. Young's modulus for compression and the compressive strength of a given material may or may not be the same as for stretching.

Shear and Bulk Moduli A sidewise deformation resulting from forces that act parallel to the surface of a material is called a *shear* (see Figure 10-1). Shearing stress and shearing strain are proportional within the elastic limit of a material; the constant of proportionality is the *shear modulus*.

Figure 10-1

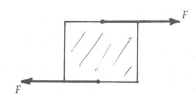

The distinction between fluids and solids is that fluids are unable to support any shear stress and so fill up a container of any shape. An increase in the pressure on a material tends to compress it in all directions at once. The *bulk modulus* is the ratio of the increase in pressure to the resulting fractional decrease in volume; its inverse is the *compressibility* of the material.

Pressure in a Fluid In a fluid, the force per unit area on a surface at a given depth, oriented anyhow, is the same. This force per unit area is called the *pressure*. A variety of units of pressure are in common use. The pressure in a fluid increases with increasing depth because of the weight of the fluid above; the pressure is the same at all points that are at the same depth. Any increase in pressure is the same throughout the fluid (this is *Pascal's principle*). In many applications, the *gauge pressure*—the difference from atmospheric pressure—is used. Liquids are roughly incompressible; as a result, the pressure change with height in a liquid is linear. Gases, on the other hand, are highly compressible, with a density that is approximately proportional to the pressure. The pressure decrease with altitude in the atmosphere, therefore, is approximately *exponential*—that is, it decreases by a constant fraction for a given increase in height.

Archimedes' Principle There is an upward, buoyant force on an object submerged in a fluid that is equal to the weight of the fluid displaced by the object. An object that is less dense than water will float (in equilibrium) such that the water it displaces weighs the same as the object.

Surface Tension The boundary surface of a liquid is under tension due to the mutual attraction of the molecules of the liquid and thus exerts a force that resists any stretching or breaking of the surface. This force per unit length is the *coefficient of surface tension* of the liquid. Surface tension is the reason that free small droplets of a liquid tend to assume a spherical shape. The behavior of a fluid at a surface of contact with another material depends on the relative strength of the cohesive forces in the fluid and the adhesive forces between the fluid and the second material. Thus, water "wets" glass because the glass-water adhesive forces are much stronger than the cohesive forces within the water.

Capillary Action Adhesive forces will cause a liquid to rise in a small tube above the level of the surrounding fluid if the liquid is one that wets the material of the tube. This is spoken of as *capillary action*.

Bernoulli's Equation The general motion of a fluid can be very complicated, and we confine our study of fluid motion to steady, nonturbulent flow. If we assume the fluid is incompressible, then the volume flow rate in a tube or pipe is the same throughout. If the fluid flow is without internal energy dissipation (we say the fluid is *nonviscous*), we can apply the work-energy theorem to get *Bernoulli's equation*, the fundamental dynamic equation for the steady flow of a nonviscous, incompressible fluid. One consequence of Bernoulli's equation is that, in a level flow, the pressure is lower where the fluid is flowing faster (the *Venturi effect*).

Viscosity In real fluids, energy is dissipated by internal friction in a flowing fluid; this property is known as *viscosity*. Because of viscosity, the pressure drops (moving along the direction of flow) in steady flow through a long horizontal pipe (contrary to Bernoulli's equation) by an amount proportional to the flow rate. The *coefficient of viscosity* is a measure of the fluid's resistance to flow; it is defined in terms of the force required to move parallel plates in the fluid.

Reynolds Number All our discussion of fluid dynamics—Poiseuille's law, Bernoulli's equation, and so forth—applies only to steady, nonturbulent flow. In a real situation, steady flow often breaks down and becomes turbulent. The *Reynolds number* is a dimensionless parameter that characterizes the flow of a fluid; for values of the Reynolds number above a few thousand, the flow becomes turbulent.

II.   Numbers and Key Equations

Numbers

1 pascal (Pa) = 1 N/m^2

1 atm = 1.01325 × 10^5 Pa

1 bar = 1000 millibar = 10^5 Pa

1 mmHg = 1 torr = 133.4 Pa

1 ft H$_2$O = 2991 Pa

1 poise = 0.100 Pa·s = 0.100 N·s/m^2

Key Equations

Density

$$\rho = \frac{m}{V}$$

Weight density

$$\rho g = \frac{mg}{V}$$

Stress

$$\text{Stress} = \frac{F}{A}$$

Tensile strain

$$\text{Strain} = \frac{\Delta L}{L}$$

Young's modulus

$$Y = \frac{\text{stress}}{\text{strain}} = \frac{F/A}{\Delta L/L}$$

Shear modulus

$$M_s = \frac{\text{shear stress}}{\text{shear strain}} = \frac{F_s/A}{\Delta X/L} = \frac{F_s/A}{\tan \theta}$$

Bulk modulus

$$B = -\frac{\Delta P}{\Delta V/V}$$

Pressure

$$P = \frac{F}{A}$$

Pressure variation with height in a liquid

$$P = P_0 + \rho g h$$

Gauge pressure

$$P_{\text{gauge}} = P - P_{\text{at}}$$

Archimedes' principle

$$B = W_{\text{fl}}$$

Surface-tension force on wire

$$F = (2L)\gamma$$

Height in capillary tube

$$h = \frac{2\gamma \cos \theta}{\rho g r}$$

Continuity equation

$$I_v = Av = \text{constant}$$

Bernoulli's equation

$$P + \rho g y + \tfrac{1}{2}\rho v^2 = \text{constant}$$

Viscous fluid flow

$$\Delta P = P_1 - P_2 = I_v R$$

Coefficient of viscosity

$$F = \eta \frac{vA}{z}$$

Poiseuille's law

$$\Delta P = \frac{8\eta L}{\pi r^4} I_v$$

Reynolds number

$$N_R = \frac{2r\rho v}{\eta}$$

III.   Possible Pitfalls

The elastic moduli are constants *only* for stresses that are less than the elastic limits of a material. In particular, the stress-strain ratio is not at all constant for stresses approaching the tensile or compressive strength of the material.

It's easy to mix up density, weight density, and specific gravity; don't. Density (or mass density) is the mass per unit volume of a substance; multiplying it by g gives the weight per unit volume or weight density. Specific gravity is a dimensionless number; it is the ratio of the density of a substance to that of water. It is thus numerically the same as density in grams per cubic centimetre.

Don't forget that the decrease of pressure with height (or the increase with depth) is *linear* only for a fluid that is incompressible and therefore has a constant density. The decrease of atmospheric pressure with height, for example, is not at all linear, as air is very compressible and its density is therefore not constant.

The density of water is defined to be 1—but *not* in SI units. It is 1 g/cm³ (or 1 kg/L) but 1000 kg/m³.

Contrary to the way it may sound, specific gravity has nothing to do with weight. It is the density of a substance *relative* to that of water.

In doing problems that deal with the pressure at some depth in a fluid, don't forget that ρgh is the pressure *difference* over a vertical height h; it's not necessarily the pressure at either the top or bottom.

The equilibrium condition for an object floating in a fluid is that the weight of the object is equal to the weight of the volume of water that the submerged portion of the object displaces; it is *not* that the volume of the water displaced is equal to the volume of the object.

Just because an object sinks in a liquid rather than floating, don't think that there is no buoyant force acting on it. This only means that the buoyant force is less than the object's weight.

The continuity equation for fluid flow and Bernoulli's equation are strictly correct only for a fluid of constant density. Thus, they apply pretty well to most liquids, but not (at least without modification) to gases.

You can really mess up calculations with Bernoulli's equation if you are not careful to use *consistent* units for all the quantities. It's preferable to put everything into SI units.

IV.   Questions and Answers

Questions

1. Assume that you float in freshwater with only about 5 percent of your body above water. From this, estimate your body volume.

2. The "fog" produced by evaporating "dry ice" (frozen CO_2) stays at ground level and spreads along the ground. Why is this?

3. Water bugs can walk on the surface of a pond or lake. Could a monster water bug (probably in a horror movie) 10 or 20 times larger in all its dimensions pull the same trick? Why or why not?

4. A long steel rod is squeezed along its length. Is it Young's modulus or the bulk modulus of steel that you would need to know to calculate the resulting strain?

5.   When a real wire is stretched by a tensile stress, its diameter is decreased. We speak of the fractional change in the diameter as the *transverse strain*. For nickel wire, it is found that the transverse strain is 0.36 times the longitudinal strain. Does the density of nickel wire increase or decrease as it is stretched?

6.   Figure 10-2 represents a small vessel upended in water (which is contained in a larger vessel) so as to trap some air inside. This gadget is known as a "Cartesian diver." If the air pressure above the water is increased (for example, by pressing on a flexible lid on the larger container), the small vessel sinks to the bottom. Why does it sink? How can it be made to rise again?

Figure 10-2

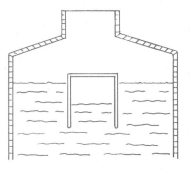

7.   Once, in the course of an experiment, I needed to know the exact volume of a large, irregular, thin-walled aluminum container. The volume was of the order of 1 m³. Someone suggested the obvious, which was to fill it with water from vessels of known volume. The only drawback was that a cubic metre of water weighs about a ton and would have crushed the 1.5-mm aluminum walls of the container. Can you suggest a way around this?

8.   A colleague of mine likes to do the following demonstration. He pours some red-dyed water into a glass U-tube that is open at both ends to show that the fluid "seeks its own level" (that is, it rises to the same height on both sides of the U-tube). He then pours in some more red liquid, and it comes to rest as shown in Figure 10-3. What's the trick here?

Figure 10-3

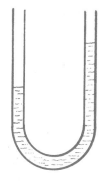

9.   If the height to which capillary action would raise water in a tube is greater than the tube's actual height, will the water rise and flow out the top of the tube like a fountain?

10. An ice cube floats in a glass of water. As the ice melts, does the water level rise or fall? Explain.

11. Aircraft carriers steam into the wind while launching planes. Why? In what direction should they move while recovering planes?

12. In a real pipe, the fluid pressure decreases as you go downstream, even though the pipe is level and the fluid incompressible. Why?

13. No pump can raise water in a pipe by suction higher than about 34 feet. Why not?

14. You can't breathe underwater by drawing air through a tube from the water surface (a "snorkel") at a depth of more than a foot or so. Why not?

Answers

1. Of course it depends on your mass. Suppose your mass is 70 kg. In order for you to float, the buoyant force of the water must equal your weight; thus, by Archimedes' principle, you must displace 70 kg of water. A 70-kg mass of water has a volume of 70 L or 0.070 m^3. Thus, the submerged 95 percent of your body has a volume of 0.070 m^3, and your body volume is 0.070 m^3/0.95 = 0.074 m^3.

2. The "fog" that you see is water vapor condensing out of the air locally due to the presence of the cold CO_2 vapor. Although the CO_2 vapor itself is transparent, you can see where it is by means of the condensed water vapor. Of course, it stays down or sinks down because CO_2 vapor is substantially heavier than air. The fog that forms over ponds and streams at night appears to do the same thing, but we need a different explanation; water vapor is lighter than air. What's happening in this case?

3. It seems to me I saw an old black and white second feature a long time ago in which this happened. In any case, it couldn't really happen. A scaled-up water bug 20 times bigger in every direction would weigh 8000 times as much as one of normal size. But the supporting force due to surface tension would increase proportionally to the circumference of his foot, so it would be only 20 times greater. The force would be far too small to hold him up.

4. You must use Young's modulus. The bulk modulus is used only when there is a uniform pressure on the object in all directions—as at the bottom of a body of water, say. When you squeeze the rod along its length, it actually expands laterally (see Question 5).

5. Suppose the wire is stretched such that its length increases by 0.1 percent (a tensile strain of 0.001). This effect alone would increase its volume by a factor of 1.001. But at the same time, the diameter of the wire decreases by 0.36 as much. The radius is thus 0.99964 of its original value, and the cross-sectional area is $(0.99964)^2$ = 0.99928 of its original value. The volume of the stretched wire is therefore (1.001)(0.99928) = 1.00028 times its original value. Since the volume has increased slightly, the density of the stretched wire is slightly lower.

6. The increased pressure on the water surface is transmitted to the trapped air inside the small vessel; this is Pascal's principle. The trapped air is compressed and displaces less water, so the buoyant force holding the small vessel up is reduced. Increase the pressure enough, and the small vessel sinks. Decrease it enough, and the trapped air expands enough to raise the sunken vessel again.

7. We floated the container in a swimming pool and filled it with water from a calibrated vessel. Except for the very small weight of the aluminum shell, the water pressure inside and outside the container was then the same, so there was little or no stress on its fragile walls.

8. The fluid he added the second time wasn't water but some liquid that is less dense than water and doesn't mix readily with it. Thus, a taller column of the second fluid was required to produce the same pressure at a given height in the water below.

9. No. The only reason the water rises is because of the adhesive force of the vertical wall of the tube pulling it upward. Thus, it can only rise as high as the vertical wall of the tube.

10. When the ice is floating, the part of it that is underwater is displacing a volume of water that is equal in weight to the whole weight of the ice. But that is exactly the volume of the water produced by the melting ice, so the water level goes neither up nor down but stays the same.

11. To handle both launch and recovery most easily, you want the planes to be moving as slowly as possible relative to the flight deck of the aircraft carrier. By steaming into the wind, you maximize the velocity of the air relative to the flight deck. Since lift is determined by the air speed relative to the wings of the plane, this reduces the speed necessary for takeoff relative to the deck. For the same reason, the carrier will steam into the wind to recover planes.

12. The pressure decreases because any real fluid has some viscosity, and thus energy is being dissipated in the fluid as it flows. If a steady flow is to be maintained, there must be a pressure difference to do work on the fluid.

13. A suction pump raises water by creating a pressure lower than atmospheric pressure; it is the pressure difference that raises the fluid. But atmospheric pressure corresponds to a height of 34 ft of water, and—since the pressure at the pump cannot be less than zero—the largest pressure difference that can be created is equal to atmospheric pressure.

14. Because to draw air into your lungs, your diaphragm must lower the pressure in your lungs to less than atmospheric pressure. At a depth of a foot, the water outside your body is already at a pressure around 25 torr higher than atmospheric. Because 30 to 50 torr is as much external pressure difference as your diaphragm can handle, you can't suck air through a pipe at a depth of more than a foot or two.

V. Problems and Solutions

Problems

1. The density of water is 1.00 g/cm^3 at 4°C. Suppose that a 500-mL flask is filled exactly full of water at a temperature of 60°C, where the density of water is (say) 0.980 g/cm^3. When the flask of water is cooled to 4°C, how much more water must be added in order to fill the flask? Neglect the thermal expansion or contraction of the flask itself.

How to Solve It
* How many grams of water are in the flask at 60°C?

* How many cubic centimetres does this much water occupy at 4°C?

2. The mass of a small analytic flask is 15.2 g. When the flask is filled with water, its mass is 119.0 g; when it is filled with an unknown fluid, its mass is 96.7 g. What is the specific gravity of the unknown fluid?

How to Solve It
* The total mass of the filled vessel is the mass of the fluid in the flask plus the mass of the flask itself.

* Thus, you know the mass of the same (unknown) volume of water and the unknown fluid.

* Specific gravity is defined as the ratio of the density of a substance to that of water. Since equal volumes are involved here, the ratio of the mass of the unknown fluid to that of water gives you the specific gravity of the unknown fluid.

3. Young's modulus for steel is 2.0 × 10^{11} N/m^2; for tungsten, it's 3.6 × 10^{11} N/m^2. A piece of tungsten wire 0.8 m long and a 0.5-m length of steel wire, each of diameter 1.0 mm, are joined end to end, making a total length of 1.3 m. If a mass of 22 kg is hung vertically from one end of this wire, how much does the wire stretch?

How to Solve It
- The tension is the same throughout both wires if they have the same diameter and are connected end-to-end.

- The tensile stress is thus the same for each wire.

- Calculate the strain in each wire using the appropriate Young's modulus.

- From the strains, calculate how much each wire stretches and the total stretch.

4. Young's modulus for copper is 1.1×10^{11} N/m². If a load of mass 40 kg is hung from a 2-m copper wire of diameter 1.5 mm, by how much does the wire stretch?

How to Solve It
- Calculate the cross-sectional area of the wire.

- From this, calculate the tensile stress in the wire.

- The stress divided by Young's modulus gives the fractional elongation of the wire. From this, calculate how much the wire stretches.

5. A lead brick measuring $5.0 \times 10.0 \times 20.0$ cm is dropped into a swimming pool 3.5 m deep. By how much does the volume of the brick change?

How to Solve It
- What is the pressure at the bottom of 3.5 m of water? Be careful: should you include the pressure of the atmosphere in this calculation?

- The bulk modulus of lead is given in Table 10-3 on page 226 in the text.

- The pressure on the brick divided by the bulk modulus gives the fractional decrease in the brick's volume.

6. Blood pressure is normally measured on a patient's arm at approximately the level of the heart. If it were measured instead on the leg of a standing patient, how significantly would the reading be affected? Normal blood pressure is in the range of 70 to 140 torr, and the specific gravity of normal blood is 1.06.

How to Solve It

- The question is just, how much does pressure change with height in blood?

- You'll have to pick some appropriate height between arm and leg— say, one metre to be simple.

- Calculating the pressure difference over this height won't give you the correct pressure in the leg because the actual blood is flowing and it is viscous. It will give you an idea of how significant the difference is, however.

7. Some colleagues and I once had to carry out an experiment on a lake. To do this we built a raft out of Styrofoam, 1 ft (0.305 m) thick and 8 ft (2.44 m) square. Take the specific gravity of Styrofoam to be 0.035. (*a*) How deep in the water did the unloaded raft float? (*b*) With three men (whose average mass was 88 kg each) and 120 kg of experimental equipment on the raft, how deep did it float?

How to Solve It

- Calculate the mass of the Styrofoam raft, with and without the load.

- When it is floating in equilibrium, this is the mass of water it will displace.

- The volume of water displaced is the area of the raft multiplied by the depth to which it sinks.

8. The density of air under "ordinary" (not standard) conditions is about 1.2 kg/m^3, whereas that of helium is 0.17 kg/m^3. What must the radius of a spherical helium-filled balloon be if it is to lift a total load of 350 kg?

How to Solve It

- The balloon will rise if its mass (including the payload) is less than that of the air it displaces.

- What must be the radius of a sphere of air in order for its mass to be 350 kg?

9. In Figure 10-4, oil, which has a specific gravity of 0.68, lies on top of water in a container. A wooden object is floating at the fluid boundary such that one third of its volume lies below the boundary. What is the density of the wood?

Figure 10-4

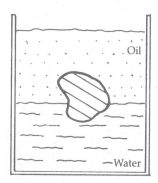

How to Solve It

- The wooden object will float such that the weight of the fluid it displaces is equal to its weight.

- The fluid it displaces is 1/3 water and 2/3 oil, by volume.

- What is the total mass of the displaced fluids? This must equal the mass of the wooden object.

10. A water bug stands on the surface of a pond on six legs and does not sink. The surface tension of water is about 0.073 N/m. Assuming that the diameter of the end of each leg is 0.107 mm, what is the maximum possible mass of the bug?

How to Solve It
- Calculate the circumference of one of the bug's feet.

- This distance multiplied by the coefficient of surface tension of water gives the maximum support force that the water can provide for each foot.

- But the bug has six feet.

11. The surface tension of water is about 0.073 N/m. A vertical capillary tube of inside diameter 0.8 mm extends 15 cm below the surface of a body of water. What is the minimum (gauge) pressure required to blow air through the tube and out its bottom? You can take the glass-water contact angle to be 0°.

How to Solve It
- The applied pressure must shove the water in the tube down farther than 15 cm because, left to itself, the water would rise in the tube due to capillary action.

- How high would the water level be with atmospheric pressure at the top of the tube?

- Calculate the excess pressure needed to push the water level down from this height to the bottom of the tube.

12. Water is flowing smoothly at 15 ft/s in a horizontal pipe of 2 in inside diameter at an absolute pressure of 40 lb/in². Neglect the viscosity of water. At a certain point the diameter of the pipe necks down to 1.05 inches. (*a*) How fast does the water flow in the narrow section? (*b*) What is its pressure there (in lb/in²)?

How to Solve It

- Before you do anything else, convert the flow velocity, pressure, and pipe diameters to SI units. It's hopeless to try to do a complicated problem without putting all quantities in a consistent system of units.

- Use the continuity equation to find the water speed in the narrow section.

- Since the pipe is horizontal, there are only two terms (the pressure and velocity terms) in Bernoulli's equation.

- Use Bernoulli's equation to find the pressure difference required to cause this change in speed.

13. A town water tank stands 10 m off the ground. Its diameter is 18 m, and the water in the tank is 7.5 m deep; the top is open to the atmosphere. If a hole is punched in the bottom of the tank and the water flows out in a stream 1 cm in diameter, how long does it take for the water level to drop by 10 cm? Neglect the viscosity of the water.

How to Solve It

- Compare the areas of the hole and the top of the water. Is the speed at which the water level at the top of the tank falls significant?

- You know enough to calculate all three terms in Bernoulli's equation at the top surface of the water.

- Use Bernoulli's equation to calculate the speed at which water leaves the hole.

- You now know the velocity and cross-sectional area of the water stream; calculate the volume flow rate.

14. At a certain point in a pipe 2.5 cm in diameter, water is flowing at 10 m/s. The pipe gradually descends vertically by 7 m, and over the same distance its diameter increases to 3.5 cm. What is the pressure difference between the upper and lower points?

How to Solve It

- The continuity equation tells you how fast the water is flowing when the pipe diameter is 3.5 cm.

- Use Bernoulli's equation to calculate the pressure difference between these two points.

15. Water is supplied to an outlet from a pumping station 5 km away. From the pumping station to the outlet there is a net vertical rise of 19 m. Take the coefficient of viscosity of water to be 0.01 poise. The pipe leading from the pumping station to the outlet is 1 cm in diameter, and the gauge pressure in the pipe at the point where it exits the pumping station is 520 kPa. At what (volume) rate does water flow from the outlet?

How to Solve It

- Bernoulli's equation alone will not work here since the water is viscous.

- There will be a pressure drop due to viscous flow in the pipe and a drop due to the 19-m vertical rise.

- You know the total pressure drop from the pumping station to the outlet and can calculate the drop due to the vertical rise. The difference is the pressure drop due to viscous drag in the pipe.

- Use Poiseuille's law to find the volume flow rate at the outlet.

16. Blood flows in the finer capillaries of the body at a rate of around 1 mm/s. If blood flows at this rate through a capillary 8 μm in diameter, what is the pressure difference required to move blood at this rate through a capillary 1.8 mm long? Assume that the coefficient of viscosity of blood is 4.0 mPa·s.

How to Solve It

- Calculate the volume flow rate in the capillary.

- If you get the units all consistent, this is just a plug into Poiseuille's law.

- It may interest you to put the result in torr to compare it to ordinary blood pressure values.

17. Water is pumped into one end of a pipe 1 inch in diameter and 50 ft long at a gauge pressure of 40 lb/in. (*a*) Assuming that the flow is nonturbulent, how much water can be pumped through the pipe in 1 min to an outlet that is open to the atmosphere? (*b*) Calculate the Reynolds number for this flow to see whether or not your assumption was justified.

How to Solve It

- Use Poiseuille's law to get the volume flow rate in the pipe.

- From the volume flow rate, calculate the velocity at which the water is flowing.

- From the velocity and the other data given, calculate the Reynolds number.

Solutions

1. The density of water at 60°C is given as 0.980 g/cm³. The mass of the water at that temperature is thus

$$m = \rho V = (0.980 \text{ g/cm}^3)(500 \text{ cm}^3) = 490 \text{ g}$$

The density of water at 4°C is 1.00 g/cm³, so the 490 g of water occupies 490 cm³ at 4°C. Thus, 10 cm³ must be added to fill the flask again.

2. 0.785

3. The cross-sectional area of each wire is

$$\pi(0.5 \times 10^{-4} \text{ m})^2 = 7.85 \times 10^{-7} \text{ m}^2$$

If a load of 22 kg is hung by this wire, then the tensile stress is

$$\text{Stress} = \frac{F}{A} = \frac{(22 \text{ kg})(9.81 \text{ m/s}^2)}{7.85 \times 10^{-7} \text{ m}^2} = 2.75 \times 10^8 \text{ N/m}^2$$

The strain that results from this stress in steel is

$$\text{Strain} = \frac{\Delta L}{L} = \frac{\text{stress}}{Y} = \frac{2.75 \times 10^8 \text{ N/m}^2}{2.0 \times 10^{11} \text{ N/m}} = 1.37 \times 10^{-3}$$

The extension of the steel wire is thus

$$\Delta L = (\text{strain})(L) = (1.37 \times 10^{-3})(0.5 \text{ m}) = 6.9 \times 10^{-4} \text{ m}$$

By the same method, the extension of the tungsten wire is 6.1×10^{-4} m, and the total extension of the compound wire is thus 6.9×10^{-4} m + 6.1×10^{-4} m = 13.0 $\times 10^{-4}$ m = 1.30 mm.

4. 4.03 mm

5. The brick's initial dimensions were presumably measured in air; if this is so, it is the difference in pressure that compresses the brick, so we do not include atmospheric pressure. The pressure at a depth of 3.5 m in water is

$$\Delta P = \rho gh = (1000 \text{ kg/m}^3)(9.81 \text{ m/s}^2)(3.5 \text{ m}) = 34{,}300 \text{ N/m}^2$$

The bulk modulus of lead is 7.7×10^9 N/m², so the fractional decrease in the brick's volume is

$$\frac{\Delta V}{V} = \frac{\Delta P}{B} = \frac{34{,}300 \text{ N/m}^2}{7.7 \times 10^9 \text{ N/m}^2} = 4.46 \times 10^{-6}$$

The brick's volume is (5 cm)(10 cm)(20 cm) = 1000 cm³, so the brick is compressed by $(4.46 \times 10^{-6})(10^3 \text{ cm}^3) = \underline{4.46 \times 10^{-3} \text{ cm}^3}$. This is not a noticeable amount.

6.   The excess pressure at a depth of 1 m in blood would be <u>10.4 kPa</u> or <u>78 torr</u>, so the difference is significant.

7.   (*a*) The volume of the raft is

$$V = (2.44 \text{ m})(2.44 \text{ m})(0.305 \text{ m}) = 1.82 \text{ m}^3$$

and its mass is therefore

$$m = \rho V = (0.035)(10^3 \text{ kg/m}^3)(1.82 \text{ m}^3) = 63.7 \text{ kg}$$

When the raft floats in equilibrium, therefore, by Archimedes' principle it must be displacing 63.7 kg of water. Suppose the raft floats to a depth d in the water. The mass of water it displaces is then

$$m = \rho V = \rho A d = (10^3 \text{ kg/m}^3)(2.44 \text{ m})(2.44 \text{ m})(d)$$
$$= (5954 \text{ kg/m})(d)$$

In order for this to equal 63.7 kg

$$d = \frac{63.7 \text{ kg}}{5954 \text{ kg/m}} = 0.0107 \text{ m} = \underline{1.07 \text{ cm}}$$

The thing sat on top of the water—one of the sillier sights I ever saw.

   (*b*) Its total mass when loaded was

$$63.7 \text{ kg} + (3)(88 \text{ kg}) + 120 \text{ kg} = 447.7 \text{ kg}$$

so it floated at a depth

$$d = \frac{447.7 \text{ kg}}{5954 \text{ kg/m}} = 0.0752 \text{ m} = \underline{7.52 \text{ cm}}$$

It still looked pretty silly.

8.   The radius must be <u>4.32 m</u>.

9.   Let V be the volume of the wooden object. Then the total mass of fluid displaced is

$$m = \rho V_{\text{water}} + \rho V_{\text{oil}}$$

$$= (1000 \text{ kg/m}^3)(V/3) + (0.68)(1000 \text{ kg/m}^3)(2V/3)$$

$$= (787 \text{ kg/m}^3)V$$

This must equal the mass of the wooden object if it floats in equilibrium. Thus, the density of the wood is <u>787 kg/m³</u> or <u>0.787 g/cm³</u>.

10.   <u>0.015 g</u>

11. The height to which water would rise in the tube if it were open to the atmosphere at the top is given by

$$h = \frac{2\gamma \cos \theta}{\rho g r} = \frac{(2)(0.073 \text{ N/m})(\cos 0°)}{(1000 \text{ kg/m}^3)(9.81 \text{ m/s}^2)(4 \times 10^{-4} \text{ m})}$$

$$= 0.0372 \text{ m} = 3.72 \text{ cm}$$

Thus, the applied pressure must be equivalent to 15.0 cm + 3.72 cm = 18.72 cm of water:

$$P = \rho g h = (1000 \text{ kg/m}^3)(9.81 \text{ m/s}^2)(0.1872 \text{ m}) = \underline{1840 \text{ Pa}}$$

12.   (*a*) <u>54.4 ft/s</u>   (*b*) <u>21.6 lb/in²</u>

13. Using the continuity equation (Av = constant) shows that the velocity of the water in the tank at its top surface is extremely small since the area of this surface is very large compared to that of the hole. Thus, we can neglect this velocity. From Bernoulli's equation, the quantity $P + \frac{1}{2}\rho v^2 + \rho g h$ is constant for the steady flow of a nonviscous, incompressible fluid. Since the fluid is open to the atmosphere at the top, this quantity has the value

$$(1 \text{ atm}) + (0.5)(1000 \text{ kg/m}^3)(0)$$

$$+ (1000 \text{ kg/m}^3)(9.81 \text{ m/s}^2)(17.5 \text{ m})$$

$$= 1.013 \times 10^5 \text{ N/m}^2 + 1.717 \times 10^5 \text{ N/m}^2$$

$$= 2.730 \times 10^5 \text{ N/m}^2$$

there. At the point where the fluid flows out the hole, the pressure
is likewise atmospheric, so the Bernoulli's-equation quantity is

$$P + \tfrac{1}{2}\rho v^2 + \rho gh = 1.013 \times 10^5 \text{ N/m}^2 + \tfrac{1}{2}(1000 \text{ kg/m}^3)v^2$$

$$+ (1000 \text{ kg/m}^3)(9.81 \text{ m/s}^2)(10 \text{ m})$$

$$= 1.994 \times 10^5 \text{ N/m}^2 + (500 \text{ kg/m}^3)v^2$$

But this must be equal to 2.730×10^5 N/m^2 since this quantity remains constant. Thus,

$$(500 \text{ kg/m}^3)v^2 = 2.730 \times 10^5 \text{ N/m}^2 - 1.994 \times 10^5 \text{ N/m}^2$$

$$= 0.736 \times 10^5 \text{ N/m}^2$$

so

$$v^2 = \frac{7.36 \times 10^4 \text{ N/m}^2}{500 \text{ kg/m}^3} = 147.2 \text{ m}^2/\text{s}^2$$

and

$$v = 12.1 \text{ m/s}$$

The volume flow rate out the hole is thus

$$I_v = Av = (\pi)(0.005 \text{ m})^2(12.1 \text{ m/s}) = 9.53 \times 10^{-4} \text{ m}^3/\text{s}$$

When the water level has dropped by 0.1 m, the volume that has
flowed out of the tank is $(\pi)(9 \text{ m})^2(0.1 \text{ m}) = 25.4$ m^3. For this much
water to flow out the hole it takes time

$$t = \frac{V}{I_v} = \frac{25.4 \text{ m}^3}{9.53 \times 10^{-4} \text{ m}^3/\text{s}} = 2.67 \times 10^4 \text{ s} = \underline{7.42 \text{ h}}$$

14. 105.7 kPa

15. There will be a pressure drop due to viscous flow in the pipe
(given by Poiseuille's law) and a pressure drop due to the 19-m
vertical rise. Since the outlet is open to the atmosphere, the total
pressure difference is the 520 kPa given. The pressure drop due to
the vertical rise is

$$P = \rho g \,\Delta y = (1000 \text{ kg/m}^3)(9.81 \text{ m/s}^2)(19 \text{ m})$$

$$= 1.86 \times 10^5 \text{ Pa}$$

The remaining pressure drop of 520 kPa $-$ 186 kPa $=$ 334 kPa
must be due to viscosity in the pipe. From Poiseuille's law,

$$\Delta P = \frac{8\eta L}{\pi r^4} I_v$$

so

$$I_v = \frac{\pi r^4}{8\eta L} \Delta P$$

$$= \frac{(\pi)(0.005 \text{ m})^4}{(8)(0.001 \text{ N·s/m}^2)(5 \times 10^3 \text{ m})} (3.34 \times 10^5 \text{ N/m}^2)$$

$$= \underline{1.64 \times 10^{-5} \text{ m}^3/\text{s}} = \underline{16.4 \text{ cm}^3/\text{s}}$$

16. It takes $\underline{3.6 \text{ kPa}}$ or around $\underline{27 \text{ torr}}$.

17.   (a) We get the flow rate from Poiseuille's law. The pressure drop is 40 lb/in^2 = 276 kPa, the pipe's diameter is 1 in = 0.0254 m, and its length is 50 ft = 15.2 m. Thus,

$$I_v = \frac{\pi r^4}{8\eta L} \Delta P = \frac{(\pi)(1.27 \times 10^{-2} \text{ m})^4}{(8)(0.001 \text{ Pa·s})(15.2 \text{ m})} (2.76 \times 10^5 \text{ Pa})$$

$$= 0.185 \text{ m}^3/\text{s} = \underline{11.1 \text{ m}^3/\text{min}}$$

(b) The velocity of the water leaving the pipe outlet is

$$v = \frac{I_v}{A} = \frac{0.185 \text{ m}^3/\text{s}}{(\pi)(1.27 \times 10^{-2} \text{ m})^2} = 365 \text{ m/s}$$

The Reynolds number is therefore

$$N_R = \frac{2r\rho v}{\eta} = \frac{(2)(0.0127 \text{ m})(1000 \text{ kg/m})(365 \text{ m/s})}{0.001 \text{ Pa·s}} = \underline{9.3 \times 10^6}$$

The conditions given describe a flow that, if possible at all, would be very highly turbulent. Thus, the flow rate through the pipe would in fact be much slower than that calculated in part (a).

CHAPTER 11

Temperature

I. Key Ideas

Thermodynamics *Thermodynamics* is the study of temperature, heat, and energy exchange. What we experience as the *temperature* of an object or a substance is a measure of its average internal (molecular) kinetic energy, and *heat flow* is just an exchange of energy between two objects because of a difference in temperature.

Thermometers and Temperature Various physical properties of an object change when its temperature is changed, and any such property can be used as the basis of a thermometer; a very familiar example is a liquid sealed in glass. A temperature scale is established by assigning values when the thermometer is in thermal equilibrium with some reproducible system, such as water at its boiling point.

Temperature Scales The Celsius scale assigns values of zero to the freezing point of water and 100 "degrees" to its boiling point at normal atmospheric pressure. The Fahrenheit scale is still in common use in the United States. The drawback that most thermometers have is that the temperature readings of different thermometers do not agree except at the defining temperatures because the different thermometric properties do not change with temperature in the same way.

Gas Thermometers One class of thermometers, however, can be made to agree at all temperatures. These are the thermometers that depend on the properties of a gas. If a gas is confined to a constant volume, its pressure increases with increasing temperature; likewise, if it is held at constant pressure, its volume increases when it is heated. In the limit of very low gas pressure, all such "gas thermometers" agree at all temperatures.

Kelvin Temperature Scale If the pressure in a constant-volume gas thermometer is extrapolated to zero, it goes to zero at $-273.15°C$. The Kelvin scale defines temperature as proportional to pressure. The ice and steam points of water are less easy to reproduce very precisely than is the *triple point*, which is the single state (at a temperature of $0.01°C$) at which ice, water, and steam can coexist in equilibrium. The *Kelvin scale* is defined by making the triple-point temperature 273.16 K. It will turn out to be identical with an *absolute temperature scale*, which can be defined independent of any particular measuring device.

Ideal-Gas Law A gas expands when the pressure on it is decreased and vice versa. Experimentally, at constant temperature, the product PV remains constant. This is *Boyle's law*; it is approximately true for all gases in the limit of low pressures. Since, by definition, the Kelvin temperature of a gas is proportional to its pressure at constant volume, the combination PV/T is constant. This is the *ideal-gas law*; it is an *equation of state* for a gas that is valid for all gases in the low-pressure limit.

Avogadro's Number The amount of gas in the ideal-gas law is most conveniently expressed as the number of moles n. A mole is the quantity of a substance having a mass in grams equal to the molecular weight of the substance; it is Avogadro's number of atoms or molecules.

Kinetic Theory of a Gas The properties of an ideal gas can be understood in terms of a simple *kinetic* model. We can picture the gas as consisting of noninteracting molecules in motion and the pressure as the result of collisions of the gas molecules with the walls of the container. This argument shows that the temperature of the gas is a measure of the average translational kinetic energy of the molecules.

II. Numbers and Key Equations

Numbers

Avogadro's number

$$N_A = 6.02 \times 10^{23}$$

Gas constant

$$R = 8.31 \text{ J/mol·K}$$
$$= 0.0821 \text{ L·atm/mol·K}$$

Boltzmann's constant

$$k = 1.38 \times 10^{-23} \text{ J/K}$$

Triple-point temperature of water

$$T_3 = 273.16 \text{ K}$$

Degree equivalents

$$1 \text{ K} = 1 \text{ C}° = 1.80 \text{ F}°$$

Key Equations

Celsius-Fahrenheit conversion

$$t_C = \frac{5}{9}(t_F - 32 \text{ F}°)$$

$$t_F = \frac{9}{5}t_C + 32 \text{ F}°$$

Absolute temperature

$$T = \frac{273.16 \text{ K}}{P_3}P$$

$$T = t_C + 273.15$$

Ideal-gas law

$$PV = NkT = nRT \qquad n = N/N_A$$

Kinetic theory

$$PV = \tfrac{1}{3}Nmv^2 \qquad \tfrac{1}{2}(mv^2)_{av} = \tfrac{3}{2}kT$$

Molecular energy of a gas

$$E_k = \tfrac{3}{2}NkT = \tfrac{3}{2}nRT$$

Root-mean-square speed of a gas molecule

$$v_{rms} = \left(\frac{3RT}{m}\right)^{1/2}$$

III. Possible Pitfalls

Be careful not to confuse a value of temperature with a change or interval of temperature. The text makes the distinction by writing, for example, °C and C°, but there is no corresponding distinction for kelvins.

Many of the equations you will use in this chapter and the next couple are valid *only* if temperature is expressed in kelvins. Be sure you know for which equations this is true. Of course, if it's a temperature *difference* that's referred to, a kelvin and a Celsius degree are equal.

At a given temperature, molecules of different gases have the same (average) *kinetic energy*. Their speeds will *not* be the same because the atomic masses of different gases are not the same.

In applying the ideal-gas law, be sure that all quantities, including the gas constant R, are in consistent units. Trying to use pressure in torr,

volume in cubic metres, and R in L·atm/mol·K will only give you a headache.

If the ideal-gas law is used to calculate *proportionally*, as, for example, with Boyle's law $P_1V_1 = P_2V_2$, you can use any units you like. (Of course, they have to be the same on both sides of the equation.)

IV. Questions and Answers

Questions

1. Distinguish between 1°C and 1 C°. What is the corresponding distinction on the Kelvin scale?

2. In what way is the Celsius scale more convenient than the Kelvin scale for everyday use? In what way is the Kelvin scale more suitable than the Celsius scale for scientific use?

3. What change must be made in the temperature of a gas in order to have the average speed of its molecules?

4. Two different gases are at the same temperature. How do the average speeds of their molecules compare? What about their average molecular kinetic changes?

5. Negative Celsius temperatures are possible—distressingly common, in fact, where I live. Negative Kelvin temperatures are not. Why not?

6. What properties should a physical system possess in order to serve as a good thermometer?

7. Is it meaningful to speak of the temperature of a "gas" consisting of only one molecule? What about two molecules?

8. For an ideal gas, how does the density of the gas depend on the temperature? On the pressure?

9. How many grams are there in a mole of carbon monoxide gas (molecular formula CO)? How many molecules?

Answers

1. The expression 1°C refers to a temperature: 34°F, a little above freezing, a nice December day. The expression 1 C° is a temperature difference; say, the amount that the temperature of a pan of water rises if you deliver a certain amount of heat to it. There is no corresponding distinction on the Kelvin scale; 1 K can be either a temperature or a temperature difference.

2. On the Celsius scale, the temperatures that we deal with in ordinary matters tend to be small numbers, which are more convenient to use and remember than temperatures of several hundred kelvins would be. The Kelvin scale is more suitable for scientific use in that its zero corresponds to a significant physical limit, and it is identical with the absolute temperature scale, which is independent of any particular thermometric device.

3. The absolute temperature of the gas must be decreased by a factor of four since it is proportional to kinetic energy and thus to speed squared.

4. If two gases are at the same temperature, they have the same average kinetic energy per molecule, by definition. If they do not happen to have the same molecular mass, however, then the molecules of the lighter gas will be moving at higher speeds on the average.

5. A negative Celsius temperature just means "colder than the freezing point of water." The Kelvin temperature, however, is proportional to the kinetic energy per molecule, and kinetic energy can never be negative.

6. It depends a lot on what the thermometer is supposed to be good for, but there are some general requirements that apply to most cases. Whatever thermometric property is to be used—length, electrical resistance, pressure, or whatever—should be easily measurable itself and should vary linearly, or nearly so, with absolute temperature; otherwise, the scale will be usable only in a few narrow ranges of temperature. In most cases, the thermometer will need to be physically small so that it will come quickly to equilibrium with whatever system is to be measured and also so that it will affect that system as little as possible. Other considerations will arise in each particular situation.

7. Temperature is an intrinsic property of a system; to use the language of mechanics, it measures the kinetic energy of motion relative to the center of mass of the system, not motion of the system as a whole. For a "system" of only one particle, there is no internal motion, and the concept of the temperature of such a system is meaningless. Strictly speaking, temperature could have meaning for a system of two particles, although we usually use it only when referring to a system containing many particles.

8. Density ρ is mass per unit volume. If M is the molar mass of the gas,

$$PV = nRT = \frac{m}{M} RT \qquad \rho = \frac{m}{V} = \frac{MP}{RT}$$

Thus, the density is inversely proportional to the absolute temperature T of the gas and directly proportional to the pressure P.

9. The molar mass of CO is $12 + 16 = 28$, so a mole of CO has a mass of 28 g. There are Avogadro's number (6.02×10^{23}) of molecules in a mole of anything.

V. **Problems and Solutions**

Problems

In certain of these problems you will need the molar mass of some gas. These values can be calculated from the atomic mass values given in Appendix F of the text.

1. A certain constant-volume gas thermometer reads a pressure of 88 torr at the temperature of the triple point of water. (*a*) What will the pressure in the thermometer be at a temperature of 310 K? (*b*) What is the thermometer temperature when the pressure reads 100 torr?

How to Solve It
• For the constant-volume gas thermometer, the absolute temperature is proportional to the pressure reading.

• Part (*b*) is just the same calculation in reverse.

2. Mercury freezes at $-39°C$. Express this temperature on the Fahrenheit and Kelvin scales.

How to Solve It
• To convert from C° to kelvins you need only add 273.15 to the Celsius temperature; a kelvin and a Celsius degree are the same.

- There is a formula in the text for relating the Celsius and Fahrenheit scales.

3. You inflate your bicycle tires to 75 lb/in^2 on the gas station gauge the first thing on a cool morning when the thermometer reads 52°F. You then spend the day riding on hot pavement, raising the temperature of your tires to 125°F. Assuming that the volume of the tires hasn't changed, what is the pressure in them now?

How to Solve It
- Assume that the air in your tires obeys the ideal-gas law.

- Notice that what the gauge reads is (reasonably enough) gauge pressure. To get absolute pressure, add the value of atmospheric pressure, 14.7 lb/in^2.

- Since the volume of the air remains constant, the pressure and the absolute temperature are inversely proportional.

- Remember to convert the Fahrenheit temperatures to kelvins.

4. (*a*) If 3 mol of an ideal gas occupy a volume of 40 L at a pressure of 1 atm, what is the temperature of the gas? (*b*) If the gas is heated at constant pressure to 273 K, what volume does it occupy now?

How to Solve It
- Use the ideal-gas law directly to find the temperature in part (*a*). Be careful of units!

- In part (*b*), use the same equation to find the volume, given the new temperature.

5. The gas in intergalactic space is mostly atomic hydrogen at a temperature of 3 K and a pressure of the order of 10^{-21} atm. How many H atoms are there per cubic metre of space? (*Note:* It's interesting to compare this result with that of the next example, which deals with a very good laboratory vacuum.)

How to Solve It
- The number of moles in a sample of gas can be calculated from the ideal-gas law.

- Each mole of hydrogen corresponds to Avogadro's number of atoms.

6. A high-vacuum pump reduces the pressure in a container to 10^{-8} torr. If the temperature is 20°C, how many gas molecules per cubic metre are there in the container?

How to Solve It

- Put P, T, and the gas constant R in consistent units. Unit conversions play an important role in this sort of problem.

- Solve the ideal-gas law for n/V, the number of moles per unit volume. Each mole corresponds to Avogadro's number of molecules.

- Convert the result to molecules per cubic metre.

7. The mass of a certain gas sample that occupies 3 L at 20°C and 10 atm pressure of is found to be 55 g. What gas is this?

How to Solve It

- Again you must put all the quantities you will use into consistent units.

- Then you can solve the ideal-gas law for the number of moles of gas in the sample.

- Knowing the mass of this many moles, you can calculate the molar mass, after which there aren't many possibilities for what the gas is.

8. A high-pressure gas cylinder has a mass of 21.22 kg when empty. Its interior volume is 1.33 L. It is filled at room temperature (20°C) with nitrogen gas, after which its mass is 21.61 kg. What is the gas pressure in the filled cylinder?

How to Solve It

- From the data given, find the mass of gas in the container.

- Since you know the molecular weight of nitrogen gas, this tells you the number of moles of nitrogen in the cylinder. Don't forget that nitrogen gas is diatomic!

- Now you can solve the ideal-gas law for the pressure in the cylinder.

9. Gas confined in a cylinder at constant volume is initially at a pressure of 10^6 Pa and a temperature of 10°C. The cylinder is immersed in boiling water and allowed to come to thermal equilibrium. While the cylinder is still in the water, a valve is opened and gas is vented from it until the pressure returns to 10^6 Pa. The valve is then closed, and the cylinder, with the remaining gas, is removed from the water and cooled back to 10°C. What is the final pressure in the cylinder?

How to Solve It

- The problem is best treated in three pieces.

- We can assume that the cylinder's volume is constant. Thus, while the cylinder is being heated, its pressure is proportional to its absolute temperature.

- When the gas is vented at constant volume and temperature, the pressure changes proportional to the number of moles n of gas in the cylinder.

- Finally, the gas is cooled again at constant volume.

10. Two gases present in the atmosphere are water vapor (H_2O) and argon. What is the ratio of their rms speeds?

How to Solve It

- In thermal equilibrium, the mean kinetic energy per molecule is the same for both species.

- Since you also know the ratio of their molecular masses, you can solve for the ratio of their velocities.

11. The temperature at the surface of the sun is about 6000 K, and the sun consists primarily of atomic hydrogen (not H_2). Calculate the rms speed of a hydrogen atom at the surface of the sun, and compare it to the escape speed from the sun's surface. The sun's radius is 6.96×10^8 m and its mass is 1.99×10^{30} kg.

How to Solve It

- Calculate the rms speed of a hydrogen atom at 6000 K.

- Find the escape velocity at the surface of the sun from the formulas in Chapter 9.

- The ratio of these speeds is significant in astrophysics. The sun consists mostly of hydrogen; this calculation tells us whether or not gravity is strong enough to hold it together.

12. (*a*) What is the total kinetic energy of the atoms in 1 L of neon gas (atomic weight 22) at 20°C and 1 atm pressure. (*b*) If the gas is expanded at constant temperature to a volume of 2 L, how much does this total kinetic energy change?

How to Solve It
- The average kinetic energy of a gas molecule is $\frac{3}{2}kT$.

- Use the ideal-gas law to calculate the number of neon atoms present.

- The number of atoms multiplied by the average kinetic energy of each one gives the total kinetic energy.

- Given constant temperature, how will expansion change this result?

Solutions

1. If we call the temperature of the triple point of water T_3, the temperature T or pressure P read by the thermometer can be found using

$$\frac{P}{T} = \frac{P_3}{T_3} = \frac{88 \text{ torr}}{273.16 \text{ K}}$$

where $P_3 = 88$ torr.
  (*a*) When $T = 310$ K,

$$P = \frac{(88 \text{ torr})(310 \text{ K})}{273.16 \text{ K}} = \underline{99.87 \text{ torr}}$$

  (*b*) When $P = 100$ torr,

$$T = \frac{(273.16 \text{ K})(100 \text{ torr})}{88 \text{ torr}} = \underline{310.4 \text{ K}}$$

2. $\underline{-38.2°F; \quad 234.2 \text{ K}}$

3. Apply the ideal-gas law in the form

$$\frac{P_1 V_1}{T_1} = \frac{P_2 V_2}{T_2}$$

The temperatures must be converted to an absolute scale. When the Fahrenheit temperature is 52°,

$$t_{C1} = \frac{5}{9}(52 - 32) = 11.1°C$$

$$T_1 = 11.1 + 273.15 = 284.3 \text{ K}$$

In just the same way, 125°F corresponds to $T_2 = 324.8$ K, and the gauge pressure of 75 lb/in² corresponds to an absolute pressure of

$$P_1 = 75 \text{ lb/in}^2 + 14.7 \text{ lb/in}^2 = 89.7 \text{ lb/in}^2$$

Since we assume that the volume hasn't changed,

$$P_1/T_1 = P_2/T_2$$

$$\frac{89.7 \text{ lb/in}^2}{284.3 \text{ K}} = \frac{P_2}{324.8 \text{ K}}$$

so

$$P_2 = 102.5 \text{ lb/in}^2 \text{ absolute} = \underline{85.8 \text{ lb/in}^2 \text{ gauge}}$$

4.   (a) 162.5 K   (b) 67.2 L

5.   The ideal-gas law

$$PV = nRT$$

gives

$$\frac{n}{V} = \frac{P}{RT}$$

$$= \frac{10^{-21} \text{ atm}}{(0.0821 \text{ atm·L/mol·K})(3 \text{ K})}$$

$$\approx 4.1 \times 10^{-21} \text{ mol/L}$$

Thus, the number of atoms in one cubic metre is

$$\frac{N}{V} = (4.1 \times 10^{-21} \text{ mol/L})(10^3 \text{ L/m}^3)(6.02 \times 10^{23} \text{ atoms/mol})$$

$$= \underline{2.4 \times 10^6 \text{ H atoms/m}^3}$$

6. $\underline{3.3 \times 10^{14} \text{ molecules/m}^3}$

7.   What you can get from the data given is the molar mass of the gas. From the ideal-gas law,

$$PV = nRT$$

so

$$n = \frac{PV}{RT} = \frac{(10 \text{ atm})(3 \text{ L})}{(0.0821 \text{ L·atm/mol·K})(293 \text{ K})} = 1.25 \text{ mol}$$

Since we know that this much gas has a mass of 55 g,

$$M = \frac{55 \text{ g}}{1.25 \text{ mol}} = \underline{44.0 \text{ g/mol}}$$

The gas may be anything with a molar mass of 44.0. I had CO_2 (12 + 16 + 16 = 44) in mind when I wrote the problem, but it could just as well be N_2O (14 + 14 + 16 = 44).

8. $\underline{252 \text{ atm}}$ or $\underline{255 \times 10^7 \text{ Pa}}$

9. For any ideal gas, the quantity PV/nT is constant. In this case, the volume of the gas remains constant throughout, so P is directly proportional to nT; that is, P/nT is constant. During the heating, T increases from 283 K (10°C) to 373 K (100°C), the temperature of boiling water:

$$\frac{10^6 \text{ Pa}}{(n_1)(283 \text{ K})} = \frac{P_2}{(n_1)(373 \text{ K})}$$

giving

$$P_2 = 1.32 \times 10^6 \text{ Pa}$$

Then the gas is vented, reducing the number of moles of gas in the cylinder:

$$\frac{1.32 \times 10^6 \text{ Pa}}{(n_1)(373 \text{ K})} = \frac{10^6 \text{ Pa}}{(n_2)(373 \text{ K})}$$

giving

$$n_2 = 0.759 n_1$$

Finally, the gas is cooled back to 10°C:

$$\frac{10^6 \text{ Pa}}{(n_2)(373 \text{ K})} = \frac{P_3}{(n_2)(283 \text{ K})}$$

which gives

$$P_3 = \underline{7.59 \times 10^5 \text{ Pa}}$$

Actually, the way the problem is stated, we could have skipped the second step, as we never needed to know n_2.

10. The ratio is $\underline{1.49 \text{ to } 1}$.

11. The escape speed at the surface of the sun is given by

$$v_e^2 = \frac{2GM}{R} = \frac{(2)(6.67 \times 10^{-11} \text{ N·m}^2/\text{kg}^2)(1.99 \times 10^{30} \text{ kg})}{6.96 \times 10^8 \text{ m}}$$

$$= 3.81 \times 10^{11} \text{ m}^2/\text{s}^2$$

so

$$v_e = 6.17 \times 10^5 \text{ m/s}$$

Now

$$\tfrac{1}{2}(mv^2)_{av} = \tfrac{3}{2}kT$$

so

$$(v^2)_{av} = \frac{3kT}{m}$$

The mass of a hydrogen atom is

$$m = \frac{M}{N_A} = \frac{1.008 \text{ g/mol}}{6.02 \times 10^{23} \text{ atoms/mol}} = 1.67 \times 10^{-24} \text{ g}$$

$$= 1.67 \times 10^{-27} \text{ kg}$$

so

$$(v^2)_{av} = \frac{(3)(1.38 \times 10^{-23} \text{ J/K})(6000 \text{ K})}{1.67 \times 10^{-27} \text{ kg}} = 1.49 \times 10^8 \text{ m/s}$$

or

$$v_{rms} = \underline{1.22 \times 10^5 \text{ m/s}}$$

Since this is only about 2 percent of the escape speed, hydrogen at this temperature will not readily escape from the sun's surface. Some hydrogen nuclei (protons) are driven off the surface by radiation pressure, which we don't take into account here.

12.   (*a*) 152 J   (*b*) It does not change.

CHAPTER 12

Heat and the First Law of Thermodynamics

I. Key Ideas

Heat *Heat* is energy that is transferred from one object to another because of a temperature difference between them. Heat drawn from or added to a body represents a change in the internal molecular energy of the body. Heat, of itself, can't be regarded as a conserved quantity, but heat exchange is a part of the conservation of energy when all forms of energy are accounted for.

Heat Capacity The transfer of heat is reflected by changes in temperature. The amount of heat needed to raise the temperature of an object by one degree is called its *heat capacity*. The heat capacity per unit mass of a substance is its *specific heat*. The specific heat of water is quite large compared to that of other ordinary materials; this accounts for some of the special properties of water in nature. The *molar heat capacities* of metals are all about the same, just about 3*R*.

The Calorie The specific heat of water is nearly constant, and the heat capacities of other materials are measured by comparison with water. The *calorie*, now defined as 4.184 J, is approximately the amount of heat required to raise the temperature of 1 gram of water by 1 K. The calorie remains in common use although it is just an alternate unit of energy. The "calorie" used in measuring the nutritional value of food is actually a kilocalorie.

Heat Capacity and Work If a material is allowed to expand as it is heated (as nearly all materials tend to do), *more* heat input is required for a given temperature change than if the volume of the material is held constant because the expanding material must do work against its surroundings. For solids and liquids, the difference is very small, and the specific heat at constant pressure is normally quoted since maintaining constant volume is difficult. For gases, however, the heat capacity at constant volume is substantially less than that at constant pressure because gases expand very freely.

The First Law of Thermodynamics The *first law of thermodynamics* is simply a statement of the conservation of energy that includes heat transfer. The heat delivered to a system is equal to the increase in the system's internal energy plus the amount of work done *by* the system on its surroundings. *Joule's experiment* measured the amount of mechanical work required to produce a given amount of heat.

Internal Energy In a process in which zero work is done by (or on) a system, the change in the system's internal energy is equal to the heat input. The internal energy of a system is a definite property of the state of the system, but heat and work are not, as they depend on *how* the change of state is made.

The Ideal Gas An ideal gas is a particularly simple system; its state can be specified by just two variables, like pressure and volume, and thus as a point on a graph of pressure versus volume (a *PV diagram*). Suppose the gas expands slowly at constant pressure against a piston; the work it does is $P \Delta V$. (The work done in an arbitrary change of state is equal to the area under the *PV* curve.) The internal energy (and temperature) of the gas will decrease as it expands unless the energy expended doing work is made up by heating the gas.

Internal Energy of an Ideal Gas The internal energy of a material depends upon its molecular structure. The molar heat capacity of a gas at constant pressure is larger than that at constant volume because of the work done by the expanding gas at constant pressure; for an ideal gas, they differ by the gas constant R. If the translational kinetic energy of a gas ($\frac{3}{2}kT$ per molecule) is its only internal energy, then its molar heat capacity at constant volume should be just $\frac{3}{2}R$. Empirically, this holds for monatomic gases but not for more complex molecules.

The Equipartition Theorem The average kinetic energy per molecule of a monatomic gas is $\frac{3}{2}kT$, because there are three equivalent, independent directions of motion, and a kinetic energy $\frac{1}{2}kT$ is associated with each. For gases with complex molecules, other types of motion are also possible—for example, rotation and vibration. The theorem of equipartition of energy states that, on the average, each such *degree of freedom* has an equal share, $\frac{1}{2}kT$ per molecule, of the total energy. Diatomic molecules can rotate in either of two ways, so there are five degrees of freedom per molecule, and the internal energy is $\frac{5}{2}nRT$.

Solids The molecules in a solid are fixed in place but each can vibrate in three dimensions. Solids thus have a total of six degrees of freedom, and the equipartition theorem predicts a molar heat capacity of $3R$ for every crystalline solid. This is the *Dulong-Petit law*.

II.   Numbers and Key Equations

Numbers

$1 \text{ cal} = 4.184 \text{ J}$

$1 \text{ Btu} = 252 \text{ cal} = 1054 \text{ J}$

$1 \text{ L·atm} = 101.3 \text{ J}$

Gas constant

$R = 8.31 \text{ J/mol·K} = 1.987 \text{ cal/mol·K}$

Key Equations

Heat and temperature

$$Q = C \Delta T = mc \Delta T$$

Molar heat capacity

$$C_m = Mc \qquad C = nC_m$$

Heat capacity of a solid (Dulong-Petit law)

$$C_m \approx 3R$$

First law of thermodynamics

$$Q = \Delta U + W$$

Internal energy change

$$U = mc_v \Delta T$$

Work done in expansion

$$W = P \Delta V$$

Isothermal work by an ideal gas

$$W_{iso} = nRT \ln(V_2/V_1)$$

Ideal-gas heat capacities

$$C_p = C_v + nR$$

Internal energy of an ideal gas

$$U = \tfrac{3}{2}nRT \qquad \text{(monatomic)}$$

$$U = \tfrac{5}{2}nRT \qquad \text{(diatomic)}$$

Internal energy of a solid

$$U = 3nRT$$

III.   Possible Pitfalls

When the number of moles n doesn't appear in the heat transfer formulas, work formulas, and so forth for a gas, the quantities being calculated are heat or work *per mole*. Conversely, when you're dealing with an arbitrary amount of gas, there should be an n in the formula.

Be very careful of *signs* in first-law problems, heat exchange problems, and such; it's awfully easy to get mixed up. The convention followed here is that Q is positive when heat is *added* to a system and W is positive when work is done *by* the system.

In doing first-law problems, don't assume that heat flow and temperature change always go together. There can be a temperature increase without any heat transfer, and there can be a heat transfer without any temperature change.

The only type of system for which the internal energy depends only on temperature is an ideal gas. Don't try to use this idea in other kinds of systems.

The formula $W = P \, \Delta V$ holds only if the volume change ΔV is carried out at constant pressure. Otherwise, the work must be calculated graphically or by using calculus.

When a gas expands, it does positive work on its surroundings, pushing its surroundings out of its way. When a gas is compressed—that is, when its volume is decreased—the work done *by* the gas is always negative.

The ideal-gas law relates initial and final states of some process, but (regardless of temperature change) it doesn't say whether heat flowed to or from the gas in the process. The amount of heat exchanged depends on how the process was carried out.

The formula $Q = mc \, \Delta T$ assumes that c is constant over the temperature interval ΔT. In fact, this is only an approximation (but an acceptable one in most of the problems you'll be doing); actually, there is some dependence of heat capacity on temperature for all materials.

IV.   Questions and Answers

Questions

1. Can a system absorb heat without its temperature increasing? Can a system absorb heat without its internal energy increasing?

2. The temperature of 2 L of water is increased from 20 to 30°C. Can this be done without any work being done on the water? Explain.

3. For solids and liquids, in practice, you usually don't distinguish between specific heats at constant volume and at constant pressure. Why not? If you did want to make this distinction, which one would be larger? Which specific heat are you most likely to find tabulated?

4. A gas expands at constant temperature and does work on a piston, say. Does the internal energy of the gas decrease? If not, where does the energy to do the work come from?

5. Give an example of a process in which heat is added to a system without changing its temperature. Give an example of a process in which the temperature of a system is increased without any heat input or output.

6. A gas expands slowly to twice its initial volume (*a*) at constant pressure and (*b*) at constant temperature. In which case does it do more work on its surroundings? Why? Answer in terms of the first law of thermodynamics.

7. Consider nitrogen (N_2) and helium (He) gases. For which gas is the internal energy of 1 mol greater at a given temperature? For which gas is the internal energy of 1 kg greater at a given temperature?

8. The ratio of molar heat capacities (C_{mp}/C_{mv}) for a monatomic ideal gas is 5/3; for a diatomic gas it is 7/5. Would you expect the ratio to be higher or lower for a gas having three or more atoms per molecule?

9. A body of water is heated at constant pressure from 20 to 40°C. Explain why it is incorrect to say that the water "contains more heat" than it did initially.

Answers

1. According to the first law of thermodynamics, if heat is added to a system and work is done by the system in the same process, the internal energy (and also the temperature) of the system may increase, decrease, or stay the same.

2. If during the temperature increase, the volume of the water is kept the same (which by the way is very difficult to do), then no work has been done.

3. There is a distinction, and for any material that expands when its temperature is increased, c_p is greater than c_v, just as for a gas. You pay no attention to the distinction because the volume change is very small. What one ordinarily measures and tabulates is c_p.

4. If the gas is ideal, its internal energy doesn't change in an isothermal process since the internal energy of an ideal gas depends only on its temperature. The work that is done must be equal to the heat that was added to the gas during the expansion to maintain the constant temperature. This is just the first law of thermodynamics.

5. An ideal gas that expands isothermally absorbs heat even though its temperature remains constant. If a gas in a thermally insulated container expands (without heat transfer), it cools; the work it has done in expanding must have come from the decrease in its internal energy.

6. If the gas expands at constant pressure, the work it does is $P \Delta V$. When the gas undergoes the same expansion at constant temperature, its pressure drops as it expands. Thus, in the second case, the pressure is lower during the expansion, and, consequently, the work the gas does is less—it "shoves its surroundings out of the way" with a greater force.

7. The molar masses of N_2 and He are 28 g/mol and 4 g/mol, respectively. Since nitrogen is diatomic, its internal energy is $\frac{5}{2}RT$ per mole and therefore $\frac{25}{28} \approx 0.089RT$ per gram. Helium is monatomic, so its internal energy is $\frac{3}{2}RT$ per mole and therefore $0.375RT$ per gram. Thus, nitrogen has the greater internal energy per mole, but helium has the larger internal energy per unit mass.

8. The ratio of heat capacities is

$$\frac{C_{mp}}{C_{mv}} = \frac{C_{mv} + R}{C_{mv}} = 1 + \frac{R}{C_{mv}}$$

which is smaller when the number of degrees of freedom is larger (because C_{mv} is then larger). Thus, for polyatomic gases, the ratio is even smaller than it is for diatomic gases; empirically, many such gases have a ratio of around 1.35.

9. Heat is not a function of state. It is incorrect here to say that the water at 40° "contains more heat" because the same state change could be brought about in a different way with less heat input or even with none.

V. Problems and Solutions

Problems

1. A 160-g mass of a certain metal at an initial temperature of 88°C is dropped into 140 g of water in an insulated container. The water is initially at 10°C. The system finally comes to equilibrium at

18.4°C. (*a*) If heat losses to the container and the surroundings are negligible, what is the specific heat of the metal? (*b*) If the Dulong-Petit law holds, what is the molar (atomic) mass of the metal?

How to Solve It
- You know the heat capacity, temperature change, and mass of the water, so you can calculate the heat transferred from the unknown metal to the water. Use this together with the temperature drop and mass of the metal to calculate the specific heat of the metal.

- The specific heat times the molar mass gives the heat capacity per mole.

- According to the Dulong-Petit law, the heat capacity per mole is approximately equal to $3R$.

- Use the two expressions for heat capacity per mole to find the molar mass.

- Knowing the molar mass, you can use Appendix F in the text to find out what the metal is.

2. The specific heat of copper is 0.0923 kcal/kg·C°. If 180 g of copper at 200°C is dropped into an insulated container containing 280 g of water at 20°C, what is the final equilibrium temperature of the copper and water?

How to Solve It
- If the heat losses to the container and the surroundings are negligible, the heat lost by the copper is equal to that gained by the water.

- You know the heat capacity of each material, so you can set up equations for Q_{in} and Q_{out}, which you can solve to find the final temperature.

- The answer you get will be higher than would actually occur since, when you first drop the copper in, a lot of its heat will go into boiling water away until its surface temperature falls below 100°C. Real problems are mostly hard.

3. A piece of iron (specific heat 0.103 kcal/kg·C°) of mass 80 g at a temperature of 98°C is dropped into an insulated vessel containing 120 g of water at 20°C. At what final temperature does the system come to equilibrium?

How to Solve It
- If the heat losses to the container and the surroundings are negligible, the heat lost by the iron is equal to that gained by the water.

- You know the heat capacity of each material, so you can set up equations for Q_{in} and Q_{out}.

- Solve these to find the final temperature of the iron and water.

4. Imagine that you want to take a warm bath, but there's no hot water. You draw 40 kg of tap water at 18°C in the bathtub and heat water on your stove to warm the bath water up. If you heat the water to 100°C in a 2-L saucepan, how many panfuls must you add to raise the temperature of the bath to 40°C?

How to Solve It

- Forty kilograms of water is to be warmed from 18°C to 40°C while an unknown amount of water is to be cooled from 100°C to 40°C.

- If heat losses to the surroundings can be neglected, this tells you what the unknown amount of water must be.

- How many 2-L panfuls is this?

5. A lead bullet strikes and is stopped in a block of wood. The bullet's temperature rises from 18°C to 85°C in the process. Assuming that all of the bullet's initial kinetic energy goes into heating the lead, how fast was it going when it hit the block? The specific heat of lead is 0.0305 cal/g·C°.

How to Solve It

- You don't know the mass of the bullet—but you don't need it. Calculate the internal energy increase of the lead per unit mass from the temperature change given.

- Since we assume all the bullet's initial kinetic energy is transformed into internal energy of the bullet, we know the initial kinetic energy per unit mass of the bullet.

- Calculate the speed of the bullet.

6. A hydrotherapy unit contains 40 kg of water and is heated by a 1500-W heater. In a 10-min period, 4.00×10^5 J of work is done on the water by bubbling high-pressure air through it, and 88 kcal of heat is lost to the surroundings. (*a*) By how much was the internal energy of the water increased in this period? (*b*) If the water was initially at 16°C, what is its final temperature?

How to Solve It

- Both the heat supplied by the resistance heater and the heat loss to the surroundings contribute to Q.

- Since the work done on the water is given, you can calculate the change in its internal energy from the first law.

- If this net increase in internal energy had been delivered as heat, by how much would the temperature of the water have increased?

7. Ice expands as its temperature increases, as do most materials. (We'll discuss thermal expansion more fully in the next chapter.) For ice, the volume increase per K is $\beta = 1.53 \times 10^{-4}$. The specific gravity of ice is 0.92 and its molar mass is 18.02 g. Calculate the difference between the heat capacities per mole of ice at constant volume and at constant pressure.

How to Solve It

- The difference in the heat capacities is due to the work that the ice does on its surroundings as it expands when it is heated at constant pressure.

- Since the pressure is not specified, assume that atmospheric pressure is meant.

- What is the volume of a mole of ice? By how much does it expand when its temperature is increased by 1 K?

8. A sample of gas is initially in a 15-L container at 20°C and a pressure of 240 kPa. It is compressed at constant pressure to a volume of 6 L. How much work is done by the gas in the process?

How to Solve It

- In a change of state at constant pressure, the work that the gas does is the pressure multiplied by the increase in volume.

- If the gas is compressed, the work that it does is negative.

9. In the PV diagram shown in Figure 12-1, path A is an isothermal expansion of a monatomic ideal gas, and path B is an expansion at constant pressure followed by cooling at constant volume. For each path, calculate the heat input, the work done, and the change in internal energy for 1 mol of the gas.

How to Solve It

- There is a formula for the work done in an isothermal expansion of an ideal gas.

- In an isothermal process, the temperature and therefore the internal energy of the gas do not change, so from the first law, the heat input and the work done are equal.

Figure 12-1

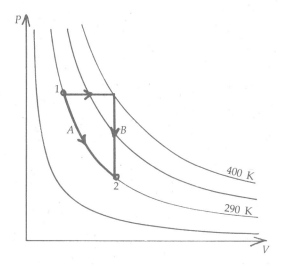

- The heat capacities for a monatomic ideal gas are known, so you can calculate the heat input in process B. What is the internal energy change in process B?

10. One-half mole of nitrogen gas is heated from room temperature (20°C) and a pressure of 1 atm to a final temperature of 120°C. (*a*) How much heat must be supplied if the volume is kept constant while the gas is heated? (*b*) How much heat must be supplied if the heating is at constant pressure? (*c*) By how much is the internal energy changed in each case?

How to Solve It
- Remember that nitrogen is a diatomic gas.

- You can find the molar heat capacities of nitrogen at constant volume and constant pressure in Table 12-2 in the text.

- What determines the change in the gas's internal energy?

11. For 195 g of a certain ideal gas, the heat capacity at constant volume is 145 J/K and the heat capacity at constant pressure is 203 J/K. (*a*) How many moles of the gas are there? (*b*) What is its molar mass?

How to Solve It
- The difference in the heat capacities per mole is the same for any ideal gas.

- From the given difference, therefore, you can calculate how many moles there are.

- That number of moles has a mass of 195 g; calculate the mass of one mole.

12. (*a*) One mole of an ideal monatomic gas is heated at constant volume from 273 K to 500 K. Find the heat added to the gas, the work done by it, and the change in its internal energy. (*b*) Repeat the calculations if the gas is heated at constant pressure.

How to Solve It
- You know what the molar heat capacity at constant volume of a monatomic ideal gas is. Use it to calculate the heat input.

- When the gas is heated at constant volume, it does no work on its surroundings. Thus, the heat input equals the increase in internal energy.

- The internal energy of an ideal gas depends only on the temperature, so the change in internal energy is the same in (*a*) and (*b*).

Solutions

1.   (*a*) If heat losses are negligible, the heat lost by the metal must equal the heat gained by the water. Thus,

$$(mc \, \Delta T)_{\text{metal}} = (mc \, \Delta T)_{\text{water}}$$

$(160 \text{ g})(c)(88°C - 18.4°C) = (140 \text{ g})(1.00 \text{ cal/g·C°})(18.4°C - 10°C)$
which we can solve to give $c = \underline{0.106 \text{ cal/g·C°}}$

   (*b*) Multiplying this by the molar mass M of the metal gives the heat capacity per mole. According to the Dulong-Petit law, the heat capacity per mole is also equal to $3R$. Thus,

$$C_m = cM = 3R$$

so

$$M = \frac{3R}{c} = \frac{(3)(1.987 \text{ cal/mol·C°})}{0.106 \text{ cal/g·C°}} = \underline{56.2 \text{ g}}$$

The unknown metal is presumably <u>iron</u>.

2.   <u>30.1°C</u>

3. The heat that has been gained by the water is

$$Q_w = m_w c_w \, \Delta T_w = (120 \text{ g})(1.00 \text{ cal/g} \cdot \text{C}°)(T_e - 20°\text{C})$$

where T_e is the final equilibrium temperature. That lost by the chunk of iron is

$$Q_i = m_i c_i \, \Delta T_i = (80 \text{ g})(0.103 \text{ cal/g} \cdot \text{C}°)(98°\text{C} - T_e)$$

If there are no heat losses, these two must be equal:

$$Q_w = (120 \text{ cal/C}°)(T_e - 20°\text{C}) = Q_i = (8.24 \text{ cal/C}°)(98°\text{C} - T_e)$$

$$(120 \text{ cal/C}°)T_e - 2400 \text{ cal} = 808 \text{ cal} - (8.24 \text{ cal/C}°)T_e$$

$$(128.2 \text{ cal/C}°)T_e = 3208 \text{ cal}$$

Thus

$$T_e = \underline{25.0°\text{C}}$$

4. $\underline{14.7 \text{ kg}}$ of hot water or $7\frac{1}{3}$ panfuls

5. The assumption of the problem is that all of the bullet's initial kinetic energy is transformed into internal energy:

$$\Delta U = \Delta E_k = \tfrac{1}{2}mv_0^2$$

The same internal energy increase delivered as heat would cause a temperature increase ΔT given by

$$\Delta U = mc \, \Delta T$$

These two must be equal—the same change in temperature and internal energy corresponds to a given state change regardless of how it's made. Thus,

$$\Delta U = mc \, \Delta T = \tfrac{1}{2}mv_0^2$$

so

$$v_0^2 = 2c \, \Delta T$$

The specific heat of lead is

$$c = (0.0305 \text{ kcal/kg} \cdot \text{C}°)(4184 \text{ J/kcal}) = 127.6 \text{ J/kg} \cdot \text{C}°$$

so

$$v_0^2 = (2)(127.6 \text{ J/kg} \cdot \text{C}°)(85°\text{C} - 18°\text{C}) = 1.71 \times 10^4 \text{ m/s}$$

or

$$v_0 = \underline{131 \text{ m/s}}$$

6.   (a) 9.32×10^5 J   (b) $21.6°C$

7.   The mass of a mole of ice is 18.02 g. How much work does it do in shoving the rest of the universe out of the way when it expands? The volume of the mole of ice is

$$V = \frac{m}{\rho} = \frac{18.02 \text{ g}}{0.92 \text{ g/cm}^3} = 19.6 \text{ cm}^3$$

For a temperature change of 1 K,

$$\Delta V = \beta V = (1.53 \times 10^{-4})(19.6 \text{ cm}^3)$$

$$= (3.00 \times 10^{-3} \text{ cm}^3) \left(\frac{1 \text{ L}}{10^3 \text{ cm}^3}\right) = 3.00 \times 10^{-6} \text{ L}$$

The work done by the expanding ice is

$$W = P \Delta V = (1 \text{ atm})(3.00 \times 10^{-6} \text{ L})$$

$$= (3.00 \times 10^{-6} \text{ L·atm}) (24.2 \text{ cal/L·atm})$$

$$= 7.26 \times 10^{-5} \text{ cal}$$

What we have calculated is the work done by 1 mol of ice expanding when its temperature is increased by 1 K at a constant pressure of 1 atm. But this is the difference in the heat capacities. Thus,

$$C_p - C_v = 7.26 \times 10^{-5} \text{ cal/mol·K}$$

In practice this very small difference can usually be neglected; we do not ordinarily worry about the difference in heat capacities of a solid.

8. -2160 J

9.   First, for path A, we use the equation for the work done by a gas expanding isothermally:

$$W = nRT \ln(V_2/V_1)$$

As we don't know the initial and final volumes, let's shelve that for a moment. On path B, the gas is first expanded at constant pressure, its temperature changing from 290 K to 400 K. Since pressure is constant, the volume of the gas is proportional to its absolute temperature, and

$$\frac{V_2}{V_1} = \frac{T_2}{T_1} = \frac{400 \text{ K}}{290 \text{ K}} = 1.379$$

Now at constant pressure

$$W = P\,\Delta V = P_1(V_2 - V_1) = nR(T_2 - T_1)$$

$$= (1\ \text{mol})(8.31\ \text{J/mol·K})(400\ \text{K} - 290\ \text{K}) = 915\ \text{J}$$

For the constant-volume part of path B, no work is done, so the total work done in process B is

$$W_B = \underline{915\ \text{J}}$$

Since the gas ends up back at its initial temperature, there has been no net change in its internal energy, so from the first law

$$\Delta U_B = \underline{0} \quad \text{and} \quad Q_B = W_B = \underline{915\ \text{J}}$$

As for path A, we now know that the volume ratio is 1.379, so

$$W_A = nRT \ln(V_2/V_1) = (1\ \text{mol})(8.31\ \text{J/mol·K})(290\ \text{K}) \ln(1.379)$$

$$= \underline{775\ \text{J}}$$

Again, there is no temperature change, so

$$\Delta U_A = \underline{0} \quad \text{and} \quad Q_A = W_A = \underline{775\ \text{J}}$$

10. (a) $\underline{248\ \text{cal}}$ (b) $\underline{348\ \text{cal}}$ (c) $\underline{248\ \text{cal}}$

11. (a) For any ideal gas, the difference in the molar heat capacities at constant pressure and constant volume is equal to $R = 8.31$ J/mol·K, so

$$C_p - C_v = nR$$

$$n = \frac{C_p - C_v}{R} = \frac{203\ \text{J/K} - 125\ \text{J/K}}{8.31\ \text{J/mol·K}} = \underline{7.0\ \text{mol}}$$

(b) Since the sample mass is 195 g, the molar mass of the gas is

$$M = \frac{m}{n} = \frac{195\ \text{g}}{7.0\ \text{mol}} = \underline{28\ \text{g/mol}}$$

The gas might be CO or N_2.

12. (a) $W = \underline{0}$, $Q = \Delta U = \underline{2830\ \text{J}}$ (b) $Q = \underline{4720\ \text{J}}$, $W = \underline{1890\ \text{J}}$, $\Delta U = \underline{2830\ \text{J}}$

CHAPTER 13

Thermal Properties and Processes

I. Key Ideas

Thermal Properties When heat is transferred to or from a body, one or another of its properties may change—it may expand or contract, liquefy or vaporize—with or without an accompanying change in its temperature.

Thermal Expansion Usually when the temperature of an object is increased, it expands. (An important exception is water, which between 0 and 4°C *contracts* as its temperature is increased.) The amount by which a material expands is proportional to the temperature change if the change is not too large. The fraction by which a material expands per unit temperature increase is called its *coefficient of linear thermal expansion*. The expansion is the same in every direction, so the *coefficient of volume expansion* is three times the linear coefficient. The expansion coefficient of a gas is not even roughly independent of temperature.

Change of Phase At certain temperatures a substance can absorb large quantities of heat without any corresponding temperature change as the physical state of the substance changes. For example, water is transformed from a solid to a liquid by the addition of heat at a constant temperature of 0°C. A specific amount of heat per unit mass, called the *latent heat* of the *phase change*, is required to cause this transition. For a given

substance there may be more than one solid or liquid phase—examples are the superfluid and ordinary fluid phases of liquid helium.

State of a Gas The ideal-gas equation of state describes the behavior of all gases in the limit of low density. At low temperatures and high pressures, however, where the density of the gas is high, its behavior is nonideal and its equation of state is more complex. The *van der Waals equation* of state takes into account the finite volume of gas molecules and the attractive forces between the molecules. It gives a good description of gas properties above a *critical temperature* T_c.

Vapor Pressure At temperatures below T_c, a gas can be liquefied by applying sufficient pressure to it. At a given temperature below T_c, there is a value of pressure at which the gas and liquid phases are in equilibrium; this is the *vapor pressure* of the liquid at that temperature. Above T_c, the liquid phase cannot exist at any pressure.

Phase Diagram The changes of phase of a pure substance can be indicated on a phase diagram, which is a graph of pressure versus temperature. The vapor-pressure curve (up to temperature T_c) displays the equilibrium between gas and liquid phases. The *normal boiling point* of the substance is the temperature at which the vapor pressure is 1 atm. Likewise, the states at which liquid and solid phases are in equilibrium form a *melting* *curve* on the diagram. Where the melting and vapor-pressure curves intersect is the *triple point* of the substance, the unique temperature and pressure at which all three phases can exist in equilibrium.

Humidity Water vapor is a normal component of the atmosphere. The partial pressure of each component of a mixture of gases is independent of the others. If the partial pressure of water vapor in the air is equal to the vapor pressure of water for the same temperature, the air is *saturated*. The *relative humidity* is the ratio of the actual partial pressure of water vapor to the saturated vapor pressure. For a given partial pressure, the temperature at which the air would be saturated is the *dew point*.

Heat Transfer Heat can be transferred by any one or more of three fundamental physical processes: conduction, convection, and radiation.

Conduction *Conduction* is the transport of heat through matter by intermolecular interactions without any actual movement of matter. The rate of heat conduction is proportional to the temperature gradient in a material and to the cross-sectional area; the proportionality constant is the *coefficient of thermal conductivity*. The rate of heat conduction can also be expressed as the temperature difference divided by the thermal "resistance." For heat flow through two materials "in series" (that is, one after the other), the thermal resistances add.

Convection *Convection* is heat transport in a fluid by actual motion of the fluid. Air is a poor conductor, but a large mass of air is a poor insulator because it readily transports heat by convection. Convective heat transfer is a difficult subject to treat precisely.

Radiation Electromagnetic radiation is emitted and absorbed by all objects and is a means of transferring heat energy from one object to another that requires no intervening matter. The rate of heat emission from a body is proportional to the fourth power of its temperature and to the *emissivity* of the surface. A body in thermal equilibrium with its surroundings radiates as much energy as it absorbs. An *ideal blackbody* absorbs all the energy incident upon it; its emissivity is 1. The blackbody is an important limiting case because the properties of its radiation can be calculated theoretically. (It was discrepancies between this calculation and experiment that first led to the quantum theory.) At high enough temperatures, radiation wavelengths reach the visible; this is why hot objects glow.

II.   Numbers and Key Equations

Numbers

Stefan's constant

$$\sigma = 5.67 \times 10^{-8} \text{ W/m}^2\cdot\text{K}^4$$

Key Equations

Thermal expansion

$$\Delta L = \alpha L \, \Delta T \qquad \beta = \frac{\Delta V/V}{\Delta T} \qquad \beta = 3\alpha$$

Expansion of an ideal gas

$$\beta = \frac{1}{T}$$

Latent heat of fusion and vaporization

$$Q = mL_f \qquad \text{or} \qquad Q = mL_v$$

The van der Waals equation of state

$$\left(P + \frac{a}{V^2}\right)(V - b) = RT$$

Relative humidity

$$\frac{\text{partial pressure}}{\text{vapor pressure}} \times 100\%$$

Thermal conductivity

$$I = \frac{Q}{t} = kA\frac{\Delta T}{\Delta x}$$

Thermal resistance

$$R = \frac{\Delta x}{kA} \qquad \Delta T = IR$$

Thermal resistances in series

$$R_{eq} = R_1 + R_2$$

Stefan-Boltzmann law

$$I = e\sigma AT^4 \qquad I_{net} = e\sigma A(T^4 - T_0^4)$$

Wien's displacement law

$$\lambda_{max}T = 2.898 \times 10^{-3} \text{ m·K}$$

III. Possible Pitfalls

In heat exchange problems that include phase changes, remember that the temperature remains fixed until all the material undergoes the phase change. If the phase change is not completed, then the final temperature is that of the phase change.

In problems involving thermal expansion, the units of temperature must be consistent with those of α (or β). On the other hand, any units may be used for L and ΔL (or V and ΔV) as long as they are the same.

In conduction and convection, keeping track of the signs of everything will tell you which way the heat exchange goes. But the heat flow is always from the hotter to the colder region. The cooler region always gains heat.

In conduction and convection problems, only temperature *differences* matter, so Celsius degrees and kelvins are the same thing. In radiation problems this isn't true; you must use absolute temperatures.

IV. Questions and Answers

Questions

1. A metal plate has a circular hole drilled in it. As the plate is heated, it expands. Does the hole get bigger or smaller?

2. The ordinary thermometers we see every day are mostly alcohol in glass. How would the use of such a thermometer be affected if alcohol and glass had the same coefficient of linear thermal expansion?

3. In high-altitude regions of the world, it is hard to cook anything by boiling it in water. Eggs take forever to become hard-boiled, and pasta turns to mush without ever getting properly cooked. Why?

4. What advantage does cooking something in a pressure cooker have over boiling it in an open pan?

5. If an ordinary (liquid-in-glass) thermometer is placed in something quite hot, the liquid column may actually drop a little before it starts to rise. What is going on here?

6. When you try to open a recalcitrant jar of food, it often helps to run hot water over the metal lid for a little while. Why?

7. Under what conditions does the van der Waals equation of state reduce to the ideal-gas law?

8. The western coast of North America has, in general, a much milder climate than the eastern coast. In fact, this is typical everywhere in the "temperate" latitudes. Why? It may help to remember that prevailing winds in these latitudes are westerly.

9. When you turn on an electric fan, it doesn't cool the air in the room; in fact, the fan's electric motor actually warms the room a little. Then what good does the fan do you?

10. Vessels like thermos bottles or Dewar flasks, which are designed to keep fluids very cold, are made with double glass walls. The inner surfaces are silvered and there is vacuum between the walls. Discuss how this design minimizes heat losses.

11. Materials used commercially for building insulation tend to have relatively little mass occupying a lot of space: they are foamy, porous materials or masses of compacted fibers or some such. Why?

Answers

1. The hole enlarges with the plate. Every dimension of the metal plate expands by the same fraction, including the diameter of the hole.

2. This would make them useless, of course; the reason that the top of the fluid column rises against the scale marked on the glass is that the fluid expands more (for a given temperature change) than does the glass.

3. It's because the atmospheric pressure at high altitude is lower, so the boiling point of water is lower. Water boils—and so boiling water cooks—at the temperature at which its saturated vapor pressure is equal to atmospheric pressure. At an altitude of 4000 m, water boils at about 86°C.

4. This is the reverse of the high-altitude situation in Question 3. The sealed lid of the pressure cooker holds the inside pressure at a value higher than atmospheric pressure and thus raises the temperature at which the water inside boils.

5. If this happens, it is because the glass envelope holding the liquid, which gets heated first, expands a little before the temperature of the liquid inside has had time to rise. Then, as glass and liquid and surroundings all come to thermal equilibrium, the liquid rises with respect to the glass tube.

6. You may just be dissolving away sticky guck that is gluing the lid shut. More relevant to our discussion here, the metal lid will expand more as you increase its temperature than will the glass jar, so the lid loosens.

7. When the volume per mole is very large and the density of the molecules is correspondingly small.

8. The prevailing winds carry air from over the ocean onto the western shores. Since water has an anomalously high specific heat, the air over the ocean, in thermal contact with all that water, changes temperature much less as external conditions vary. The presence of the ocean thus tends to moderate temperature changes over the nearby land.

9. Heat is transferred from your body by convection and by evaporation of moisture from your skin. In both cases, what the fan does is physically move air around and so improve the transfer of heat away from you. What makes you comfortable or uncomfortable is the air temperature and humidity next to your skin.

10. The vacuum between the double glass walls prevents heat transfer through the walls by conduction and convection; silvering the inner surfaces of the double wall cuts down on heat loss by radiation (it reduces the emissivity). The double walls must join at the neck, and there will be some heat loss by conduction there; this is reduced by using glass, which is a relatively poor conductor.

11. Little mass means relatively little matter to conduct heat. Air is an excellent insulator with respect to conduction; the fibers or the foam material serve mainly to divide up the air into little isolated pockets so that heat transfer by convection is minimized.

V. Problems and Solutions

Problems

1. In Figure 13-1, a steel pin exactly fits into a 5-cm slot in a brass piece when both are at a temperature of 140°C. The coefficients of linear expansion are 1.1×10^{-5} per C° for steel and 1.9×10^{-5} per C° for brass; Young's modulus for steel is 2×10^{11} N/m². What is the stress in the pin when the assembly is cooled to 20°C?

How to Solve It

- The hole and the pin are the same length at 140°C; at 20°C, separately, they would not be.

- First calculate the changes in the length of the pin and in that of the slot that would result in going from 140°C to 20°C if they were separate.

- Then calculate what stress would be required to compress the pin to the length of the slot at 20°C.

Figure 13-1

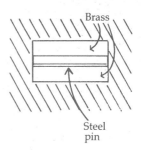

2. A surveyor's steel tape is manufactured to be accurate at 20°C. On a day when the temperature is 38°C, it is used to measure a house lot. If the area as measured with this tape was 1.2600 acres, what is the actual value of the area of the lot?

How to Solve It

- Remember that an area is being measured; length measurements and width measurements each will be affected.

- The coefficient of linear expansion of steel determines by how much each measurement is off.

- Are the measured dimensions larger or smaller than the true values?

3. A 10-L glass flask is filled to the brim with acetone of 12.4°C. To what temperature must the flask and its contents be heated or cooled for 85 cm³ of acetone to overflow the flask? Use $\alpha = 3.4 \times 10^{-6}$ per C° for glass and $\beta = 1.5 \times 10^{-3}$ per C° for the acetone.

How to Solve It

- Write an expression for the change in the volume of the glass flask in terms of the change in temperature. Remember that the volume expansion coefficient $\beta = 3\alpha$ for a solid.

- Write an expression for the change in the volume of the acetone in terms of the temperature change.

- These two changes in volumes differ by 85 cm³. Solve for the temperature at which this is true.

4. My Honda has a 40-L gas tank. If I fill it completely full of gasoline at a temperature of 12°C and then let the car sit until the noon sun raises its temperature to 30°C, how much gasoline spills out of the tank? The coefficient of volume thermal expansion for gasoline is 9.00×10^{-4} C°⁻¹; neglect the expansion of the tank.

How to Solve It

- What is the volume of the gasoline at 30°C?

- The excess gasoline (above 40 L) must overflow if the tank volume remains the same.

5. A copper bar of mass 2.5 kg at an initial temperature of 66°C is dropped into an insulated vessel containing 400 g of water and 70 g of ice at 0°C. The specific heat of copper is 0.092 kcal/kg·C°. At what final temperature does the system come to equilibrium?

How to Solve It

- Before the temperature of the water-ice mixture will change, all the ice must melt. How much heat does this require?

- By how much does the loss of the heat needed to melt the ice reduce the temperature of the copper? What would it mean if the result were below 0°C?

- Once the ice is melted, if the temperature of the copper is still above 0°C, the copper and 470 g of water come to equilibrium.

6. A cup contains 240 g of fresh-brewed coffee at a temperature of 97°C. If heat losses to the surroundings are negligible, to what final temperature is the coffee cooled if you drop an 18-g ice cube into it?

How to Solve It

- The heat capacity of coffee is the same as that of water.

- How much heat is required to melt 18 g of ice?

- This much heat must be obtained from the hot coffee. Then, when the ice is melted, there are 18 g of water at 0°C to come to equilibrium with the coffee.

7. A piece of aluminum (specific heat 0.90 kJ/kg·K) of mass 135 g initially at 20°C is placed in a large container of liquid nitrogen at 77 K (its normal boiling point). If the latent heat of vaporization of nitrogen is 199 kJ/kg, what mass of nitrogen is vaporized in bringing the aluminum to equilibrium?

How to Solve It

- Since the amount of liquid nitrogen available is "large," the aluminum will be cooled all the way to 77 K.

- How much heat must be extracted from the aluminum chunk to do this?

- How much liquid nitrogen will be evaporated by that much heat?

8. Steam at 100°C is passed into an insulated flask containing 200 g of ice at -25°C. Neglecting any heat loss from the flask, what mass of water at 100°C will finally be present in the flask? The latent heat of vaporization of water is 2260 kJ/kg.

How to Solve It

- The 200 g of ice will first be raised to 0°C, then melted, and then raised to 100°C. How much heat does it take to do all this?

- This heat is extracted from some of the steam, condensing it to liquid water. How much water is produced this way?

- The total amount of water finally present at last is the 200 g of melted ice plus the condensed steam.

9. The van der Waals constants for steam are $a = 5.43$ atm·L² and $b = 0.030$ L; its molar mass is 18.02 g/mol. One mole of steam is initially in a 6-L container at 100°C. (*a*) What is its pressure? (*b*) The steam is heated to 300°C. Assuming the volume of the container stays constant, calculate the final pressure of the gas. (*c*) How do your results compare to what you would get using the ideal-gas law.

How to Solve It

- Use the van der Waals equation to calculate the initial and final pressures.

- Repeat the calculation using the ideal-gas law.

10. On top of a high mountain, where the atmospheric pressure is 70 kPa, at what temperature does water boil? (Vapor pressure values are given in Table 13.3 in the text.)

How to Solve It

- A liquid boils when its vapor pressure is equal to atmospheric pressure.

- At what temperature is the vapor pressure of water 70 kPa?

11. Figure 13-2 shows a system holding liquid helium at 4.2 K. A cylindrical can 5 cm in diameter and 7 cm long is supported by two stainless steel pins from walls at liquid-nitrogen temperature (77 K).

Figure 13-2

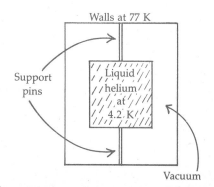

Walls at 77 K

Support pins

Liquid helium at 4.2 K

Vacuum

The space between the can and the walls is evacuated. The steel pins are each 0.1 cm in diameter and 6 cm long, with thermal conductivity 0.0032 kcal/m·s·K. The coefficient of absorption of the inner can's surface is 0.25. If the latent heat of vaporization of helium is 4.95 kcal/kg, at what rate does the liquid helium boil off?

How to Solve It
- Heat is being conducted to the liquid helium through the two support pins. At what rate? Don't forget that there are two pins.

- Calculate the surface area of the can. Heat is being absorbed by this surface from surroundings at 77 K. At what rate?

- The total rate at which heat is being delivered to the liquid helium determines the rate at which it boils away.

12. A thermos bottle with an inside diameter of 6 cm contains 150 g of water and 75 g of ice at 0°C (see Figure 13-3). Heat leakage through the walls of the bottle is negligible, but it is closed with a cork stopper 2 cm thick whose thermal conductivity is 1.2×10^{-4} kcal/m·s·K. The latent heat of fusion of water is 79.7 kcal/kg. If the surroundings are at 28°C, how long does it take for all the ice to melt?

Figure 13-3

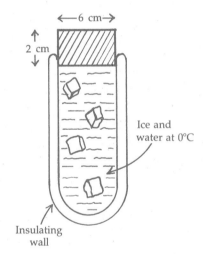

How to Solve It
- Use the equation for heat conduction to determine the rate at which heat is coming to the water through the cork.

- Calculate the heat required to melt the 75 g of ice.

- Finally calculate how long it takes for this much heat to be conducted through the stopper.

13. A home has a window area of 263 ft² of single-pane glass 0.135 in thick. (*a*) Use the value $R_f = 0.9$ h·ft²·F°/Btu (note that this is independent of the glass thickness) to calculate the rate of heat loss through the windows in Btu/h when the indoor temperature is 72°F and the outside temperature is 10°F. (*b*) For comparison, calculate the conductive heat loss rate you would get by assuming the entire temperature difference appears across the glass. (The thermal conductivity of the glass is 5.6 Btu·in/h·ft²·F°.)

How to Solve It
- Calculate the thermal resistance R by dividing R_f by the window area.

- The rate of heat loss is just the temperature difference divided by the thermal resistance.

- For part (*b*), calculate the thermal resistance directly from the thickness and the thermal conductivity of the glass. (What this amounts to is assuming that the glass surfaces are at the indoor and outdoor temperatures.)

14. The walls of a certain house consist of 160 m² of brick, 10 cm thick, with a thermal conductivity of 0.035 W/m·C°. At what rate is heat lost by conduction through the walls if the temperature of the inside wall is 16°C and that of the outside is 5°C?

How to Solve It
- Here the temperature drop across the brick is given.

- Calculate the thermal resistance of the brick directly from its thickness and its thermal conductivity.

- Calculate the rate of heat loss as the temperature drop divided by the thermal resistance.

15. Assume that the surface of a human body can be considered a perfect blackbody at infrared wavelengths with an area of 2.2 m² and a surface temperature of 31°C. (*a*) Calculate the peak wavelength of the body's radiated spectrum. (Electromagnetic radiation in this wavelength range, too long to be visible as light, is infrared radiation.) (*b*) If the body's surroundings are at a temperature of 22°C, calculate the body's total rate of heat loss by radiation.

How to Solve It
- Use Wien's displacement law. The peak wavelength is determined by the absolute temperature of the surface.

- Calculate the rate of radiation from the surface from the area of the radiator, its temperature, and the temperature of the surroundings.

- For comparison, the total energy output of a passive human body is of the order of 100 W.

Figure 13-4

16. In a certain experiment, heat is transferred from a source at 227°C to water at 30°C through a rod 3 cm in diameter. (See Figure 13-4.) The rod is made by joining an aluminum rod 12 cm long end-to-end with a copper rod 5 cm long. What is the temperature at the aluminum-copper junction? The thermal conductivities are 23 W/m·K and 401 W/m·K for Al and Cu, respectively.

How to Solve It
- Calculate the thermal resistance of each piece of the rod.

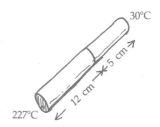

- The thermal resistance of the whole rod is the sum of the resistances of the two pieces. The temperature drop divided by this total thermal resistance gives the rate of heat flow.

- The rate of heat flow multiplied by the resistance of the aluminum rod gives the temperature drop across it—that is, from 227°C to the junction.

Solutions

1. At the initial temperature of 140°C, the pin and the slot have the same length; call this L_0. In cooling to 20°C, the length of the slot in the brass piece changes by

$$\Delta L_b = \alpha L_0 \Delta T$$

$$= (1.9 \times 10^{-5} \, C^{\circ -1})(5 \text{ cm})(20°C - 140°C)$$

$$= -0.0114 \text{ cm}$$

(I carry an extra significant figure here because I will be calculating a small difference a little later on.) In exactly the same way, I can calculate that the steel pin shrinks by 0.0066 cm. Thus, if the pin were not in the slot, it would be 0.0114 cm − 0.0066 cm = 0.0048 cm longer than the slot. Instead, it cooled in the slot and was compressed by this much. This requires a stress of

$$\text{Stress} = \frac{F}{A} = Y \frac{\Delta L}{L} = (2 \times 10^{11} \text{ N/m}^2) \frac{4.8 \times 10^{-3} \text{ cm}}{5 \text{ cm}}$$

$$= 1.9 \times 10^8 \text{ N/m}^2$$

In reality, the stress would be less than this because the brass, as well as the steel, would give way.

2. <u>1.2595 acres</u>

3. The volume expansion coefficient of the glass is

$$\beta_g = 3\alpha = (3)(3.4 \times 10^{-6} \, C^{\circ -1}) = 1.02 \times 10^{-5} \, C^{\circ -1}$$

Thus, when the flask changes temperature by ΔT, the volume of the flask changes by

$$\Delta V_g = \beta_g V_0 \Delta T$$

$$= (1.02 \times 10^{-5} \, C^{\circ -1})(10 \text{ L}) \Delta T = (1.02 \times 10^{-4} \text{ L/C}°) \Delta T$$

Likewise, the volume of the acetone changes by

$$\Delta V_a = \beta_a V_0 \Delta T = (1.5 \times 10^{-3} \, C^{\circ -1})(10 \text{ L}) \Delta T$$

$$= (1.50 \times 10^{-2} \text{ L/C}°) \Delta T$$

with the same change in temperature. The difference in these changes is the amount by which the volumes of the acetone and the container will differ at the final temperature:

$$\Delta V_a - \Delta V_g = (1.50 \times 10^{-2} \, \text{L/C°}) \, \Delta T$$

$$- (1.02 \times 10^{-4} \, \text{L/C°}) \, \Delta T$$

$$= (1.49 \times 10^{-2} \, \text{L/C°}) \, \Delta T$$

This is the 85 cm³ that overflowed, so

$$(1.49 \times 10^{-2} \, \text{L/C°}) \, \Delta T = 85 \, \text{cm}^3 = 0.085 \, \text{L}$$

or

$$\Delta T = 5.7 \, \text{C°}$$

The final temperature is therefore

$$T_f = \Delta T + T_i = 12.4°\text{C} + 5.7 \, \text{C°} = \underline{18.1°\text{C}}$$

4. $\underline{0.65 \, \text{L}}$

5. The heat required to melt 70 g of ice is

$$Q = mL_f = (70 \, \text{g})(79.7 \, \text{cal/g}) = 5580 \, \text{cal}$$

First, we need to know by how much this heat lowers the temperature of 2.5 kg of copper.

$$Q = mc \, \Delta T$$

$$\Delta T = \frac{Q}{mc} = \frac{5.58 \, \text{kcal}}{(2.50 \, \text{kg})(0.092 \, \text{kcal/kg·C°})} = 24.26 \, \text{C°}$$

Thus, when all the ice is melted, the 2.5 kg of copper is at a temperature of 66°C − 24.26 C° = 41.74°C. (If there had been more ice or less copper, this calculation might have come out negative. This would mean that there was not enough heat available from the copper to melt all the ice, and the problem would have to be reworked on different assumptions.) This copper and 400 g + 70 g = 470 g of water at 0°C (don't forget the melted ice!) all come to equilibrium at some final temperature T_f. In the process, the heat gained by the water is

$$Q_w = mc \, \Delta T = (0.470 \, \text{kg})(1.00 \, \text{kcal/kg·C°})(T_f - 0°\text{C})$$

$$= (0.470 \, \text{kcal/C°})T_f$$

while that lost by the copper is

$$Q_c = mc \, \Delta T = (2.50 \, \text{kg})(0.092 \, \text{kcal/kg·C°})(41.74°\text{C} - T_f)$$

$$= 9.60 \, \text{kcal} - (0.230 \, \text{kcal/C°})T_f$$

Equating these gives

$$Q_w = Q_c$$

$$(0.470 \text{ kcal}/C°)T_f = 9.60 \text{ kcal} - (0.230 \text{ kcal}/C°)T_f$$

so

$$T_f = \frac{9.60 \text{ kcal}}{(0.470 + 0.230) \text{ kcal}/C°} = \underline{13.71°C}$$

6. $\underline{84.7°C}$

7. The heat extracted from the piece of aluminum in cooling it to 77 K is

$$Q = mc\,\Delta T = (0.135 \text{ kg})(0.90 \text{ kJ}/\text{kg·K})(293 \text{ K} - 77 \text{ K})$$

$$= 26.24 \text{ kJ}$$

If there are no heat losses to the surroundings, all this heat has gone into evaporating liquid nitrogen. Thus,

$$Q = mL_v$$

$$m = \frac{Q}{L_v} = \frac{26.24 \text{ kJ}}{199 \text{ kJ}/\text{kg}} = \underline{0.132 \text{ kg}}$$

8. $\underline{271 \text{ g of water}}$

9. (*a*) The van der Waals equation of state for one mole of gas is

$$(P + a/V^2)(V - b) = RT$$

Solving for P gives

$$P = \frac{RT}{V - b} = \frac{a}{V^2}$$

The pressure at the steam point (100°C) is thus

$$P_1 = \frac{(8.21 \times 10^{-2} \text{ L·atm}/\text{mol·K})(373 \text{ K})}{6 \text{ L} - 0.030 \text{ L}} - \frac{5.43 \text{ atm·L}^2}{(6 \text{ L})^2}$$

$$= \underline{4.98 \text{ atm}}$$

(*b*) Likewise, at 300°C = 573 K we get $P_2 = \underline{7.73 \text{ atm.}}$
(*c*) Using the ideal-gas law amounts to setting $a = b = 0$.

Thus, at 100°C

$$P_1 = \frac{RT}{V} = \frac{(8.21 \times 10^{-2} \text{ L·atm/K})(373 \text{ K})}{6 \text{ L}}$$

$$= \underline{5.10 \text{ atm}}$$

and at 300°C, $P_2 = \underline{7.84 \text{ atm}}$.

10. $\underline{90.0 \text{ °C}}$

11. Heat is delivered to the helium and evaporates it by conduction through the support pins and by radiation from the 77-K walls. To get the rate of heating by conduction, first calculate the thermal resistance of one of the steel support pins.

$$R_{\text{pin}} = \frac{\Delta x}{kA}$$

The cross-sectional area of a pin is

$$A = \pi(d/2)^2 = (\pi)(5 \times 10^{-4} \text{ m})^2 = 7.85 \times 10^{-7} \text{ m}^2$$

so

$$R_{\text{pin}} = \frac{0.06 \text{ m}}{(3.2 \text{ cal/m·s·K})(7.85 \times 10^{-7} \text{ m}^2)} = 2.39 \times 10^4 \text{ s·K/cal}$$

There are two pins, which are in "parallel" since each has one end at 4.2 K and the other at 77 K. Thus,

$$\frac{1}{R_{\text{eq}}} = \frac{1}{R_{\text{pin}}} + \frac{1}{R_{\text{pin}}}$$

$$R_{\text{eq}} = \tfrac{1}{2}R_{\text{pin}} = 1.19 \times 10^4 \text{ s·K/cal}$$

Thus, the rate at which heat is being delivered by conduction is

$$I_{\text{cond}} = \frac{\Delta T}{R_{\text{eq}}} = \frac{77 \text{ K} - 4.2 \text{ K}}{1.19 \times 10^4 \text{ s·K/cal}} = 6.10 \times 10^{-3} \text{ cal/s}$$

In order to calculate the rate at which heat is delivered by radiation, we first need the surface area of the can:

$$A = A_{\text{wall}} + 2A_{\text{end}} = 2\pi rh + 2\pi r^2$$

$$= (\pi)(2r)(h + r) = (\pi)(0.05 \text{ m})(0.07 \text{ m} + 0.025 \text{ m})$$

$$= 1.49 \times 10^{-2} \text{ m}^2$$

The heat absorbed by the can from its surroundings (at 77 K) is

$$I_{rad} = a\sigma A(T_s^4 - T^4)$$

$$= (0.25)(5.67 \times 10^{-8} \text{ W/m}^2 \cdot \text{K}^4)$$

$$\times (1.49 \times 10^{-2} \text{ m}^2)(77^4 \text{ K}^4 - 4.2^4 \text{ K}^4)$$

$$= 7.44 \times 10^{-3} \text{ W} = 1.78 \times 10^{-3} \text{ cal/s}$$

The total rate at which heat flows to the can is thus

$$I = I_{cond} + I_{rad} = 6.10 \times 10^{-3} \text{ cal/s} + 1.78 \times 10^{-3} \text{ cal/s}$$

$$= 7.88 \times 10^{-3} \text{ cal/s}$$

The rate at which this will boil off liquid helium is

$$\frac{I}{L_v} = \frac{7.88 \times 10^{-3} \text{ cal/s}}{4.95 \text{ cal/g}} = \underline{1.59 \times 10^{-3} \text{ g/s}}$$

or about 5.7 g/h.

12. <u>3.50 h</u>

13.   (a) The thermal resistance of the walls is

$$R = \frac{R_f}{A} = \frac{0.9 \text{ h} \cdot \text{ft}^2 \cdot \text{F}^\circ / \text{Btu}}{263 \text{ ft}^2} = 3.4 \times 10^{-3} \text{ h} \cdot \text{F}^\circ / \text{Btu}$$

The temperature difference here is $72^\circ \text{F} - 10^\circ \text{F} = 62 \text{ F}^\circ$, so the heat loss rate is

$$I = \frac{\Delta T}{R} = \frac{62 \text{ F}^\circ}{3.4 \times 10^{-3} \text{ h} \cdot \text{F}^\circ / \text{Btu}} = \underline{1.81 \times 10^4 \text{ Btu/h}}$$

(b) The calculation in part (a) used a tabulated thermal resistance that takes into account (roughly) the temperature gradient in the air near the window. If we were to ignore this, we would use the equation for thermal conductivity as if all the 62 F° temperature drop appeared across the thickness of the window:

$$R = \frac{\Delta x}{kA} = \frac{0.135 \text{ in}}{(5.6 \text{ Btu} \cdot \text{in/h} \cdot \text{ft}^2 \cdot \text{F}^\circ)(263 \text{ ft}^2)} = 9.2 \times 10^{-5} \text{ h} \cdot \text{F}^\circ / \text{Btu}$$

Using this gives $I = \underline{6.7 \times 10^5 \text{ Btu/h}}$. Thus, the assumption that all the temperature drop appears across the glass causes us to overestimate greatly the heat loss through the windows.

14. <u>616 W</u>, if all the temperature gradient is in the brick. The result of Problem 13 might make us wonder if this is an overestimate, however.

15.   (*a*) Wien's displacement law is

$$\lambda_{\text{max}} = \frac{2.898 \times 10^{-3} \text{ m·K}}{T}$$

Thus, the peak radiation wavelength of the body at a temperature of 31°C = 304 K is

$$\lambda_{\text{max}} = \frac{2.898 \times 10^{-3} \text{ m·K}}{304 \text{ K}} = \underline{9.53 \times 10^{-6} \text{ m}}$$

(*b*) Taking the emissivity of the body surface as 1,

$$I_{\text{rad}} = e\sigma A(T^4 - T_0^4)$$

$$= (1)(5.67 \times 10^{-8} \text{ W/m}^2 \text{·K}^4)(2.2 \text{ m}^2)(304^4 \text{ K}^4 - 295^4 \text{ K}^4)$$

$$= \underline{121 \text{ W}}$$

This result, which estimates only radiative heat loss, is as large as the body's entire resting heat budget, so some of the assumptions made must be wrong. In fact, the emissivity is not quite 1, and it is not ordinarily considered socially acceptable to expose our entire surface area directly to the surroundings.

16. <u>68.9°C</u>

CHAPTER 14

The Availability of Energy

I. Key Ideas

The Second Law of Thermodynamics The *second law of thermodynamics* is about the possibility of putting energy to use. By the first law, total energy is conserved, but in a given case, not all the energy may be *available* for use. It is easy to transform work completely into heat, but it is not possible to extract a given amount of heat from a system and use it all to do work without some other change in the surroundings being made. This is the *Kelvin-Planck statement* of the second law. The conversion of work and mechanical energy into heat is an *irreversible* process, and the second law is fundamentally a statement about the *direction* of irreversible processes.

Heat Engines Historically, the second law was first formulated in terms of the efficiency of heat engines. An elementary *heat engine* is a device that converts heat into work in a *cyclic* process. A *working substance* absorbs heat from a source at high temperature, does work, and rejects the remaining heat to a lower-temperature reservoir. The efficiency of the engine is the ratio of the work done to the heat absorbed. In practical engines, the efficiency may be as high as 50 percent or so. An engine that is 100 percent efficient would absorb heat and convert it all to work with no heat being rejected. This is impossible according to the Kelvin-Planck statement of the second law.

Refrigerators An elementary *refrigerator* is a heat engine run backwards: work is done to extract heat from a low-temperature source and transfer it to a reservoir at higher temperature. The *coefficient of performance* of a refrigerator is the ratio of the heat extracted to the work done. Practical refrigerator cycles may have coefficients of performance of 5 to 6 or more. An ideal refrigerator would require zero work input and would transfer heat from a lower- to a higher-temperature reservoir without any other change to the surroundings. That this is impossible is the Clausius statement of the second law.

Equivalent Statements of the Second Law Although appearing to be very different, the Kelvin-Planck and Clausius statements of the second law are equivalent in that if we could make a device that violated one it would also violate the other. This can be seen from simple heat flow diagrams and is independent of any specific engine design features.

Reversibility By the Kelvin-Planck statement, the conversion of mechanical energy into internal energy (heat) is an *irreversible* process. By the Clausius statement, the conduction of heat across a temperature difference is an irreversible process. Likewise, a process in which a system departs significantly from equilibrium is irreversible.

Carnot's Theorem The ideal heat engine would

be one that works *reversibly*. Carnot's theorem states that no engine can be more efficient than a reversible engine working between the same reservoir temperatures. A corollary is that every reversible engine working between given reservoir temperatures has the same efficiency.

The Carnot Cycle A basic reversible engine is the *Carnot engine*. In a Carnot cycle, heat is extracted isothermally from a reservoir and work is done as the working substance expands without heat transfer, the remaining heat is rejected isothermally at a lower temperature, and the working substance is recompressed to its initial state. The efficiency of a Carnot cycle on an ideal gas can be simply calculated; by Carnot's theorem, this must be the efficiency of every reversible engine and the upper limit possible for any real heat engine. We find that the heat absorbed and the heat rejected are proportional to the reservoir temperatures.

The Heat Pump A practical application is the heat pump, which is essentially a refrigerator used to pump heat from a colder region (for example, outdoors) to a warmer region (for example, the interior of a building). The maximum coefficient of performance attainable is less than that of a Carnot cycle because perfectly adiabatic and isothermal processes cannot be carried out and because of frictional losses and the like. The heat pump effectively multiplies the heating effect of the energy used to run it by one plus the coefficient of performance.

Entropy The second law is related to the fact that physical processes go only in one direction. In all irreversible processes, the direction is such that the system and its surroundings, taken together, tend to a *less ordered* state. *Entropy* is a thermodynamic quantity that measures the "disorder" of a system. Its definition is: in a *reversible* process, the heat transferred to a system divided by temperature is the change in its entropy.

Reversibility and Entropy Change The second law gives the direction in which any irreversible process moves; the total entropy of the universe

never decreases and, in an irreversible process, it always increases. Viewed microscopically, the second law is a probabilistic statement; systems tend toward states of higher probability. Higher probability and so higher entropy is associated with a more disordered state of a system.

Lost Work The change in the entropy of the universe in a process multiplied by the lowest temperature available gives the amount of energy that has become unavailable for useful macroscopic work.

II. Key Equations

Key Equations

Work done by a heat engine

$$W = Q_h - Q_c$$

Efficiency

$$\epsilon = \frac{W}{Q_h} = 1 - \frac{Q_c}{Q_h}.$$

Coefficient of performance

$$COP = \frac{Q_c}{W}$$

Carnot cycle efficiency

$$\epsilon_C = 1 - \frac{T_c}{T_h}$$

Heat ratio for Carnot engine

$$\frac{Q_c}{Q_h} = \frac{T_c}{T_h}$$

Second-law efficiency

$$\epsilon_{SL} = \frac{\epsilon_{actual}}{\epsilon_{Carnot}}$$

Carnot coefficient of performance

$$COP_{max} = \frac{T_c}{T_h - T_c}$$

Second-law efficiency

$$\epsilon_{SL} = \frac{COP_{actual}}{COP_{max}}$$

Entropy change

$$\Delta S = \frac{\Delta Q_{rev}}{T}$$

Second law in terms of entropy change

$$\Delta S_u > 0$$

Lost work

$$W_{lost} = T \, \Delta S_u$$

III. Possible Pitfalls

The work done in a heat-engine cycle must be calculated as the *net* work done in the whole cycle, not just that done in the expansion part of the cycle.

Note also that the various statements of the second law refer to a device that works in a cycle and produces no other change in the surroundings. You can think of situations in which heat is converted completely to work, for example, but only with other irreversible effects on the rest of the universe.

Differences in entropy are *defined* only for *reversible changes of state*. But entropy is a function of state, and a given entropy difference is associated with a given state change *regardless* of how the change is brought about. To *calculate* the entropy difference, we must find a reversible way to make the state change.

Note that $\Delta Q/T$ does not tell you the entropy of the initial or final state in a process but only the *change* in the entropy. It is rather like potential energy in that only differences, or changes, are defined.

Entropy is *not* a form of energy, it's a separate quantity. Neither is it a *conserved* quantity.

The second law does not say that the entropy of some object or system cannot decrease; in fact, it can decrease. But the decrease will always be made up (or more than made up) by an increase elsewhere. The second law states that the *total* entropy of the universe may not decrease.

In calculating the change in the entropy of a system, you must always use temperature in kelvins. No other scale works.

If heat flows to a substance or object, its entropy is increased and vice versa. Entropy change has the same sign as the quantity of heat delivered *to* a substance because T cannot be negative.

IV. Questions and Answers

Questions

1. When we say "engine," we think of something mechanical with moving parts. In such an engine, friction always reduces the' engine's efficiency. Why is this?

2. We've all seen people try to keep cool on a hot summer day by leaving their refrigerator door open. But you can't cool your kitchen this way! Why not?

3. Why do engineers designing a steam electric generating plant always try to design for as high a feed-steam temperature as possible?

4.   The conduction of heat across a temperature difference is an ir-reversible process, but the object that lost heat can always be re-warmed, and the one that gained it can be recooled. The dissipation of mechanical energy, as in the case of an object sliding across a rough table and slowing down, is irreversible, but the ob-ject can be cooled and set moving again at its original speed. So in just what sense are these processes "irreversible"?

5.   In a slow, steady expansion of an ideal gas against a piston at constant temperature, all the heat input is turned into work. Is this consistent with the first law? How about the second law?

6.   If a gas expands freely into a larger volume in an insulated con-tainer so that no heat is added to the gas, its entropy increases. In view of the definition of ΔS, how can this be?

7.   Heat flows from a hotter to a colder body. By how much is the entropy of the universe changed? In what sense does this corre-spond to energy becoming unavailable for doing work?

8.   The frictional drag of traces of atmosphere on an orbiting satel-lite causes the satellite to move closer to the earth and its kinetic energy to increase. In what way does energy become unavailable for doing work in this irreversible process?

9.   Is a process necessarily reversible if there is no exchange of heat between the system in which the process takes place and its surroundings?

10. In discussing the Carnot cycle, we say that extracting heat from a reservoir isothermally does not change the entropy of the uni-verse. In a real process, this is a limiting situation that can never quite be reached. Why not? What is the effect on the entropy of the universe?

Answers

1.   The force of friction always dissipates mechanical energy since it is always directed against the motion. It is therefore an irreversi-ble process and so removes some of the available mechanical en-ergy from the engine and thus reduces its efficiency.

2.   The refrigerator just pumps heat from inside the box into the rest of the room. If you open the door, the heat being extracted from the box is dumped back into the room, with interest (the heat generated by the motor and condenser). There's an initial cooling effect near the door as cool air from within the refrigerator is convected out, but in the long run you lose.

3. The steam is the high-temperature heat source for a heat engine that drives the electric generator. Since the low-temperature heat sink is usually fixed by circumstances, increasing the feed-steam temperature is the only way to increase the Carnot efficiency limit for the generator.

4. To say that a process is irreversible means, essentially, that the universe as a whole won't go the other way. In the examples given, we can put each "system" back where it started, but not without making a permanent change in its "surroundings."

5. It is consistent with both; it must be or it wouldn't take place. All this says with respect to the first law is that the internal energy of the gas is not changing, as we know is the case for an isothermal process on an ideal gas. As for the second law, to the degree to which the process can be carried out strictly isothermally, the entropy increase of the gas is of the same magnitude as the entropy decrease of the surroundings.

6. This is an irreversible process, and entropy change is defined to equal $\Delta Q/T$ only for reversible processes.

7. Let T_1 be the higher and T_2 be the lower temperature. The total entropy change of the two bodies, taken together, is

$$\Delta S = \frac{Q}{T_2} - \frac{Q}{T_1}$$

If this is all that happens, this is ΔS_u. If, instead of just letting the heat flow, we had run a Carnot engine between these two bodies as temperature "reservoirs," the work we could have gotten from it is

$$W = \epsilon_C Q = (1 - T_2/T_1)Q = T_2 \, \Delta S$$

8. As the satellite moves closer to the earth, its kinetic energy increases, but its potential energy decreases more—by twice as much, in fact, if the orbit remains nearly circular. The satellite's total mechanical energy decreases, therefore, some of it being dissipated as heat in the atmosphere. The atmosphere is a great big heat sink, so essentially all this energy has become unavailable for doing work.

9. No, this is neither a necessary nor a sufficient condition.

10. In any real process, the entropy of the universe will increase. In reality, if everything is at exactly the same temperature, it's all at thermal equilibrium and there can be no heat transfer; there must be some small temperature difference to make the heat transfer go. But if there is a temperature difference, the entropy decrease on the hotter side will always be a little less than the entropy increase on the cooler side. The net entropy change of everything-put-together is always an increase. Furthermore, real-world heat reservoirs aren't infinite, and so extracting heat from something lowers its temperature, even if only sightly. The reversible process is an idealization.

V. Problems and Solutions

Problems

1. A certain engine absorbs 150 J of heat and rejects 88 J in each cycle. (*a*) What is its efficiency? (*b*) If it runs at 200 cycles/min, what is its power output?

How to Solve It
- The efficiency is the work done divided by the heat input.

- Whatever part of the heat absorbed that is not rejected to the low-temperature reservoir has been used to do work.

- The power output is the work done per cycle divided by the time each cycle takes.

2. A certain refrigerator requires 400 J of work to remove 350 cal of heat from its interior. (*a*) What is its coefficient of performance? (*b*) Suppose the refrigerator cycle is reversible and is run in reverse as a heat engine cycle. How much heat must be put into it in order for it to do 400 J of work?

How to Solve It
- The coefficient of performance of a refrigerator is the heat extracted from its interior divided by the work that must be done to extract it.

- If the cycle is reversible, then 400 J of work will be done when 350 cal of heat is rejected to a cold reservoir.

3. A certain refrigerator has a power rating of 88 W. Consider it as an ideal reversible refrigerator. If the outside temperature is 26°C, how long will the refrigerator take to freeze 2.5 kg of water that is put into it at 0°C?

How to Solve It
- The power rating is the work input per unit time.

- The heat from the interior is extracted at 0°C since it is extracted from freezing water.

- If the cycle is thermodynamically reversible, the heats to and from the hot and cold reservoirs are proportional to their (Kelvin) temperatures.

4. A certain refrigerator requires 35 J of work to remove 45 cal of heat from its interior. (*a*) What is its coefficient of performance? (*b*) How much heat is ejected to the surroundings at 22°C? (*c*) If the refrigerator cycle is reversible, what is the temperature inside the refrigerator?

How to Solve It
- The coefficient of performance of a refrigerator is the heat extracted from its interior divided by the work that must be done to extract it.

- The heat ejected is the work done plus the heat extracted from the interior.

- If the cycle is thermodynamically reversible, the heats to and from the hot and cold reservoirs are proportional to their (Kelvin) temperatures.

5. A not very clever idea for a ship's engine goes as follows: An ideal Carnot cycle extracts heat from seawater at 18°C and exhausts it to evaporating dry ice, which the ship carries with it, at −78°C. The latent heat of sublimation of dry ice is 137 kcal/kg. If the ship's engines are to develop 8000 horsepower, what is the minimum amount of dry ice it must carry for a day's running?

How to Solve It
- What you need to find first is the Carnot efficiency for the temperatures involved.

- How much work is required for a day's running? Be careful of units here—what's a horsepower?

- The heat that must be rejected to the low-temperature reservoir in the process of doing this much work determines the amount of dry ice that must be sublimated.

6. A certain electric generating plant produces electricity by using steam that enters its turbine at a temperature of 320°C and leaves it at 40°C. Over the course of a year, the plant consumes 4.4×10^{16} J of heat and produces an average electric power output of 600 MW. What is its second-law efficiency?

How to Solve It

● Calculate the Carnot efficiency of the plant from the temperatures involved.

● The actual efficiency is the work done in a year divided by the heat input.

● The ratio of these two efficiencies is the second-law efficiency of the plant.

7. A certain engine has a second-law efficiency of 85 percent. In each cycle, it takes in 120 cal from a reservoir at 300°C and rejects 75 cal of heat to a cold reservoir. (*a*) What is the temperature of the cold reservoir? (*b*) How much more work could be done by a Carnot engine working between the same two reservoirs and extracting the sames 120 cal in each cycle?

How to Solve It

● The energy figures given allow you to determine the actual efficiency; knowing the second-law efficiency, therefore, you can figure the Carnot efficiency.

● Use the Carnot efficiency to determine the cold-reservoir temperature.

● Determine the work that would be done by a reversible engine from the Carnot efficiency.

8. A Carnot engine removes 1200 J of heat from a high-temperature source and rejects 600 J to the atmosphere at 20°C. (*a*) What is the efficiency of this engine? (*b*) What is the temperature of the hot reservoir?

How to Solve It

● Determine the efficiency from the given values of the heat input and the work done.

● Since this efficiency is that of a reversible (Carnot) cycle, it can be used to find the ratio of the operating temperatures.

9. A refrigerator is rated at 370 W. Its interior is at 0°C and its surroundings are at 20°C. If the second-law efficiency of its cycle is 66 percent, how much heat can it remove from its interior in one minute?

How to Solve It

- The power rating is the work input per unit time. What is the work input in one minute?

- From the temperatures given, calculate the maximum coefficient of performance of the refrigerator.

- Use this and the second-law efficiency to calculate the actual coefficient of performance of the refrigerator.

- From this, calculate the heat extracted from the interior in one minute.

10. When 1 kg of water is frozen under standard conditions, by how much is its entropy changed?

How to Solve It

- "Standard conditions" means that the heat is extracted from the water at 0°C.

- The entropy of the water decreases as it is frozen because heat is extracted from it.

11. A heat engine works in a cycle between reservoirs at 273 K and 490 K. In each cycle, the engine absorbs 300 cal of heat from the high-temperature reservoir and does 475 J of work. (*a*) What is its efficiency? (*b*) By how much is the entropy of the universe changed when this engine goes through one full cycle? (*c*) How much energy becomes unavailable for doing work when this engine goes through one full cycle?

How to Solve It

- Calculate the actual efficiency of the engine from the heat input and the work done.

- The entropy change of the universe is the difference between the entropy increase of the cold reservoir and the entropy decrease of the hot reservoir.

- The "lost" work is the additional work that could have been done by a Carnot engine in the same circumstances.

12. The interior of my kitchen refrigerator's freezing compartment is at 10°F, and as I write, the kitchen is at 78°F. Suppose that heat leaks through the walls into the freezing compartment at a rate of 70 cal/min. In one hour, how much has the entropy of the universe been increased by this heat leakage?

How to Solve It

- The entropy of the interior increases by an amount equal to the heat delivered to it divided by its temperature. Watch out for units—the Fahrenheit scale may seem more natural for household stuff, but you'll have to convert to Kelvin temperatures.

- In the same way, calculate the entropy decrease of the kitchen outside the refrigerator.

- The entropy change of the universe is that of the interior of the icebox plus that of the outside world. Calculate this for a time of one hour.

13. In a thermos bottle, 350 g of water and 150 g of ice are in equilibrium at 0°C. The bottle is not a perfect insulator; over a period of time, its contents come to thermal equilibrium with the outside air at 25°C. By how much does the entropy of the universe increase in this process?

How to Solve It

- The universe here consists of (1) the ice and water in the thermos and (2) the outside air, which supplies the heat.

- Once the ice has melted, there are 500 g of water warming from 0°C to 25°C. This heat transfer does not take place at constant temperature; so you will have to estimate the entropy change of the water somehow; an exact calculation requires calculus. A simple way would be to assume the heat transfer takes place at a temperature halfway between 0°C and 25°C.

- The net entropy change of water and air together gives the entropy change of the universe.

14. I have a cup containing 220 g of fresh-brewed coffee which, at 75°C, is too warm to drink. If I pour 60 g of tap water at 26°C into it to cool it a little, estimate by how much I have increased the entropy of the universe.

How to Solve It

- The same snag applies here as in the last problem—the hot coffee and the cool water transfer heat at a temperature that isn't constant. Estimate the entropy changes in the same way.

- The entropy of the hot coffee decreases, and that of the cool water increases. The net change of the two is the entropy change of the universe.

Solutions

1. (a) The work done by the engine in one cycle is

$$W = Q_h - Q_c = 150 \text{ J} - 88 \text{ J} = 62 \text{ J}$$

Thus, the efficiency of the engine is

$$\epsilon = \frac{W}{Q_{in}} = \frac{62 \text{ J}}{150 \text{ J}} = \underline{0.413}$$

(b) If the engine performs 200 cycles in 1 min, then each cycle takes 0.30 s, and the power output is therefore

$$P = \frac{W}{t} = \frac{62 \text{ J}}{0.30 \text{ s}} = \underline{207 \text{ W}}$$

2. (a) $\underline{3.66}$ (b) $\underline{1864 \text{ J}}$

3. The work input is 88 J/s. Since the interior heat is extracted from freezing water, the temperature there is 0°C. Let Q_c be the heat extracted in 1 s. Then,

$$Q_h = Q_c + W = Q_c + 88 \text{ J}$$

Since the cycle is reversible and exhausts heat to the outside at 26°C,

$$\frac{Q_h}{Q_c} = \frac{T_h}{T_c} = \frac{273 \text{ K} + 26 \text{ K}}{273 \text{ K}} = \frac{299 \text{ K}}{273 \text{ K}} = 1.095$$

so

$$Q_h = 1.095 Q_c = Q_c + 88 \text{ J}$$

or

$$0.095 Q_c = 88 \text{ J}$$

So $Q_c = 924$ J and $Q_h = Q_c + 88$ J $= 1012$ J. Thus, 924 J is extracted from the freezing water each second. To freeze 2.5 kg of water at 0°C requires that heat in the amount of

$$Q = mL_f = (2.5 \text{ kg})(79.7 \text{ kcal/kg}) = 199.3 \text{ kcal}$$

by extracted from the ice. The time required is this total heat divided by the heat extracted per second:

$$t = \frac{(1.993 \times 10^5 \text{ cal})(4.184 \text{ J/cal})}{924 \text{ J/s}} = \underline{902 \text{ s}}$$

or just over 15 min.

4.   (a) 5.38   (b) 223 J   (c) 249 K or $-24°C$

5.   The efficiency of the Carnot cycle between these temperatures is

$$\epsilon_C = 1 - \frac{T_c}{T_h} = 1 - \frac{(273\ K - 78\ K)}{(273\ K + 18\ K)} = 0.330$$

The work that is to be done is

$$W = (8000\ hp)(1\ day)$$

$$= (8000\ hp \cdot day) \frac{550\ ft \cdot lb/s}{1\ hp} (86,400\ s/day)$$

$$= 3.80 \times 10^{11}\ ft \cdot lb$$

$$= (3.80 \times 10^{11}\ ft \cdot lb)(1.36\ J/ft \cdot lb)$$

$$= 5.16 \times 10^{11}\ J$$

The heat that must be extracted from the seawater to do this much work is

$$Q_h = \frac{W}{\epsilon_C} = \frac{51.6 \times 10^{11}\ J}{0.330} = 1.56 \times 10^{12}\ J$$

and so the heat that must go to evaporating the dry ice is

$$Q_c = Q_h - W = 1.56 \times 10^{12}\ J - 5.16 \times 10^{11}\ J$$

$$= 1.05 \times 10^{12}\ J$$

The mass of dry ice that will have to be evaporated is thus

$$m = \frac{Q_c}{L_v} = \frac{1.05 \times 10^{12}\ J}{(4184\ J/kcal)(137\ kcal/kg)} = 1.83 \times 10^6\ kg$$

That is about 2000 tons of dry ice for a day's sailing. Well, I said it wasn't a very good scheme.

6.   0.911

7.   (a) The actual efficiency of the engine is the work it does divided by the heat input. The work it does is 120 cal − 75 cal = 45 cal, so

$$\epsilon_{act} = \frac{45\ cal}{120\ cal} = 0.375$$

We are told that its second-law efficiency is 0.85, so we can find the Carnot efficiency from

$$\epsilon_{SL} = \frac{\epsilon_{act}}{\epsilon_C}$$

so

$$\epsilon_C = \frac{\epsilon_{act}}{\epsilon_{SL}} = \frac{0.375}{0.85} = 0.441$$

But the Carnot efficiency is

$$\epsilon_C = 1 - T_c/T_h$$

so

$$T_c/T_h = 1 - 0.441 = 0.559$$

and

$$T_c = 0.559T_h = (0.559)(273\ K + 300\ K) = \underline{320\ K}$$

(b) The reversible engine between the same two reservoirs would have an efficiency of 0.441. Thus, if the 120 cal taken from the hot reservoir could be input to a Carnot engine, it would do work

$$W = \epsilon_C Q_h = (0.441)(120\ cal) = \underline{52.9\ cal}$$

which is more than the 45 cal of work done by the real engine.

8.　(a) $\underline{0.500}$　　(b) $\underline{586\ K}$ or $\underline{313°C}$

9.　In 1 min, the refrigerator's engine does work

$$W = Pt = (370\ W)(60\ s) = 2.22 \times 10^4\ J$$

An ideal reversible refrigerator cycle based on these two temperatures would have a coefficient of performance of

$$COP_{max} = \frac{T_c}{T_h - T_c} = \frac{273\ K}{293\ K - 273\ K} = 13.65$$

The refrigerator's actual coefficient of performance is therefore

$$COP_{act} = \epsilon_{SL}COP_{max} = (0.66)(13.65) = 9.01$$

Thus, for the actual refrigerator, the heat extracted in 1 min is

$$Q_c = W(COP_{act}) = (2.22 \times 10^4\ J)(9.01) = \underline{2.00 \times 10^5\ J}$$

10. $\underline{1221\ J/K}$

11.   (a) The actual efficiency of the engine is

$$\epsilon_{act} = \frac{W}{Q_h} = \frac{475 \text{ J}}{(300 \text{ cal})(4.184 \text{ J/cal})} = \underline{0.378}$$

(b) The heat rejected to the cold reservoir in a cycle is

$$Q_c = Q_h - W = (300 \text{ cal})(4.184 \text{ J/cal}) - 475 \text{ J} = 781 \text{ J}$$

The entropy of the hot reservoir decreases (since heat has been extracted from it) by

$$\Delta S_h = \frac{Q_h}{T_h} = -\frac{(300 \text{ cal})(4.174 \text{ J/cal})}{490 \text{ K}} = -2.562 \text{ J/K}$$

while that of the cold reservoir increases by an amount

$$\Delta S_c = \frac{Q_c}{T_c} = \frac{781 \text{ J}}{273 \text{ K}} = 2.861 \text{ J/K}$$

The state of the working substance in the engine is unchanged after going through one full cycle, so its entropy is unchanged. Thus, the entropy change of the universe is

$$\Delta S_u = \Delta S_c + \Delta S_h = 2.861 \text{ J/K} - 2.562 \text{ J/K} = \underline{0.299 \text{ J/K}}$$

(c) The efficiency of a Carnot cycle between these two temperatures would be

$$\epsilon_C = 1 - \frac{T_c}{T_h} = 1 - \frac{273 \text{ K}}{490 \text{ K}} = 0.443$$

Thus, when $(300 \text{ cal})(4.184 \text{ J/cal}) = 1255 \text{ J}$ are withdrawn from the high-temperature reservoir, a reversible engine could have done work

$$W = \epsilon_C Q_h = (0.443)(1255 \text{ J}) = 556 \text{ J}$$

This means that $556 \text{ J} - 475 \text{ J} = 81 \text{ J}$ more work could have been done had the engine been reversible; since it was not, this $\underline{81 \text{ J}}$ has become unavailable for doing work.

12. $\underline{8.52 \text{ J/K}}$

13. First, calculate how much heat must be transferred to the interior of the thermos bottle to melt the ice. The amount of heat this takes is

$$Q_1 = mL_f = (150 \text{ g})(79.7 \text{ cal/g}) = 11,960 \text{ cal}$$

This heat is transferred at a constant $0°C = 273.2 \text{ K}$, so the entropy of the contents is increased by

$$\Delta S_1 = \frac{Q_1}{T_1} = \frac{11{,}960 \text{ cal}}{273.2 \text{ K}} = (43.8 \text{ cal/K})(4.184 \text{ J/cal}) = 183.2 \text{ J/K}$$

Next, the 500 g of water (including the melted ice) is heated from 273.2 K to 273.2 K + 25 K = 298.2 K. This requires heat

$$Q_2 = mc\,\Delta T_2 = (500 \text{ g})(1 \text{ cal/g·K})(25 \text{ K}) = 12{,}500 \text{ cal}$$

The temperature is constantly changing as this heat is transferred, and we can't calculate the entropy change exactly. If we make the approximation that this heat transfer takes place at the midpoint temperature of 12.5°C (285.7 K), then

$$\Delta S_2 \approx \frac{Q_2}{T_2} = \frac{12{,}500 \text{ cal}}{285.7 \text{ K}} = (43.8 \text{ cal/K})(4.184 \text{ J/cal}) = 183.1 \text{ J/K}$$

(This is only an approximation, but it's not bad; an exact calculation of ΔS_2 using calculus gives 43.79 cal/K.) All the heat is extracted from the external air at 25°C, so its entropy is decreased by

$$\Delta S_{air} = -\frac{Q}{T_n} = -\frac{11{,}955 \text{ cal} + 12{,}500 \text{ cal}}{298.15 \text{ K}}$$

$$= -82.08 \text{ cal/K} = -343.4 \text{ J/K}$$

The overall entropy change of the universe is thus

$$\Delta S_u = \Delta S_{water} + \Delta S_{air}$$

$$= 183.2 \text{ J/K} + 183.1 \text{ J/K} - 343.2 \text{ J/K} = \underline{23.1 \text{ J/K}}$$

14. Using the same approximation as in the last problem gives <u>2.17 J/K</u>.

CHAPTER 15

Oscillations

I. Key Ideas

Oscillations Oscillatory, periodic motions are very important in nature. *Oscillation* occurs when a system is disturbed from a stable equilibrium; a *restoring force* causes oscillation around the equilibrium position. Wave motion is a closely related phenomenon.

Simple Harmonic Motion Simple harmonic motion is the oscillatory motion that occurs under a restoring force that is proportional to the displacement of the system from equilibrium. This is nearly always the case provided that the displacement from equilibrium is small enough. The resulting motion has acceleration that is proportional to the displacement. The motion is a sinusoidal oscillation; that is, a graph of the displacement versus time has the form of a sine function. A typical example would be a mass oscillating on the end of a spring.

Period and Frequency The *period* of an oscillating system is the time for one complete cycle; the *frequency*, or number of cycles per unit time, is the inverse of the period. The frequency is determined by the force constant of the spring and the mass of the oscillating object. For simple harmonic motion, frequency is independent of amplitude.

Circular Motion and Simple Harmonic Motion There is a very close connection between simple harmonic motion and circular motion with constant speed. The projection on one coordinate axis of a point undergoing uniform circular motion is a simple harmonic motion, the angular velocity of the circular motion being 2π times the frequency.

Energy in Simple Harmonic Motion An object undergoing simple harmonic motion has a constant total energy, but its potential energy and its kinetic energy vary with time. At the turning points, the kinetic energy is zero and the potential energy has its maximum value; as the system passes through its equilibrium point, the reverse is true. The constant total energy is proportional to the square of the amplitude of the motion.

Simple Pendulum A familiar oscillating system is a pendulum—a mass on a string that is free to swing back and forth under gravity. The motion is not, in general, simple harmonic motion, but it approaches simple harmonic motion for small amplitudes. The period of the pendulum is independent of its mass, depending only on its length and the acceleration due to gravity. This makes the measurement of g easy.

Damped Oscillations In real oscillations, the motion is not conservative; it is always damped by frictional forces. As a result, the energy and amplitude of the motion decrease with time. If damping forces are small, the energy decreases exponentially, that is, by a constant *fraction* in a given time interval. The *quality factor Q* of a damped oscillation expresses the amount of damping; a high Q means the oscillation takes a long time to die out. If the damping is strong enough, there is no oscillation; the displacement just dies out.

Driven Oscillations Conversely, if the amplitude of a harmonic oscillator is to stay constant, energy must be supplied to make up for that dissipated by damping forces; that is, the oscillation must be *driven*. If a periodic driving force is applied, the eventual steady-state motion is at the driving frequency. If this is close to the natural frequency of the oscillator, the amplitude of the driven motion is large. This is called *resonance*. If the oscillator has high Q, the resonance response is strong, but only over a narrow frequency range.

II.   Numbers and Key Equations

Numbers

1 hertz (Hz) $= 1$ s^{-1}

Key Equations

Simple harmonic motion

$F = -kx$ $a = -(k/m)x$

Period and frequency

$$f = 1/T \quad f = \frac{1}{2\pi} \sqrt{\frac{k}{m}} \quad T = 2\pi \sqrt{\frac{m}{k}}$$

Displacement in simple harmonic motion

$x = A \cos (2\pi ft)$

Velocity in simple harmonic motion

$v = -(2\pi f)A \sin (2\pi ft)$

Acceleration in simple harmonic motion

$a = -(2\pi f)^2 A \cos (2\pi ft) = -(2\pi f)^2 x$

Angular velocity of circular motion

$\omega = 2\pi f$

Total energy of an oscillator

$E = \frac{1}{2}kA^2$

Potential energy of an oscillator

$E = \frac{1}{2}kx^2$

Period of a pendulum

$$T = 2\pi \sqrt{L/g} \quad g = \frac{4\pi^2 L}{T^2}$$

Quality factor of a damped oscillator

$$Q = 2\pi \frac{E}{|\Delta E|}$$

Quality factor and resonance

$$Q = \frac{f_0}{\Delta f}$$

III.   Possible Pitfalls

Don't be careless and use any of the constant-acceleration formulas for the motion of a harmonic oscillator! The force is proportional to the displacement from equilibrium, so it is definitely *not* constant.

In the formulas involving sine and cosine functions, the argument of the trig function (that is, the quantity whose sine or cosine is taken) must be a *dimensionless* number; if thought of as an angle, it must be an angle in *radians*.

In almost all cases, it's best to put the origin of the x axis at the equilibrium position of the oscillating particle since that is the zero point of the displacement of, and thus of the force on, the oscillating particle.

Remember that the formula for the period of a simple pendulum applies only to the *simple* pendulum. If the actual system is something other than a particle on the end of a massless string, then the period will depend on how the mass is distributed in space.

The formulas we write for simple harmonic motion apply to any motion under a restoring force that is directly proportional to the displacement from an equilibrium position. The "force constant k" used in the formulas is whatever quantity occupies that position in the force equation, $F = -kx$, whether or not there are springs involved.

The potential and kinetic energy of a harmonic oscillator both vary with time; it is only the total energy that is constant. The total energy may be all kinetic, all potential, or anything in between at different points in the cycle.

The motion of a mass on a spring hanging vertically is simple harmonic motion even though a second force—gravity—is involved. The weight simply changes the equilibrium point about which the mass oscillates.

IV. Questions and Answers

Questions

1. A particle is undergoing simple harmonic motion. How far does it move in one full period?

2. A mass oscillates on the end of a certain spring at frequency f. The spring is cut in half, and the same mass is set oscillating on one of the pieces. What is its new frequency of oscillation?

3. The mass of the string is usually neglected in treating the motion of a simple pendulum. If the mass of the string is not completely negligible, how is the motion of the pendulum affected?

4. A mass on the end of a string is hung over a nail and set swinging as a simple pendulum of length L. While it is swinging, the string is paid out until the free-swinging length is $2L$ (see Figure 15-1). What has happened to the pendulum's frequency? Can we use the conservation of energy to find out what has happened to its amplitude? Neglect friction at the nail.

5. If a mass-and-spring system is set oscillating, why does its motion eventually stop?

6. A mass-and-spring system is undergoing simple harmonic motion with amplitude A. If the mass is decreased by half while the amplitude is unchanged, how does the total energy of the oscillation change? What about the same mass oscillating with half the initial amplitude?

Figure 15-1

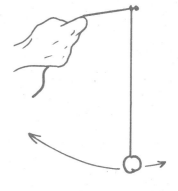

Answers

1. The particle's net displacement in one period is of course zero. The total distance it has covered, however, is from its initial position to $x = A$, from there to $x = -A$, and from $x = -A$ back to its initial position. It has covered a distance of $4A$.

2. The question really is, what is the force constant of half a spring? The same tension stretches one-half of the original spring half as much as it would the whole; so, if the force constant of the whole spring is k, that of one of the halves is $2k$. Since frequency is proportional to the square root of k, cutting the spring in half increases the frequency by $\sqrt{2}$.

3. If the string has mass, a small fraction of the total mass that is swinging is closer to the pivot than the pendulum length L, so the effective length is a little less than L. The motion, consequently, has a slightly shorter period than would a simple pendulum of the same length.

4. The period of a simple pendulum is proportional to $\sqrt{L}$, so the period is increased and the frequency is decreased by a factor of $\sqrt{2}$. The energy of the swinging pendulum is not a constant because (negative) work is done on it by whatever holds the string as it is paid out (the hand, in the figure). Since we don't know how to evaluate this work, we can't apply the conservation of energy to this problem.

5. The motion dies out because real oscillations are damped. The drag of the air and the internal forces in the material of the spring are nonconservative forces that dissipate the energy of the oscillating mass.

6. When the mass is halved, the frequency of the oscillation increases, but the total energy, $\frac{1}{2}kA^2$, is unaffected. When the amplitude is halved, the total energy is decreased by a factor of 4.

V. Problems and Solutions

Problems

1. The position of a particle is given by

$$x = 2.2 \cos (18\pi t)$$

where the quantities are measured in centimetres and seconds. What are (*a*) the frequency, (*b*) the period, and (*c*) the amplitude of the motion? (*d*) When is the first time after $t = 0$ that the particle is at $x = 0$?

How to Solve It
- The argument of the cosine function is $2\pi ft$, where f is the frequency; the inverse of the frequency is the period.

- The maximum value of the displacement x is the amplitude.

- In order that $x = 0$, clearly $\cos(18\pi t)$ must be zero. For what value of the argument is the cosine function equal to zero?

2.   The motion of a mass oscillating on the end of a spring is graphed in Figure 15-2. What is (*a*) the amplitude, (*b*) the period, and (*c*) the frequency of its motion? (*d*) If the mass is 600 g, what is the force constant of the spring?

Figure 15-2

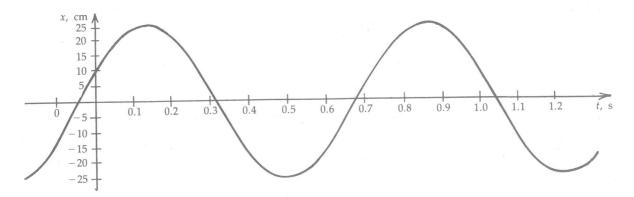

How to Solve It
- The maximum displacement is the amplitude.

- The time it takes for the motion to go through one complete cycle is the period; the inverse of the period is the frequency.

- The force constant of the spring can be calculated from the known mass and period.

3.   A certain mass hanging in equilibrium on the end of a vertical spring stretches the spring by 2 cm. (*a*) If the mass is then pulled down 5 cm further and released, at what frequency does it oscillate? (*b*) At what speed is it moving as it passes through its equilibrium position?

How to Solve It
- The period and frequency of the oscillation depend on the ratio k/m—if you know that ratio, you don't need to know k or m individually.

- You can calculate k/m from the distance the spring is stretched by the weight hanging in equilibrium.

- As it passes through equilibrium, the mass has its maximum speed.

4. A 5-kg block oscillates at a frequency of 5 Hz and an amplitude of 5 cm on the end of a spring. (*a*) What is the force constant of the spring? (*b*) What is the period of the block's motion? (*c*) What is the maximum speed at which the block moves?

How to Solve It
- The force constant can be determined from the mass of the block and the frequency of its oscillation.

- The period is the inverse of the frequency.

- The maximum speed of the oscillating block can be determined from its amplitude and frequency.

5. A platform of mass 200 g is attached to a spring of force constant 440 N/m in such a way that the system can oscillate vertically, as sketched in Figure 15-3. A 500-g block rests on the platform. The spring is compressed a distance A below equilibrium and released. What is the maximum value that A can have if the block is not to become separated from the platform?

Figure 15-3

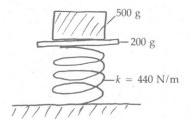

How to Solve It
- From the mass undergoing oscillation and the force constant of the spring, calculate the frequency of oscillation.

- In terms of A and the frequency, what is the maximum acceleration of the oscillating platform?

- If the acceleration exceeds g, the block will leave the platform as the platform starts downward.

6. A phonograph record 30 cm in diameter revolves once every 1.80 s. (*a*) What is the linear speed of a point on its rim? (*b*) What is its angular velocity ω? (*c*) Write an equation for the y component of the point's position as a function of time, given that it is at its maximum value on the y axis when $t = 0$.

How to Solve It
- You know the diameter of the record; how far does a point on its rim move in 1.80 s?

- The angular velocity of rotation is the linear speed divided by the radius of the disk.

- The amplitude of the oscillation is, of course, the radius of the record.

7.   An object of mass 1 kg oscillates on a spring with an amplitude of 12 cm. Its maximum acceleration is 5.0 m/s². Find its total energy.

How to Solve It
- Determine the frequency from the amplitude and the maximum acceleration.

- Use the frequency and the mass to calculate the force constant of the spring.

- Use the force constant and the amplitude of the oscillation to determine the total energy.

8.   A 400-g mass oscillates on the end of a spring with an amplitude of 2.5 cm. Its total energy is 2.0 J. What is the frequency of the oscillation?

How to Solve It
- Since you know the amplitude of the oscillation, you can use the total energy given to find the force constant of the spring.

- Use the force constant and the mass of the oscillating object to find the period and frequency of the motion.

9.   Two simple pendulums of length 1.8 m, with masses of 1.4 kg and 2.0 kg, hang side by side such that, at rest, their bobs just touch (see Figure 15-4). The pendulum on the left is pulled back 6 cm and released; it swings down and strikes the other, which is initially at rest. If the collision is elastic, to what distance does each pendulum rebound after the collision?

Figure 15-4

How to Solve It
- First, calculate the frequency of oscillation of a pendulum 1.8 m long and, from it, the velocity of the 1.4-kg pendulum when it reaches the bottom of its arc.

- Use what you learned about elastic collisions in Chapter 7 to find the velocities of the two pendulums just after they collide; remember that these are their maximum velocities.

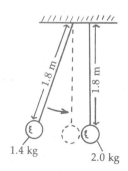

- For each pendulum, calculate the amplitude of its motion after the collision from the frequency and the maximum speed.

10. What is the natural frequency of a man of mass 75 kg swinging at the end of a rope 5 m long?

How to Solve It
- Approximate this as a simple pendulum.
- Provided that we can consider this to be a simple pendulum, the man's mass is simply window dressing. The period of a simple pendulum is determined by its length only.

11. (*a*) How long is a pendulum whose period is 2.00 s? (Such a pendulum is a common timekeeping device in mechanical household clocks.) (*b*) If this pendulum has a mass of 2.0 kg and a total energy of 0.20 J, what is the amplitude of its oscillation?

How to Solve It
- The period of a simple pendulum is determined by its length.
- The total energy is equal to the maximum value of the kinetic energy; from this, you can get the maximum speed of the pendulum.
- Knowing the maximum speed and the frequency, which you can determine from the period, you can find the amplitude of the pendulum's swing.

12. A damped harmonic oscillator of natural frequency 180 Hz loses 5 percent of its energy in each cycle. (*a*) What is the Q factor for this oscillator? (*b*) When this oscillator is driven, what is the width Δf of its resonance curve?

How to Solve It
- The fractional energy loss in a cycle is equal to $2\pi/Q$.
- Likewise, the fractional width $\Delta f/f_0$ of the resonance of the driven oscillator is equal to $1/Q$.

13. A child on a swing swings back and forth once every 3.75 s. The mass of the child (including the swing seat) is 38 kg. At the bottom of its arc, the child is moving at 1.82 m/s. (*a*) What is the total energy of the swing and child? (*b*) If the Q value is 25, what (average) power must be supplied to the swing to keep it moving with a constant amplitude?

How to Solve It
● The maximum kinetic energy of the oscillating child is equal to the total energy.

● Given that $Q = 25$, calculate the fraction of the energy that is dissipated in each cycle.

● This much energy must be supplied every 3.75 s; to what average power input does this correspond?

Solutions

1.   (a) The position of a particle undergoing simple harmonic motion is given by

$$x = A \cos (2\pi ft)$$

The argument of the cosine function (that is, what you take the cosine of) is

$$18\pi t = 2\pi ft$$

so

$$f = 9 \text{ s}^{-1} = 9 \text{ Hz}$$

(b) The period is

$$T = 1/f = \frac{1}{9 \text{ s}^{-1}} \, \underline{0.111 \text{ s}}$$

(c) Since the maximum value of the cosine is 1, the amplitude A of this motion is the maximum value of x, so $\underline{A = 2.2 \text{ cm}}$.

(d) The first value of the argument at which the cosine is zero is $\pi/2$ radians. The time at which this occurs is given by

$$18\pi t = \frac{\pi}{2}$$

$$t = \frac{\pi/2}{18\pi} \text{ s} = \frac{1}{36} \text{ s} = \underline{0.0278 \text{ s}}$$

2.   (a) <u>About 25 cm</u>   (b) <u>0.72 Hz</u>   (c) <u>1.4 s</u>   (d) <u>46 N/m</u>

3.   (a) In equilibrium, the net force on the hanging mass is

$$F_{\text{net}} = kx - mg = 0$$

Therefore,

$$k(2 \text{ cm}) = m(9.81 \text{ m/s}^2)$$

so

$$\frac{k}{m} = \frac{9.81 \text{ m/s}^2}{0.02 \text{ m}} = 490 \text{ s}^{-2}$$

Thus, the frequency is

$$f = \frac{1}{2\pi} \sqrt{\frac{k}{m}} = \frac{1}{2\pi} \sqrt{490 \text{ s}^{-2}} = \underline{3.52 \text{ Hz}}$$

(b) As the mass passes through its equilibrium position, it has its maximum speed. From the equation for velocity, the maximum speed is obviously

$$v_{max} = 2\pi f A = (2\pi)(3.52 \text{ Hz})(0.05 \text{ m}) = \underline{1.11 \text{ m/s}}$$

4.   (a) $\underline{4930 \text{ N/m}}$ (b) $\underline{0.20 \text{ s}}$ (c) $\underline{1.57 \text{ m/s}}$

5.   A total mass of 200 g + 500 g = 700 g = 0.70 kg is oscillating on a spring of force constant 440 N/m. The frequency is therefore

$$f = \frac{1}{2\pi} \sqrt{\frac{k}{m}} = \frac{1}{2\pi} \sqrt{\frac{440 \text{ N/m}}{0.7 \text{ kg}}} = 3.99 \text{ Hz}$$

From the general equation for the acceleration of a particle undergoing simple harmonic motion, the maximum acceleration of the platform is obviously

$$a_{max} = (2\pi f)^2 A = [(2\pi)(3.99 \text{ Hz})]^2 A = (629 \text{ s}^{-2}) A$$

where A is the amplitude of the motion. Since the block is not fastened to the platform, it cannot follow the platform downward with an acceleration greater than g. Thus if the block is to stay on board,

$$a_{max} < g$$
$$(629 \text{ s}^{-2}) A < 9.81 \text{ m/s}^2$$

so

$$\underline{A < 0.0156 \text{ m}}$$

6.   (a) $\underline{0.524 \text{ m/s}}$ (b) $\underline{3.49 \text{ rad/s}}$ (c) $\underline{y = (0.15 \text{ m}) \cos (3.49t)}$

7.   As in Problem 5, the maximum acceleration of the mass is

$$a_{max} = (2\pi f)^2 A$$

so

$$f = \frac{1}{2\pi}\sqrt{\frac{a_{max}}{A}} = \frac{1}{2\pi}\sqrt{\frac{5 \text{ m/s}^2}{0.12 \text{ m}}} = 1.03 \text{ Hz}$$

Use the frequency formula

$$f = \frac{1}{2\pi}\sqrt{\frac{k}{m}}$$

to give

$$k = (m)(2\pi f)^2 = (1 \text{ kg})[(2\pi)(1.03 \text{ s}^{-1})]^2 = 41.7 \text{ N/m}$$

Thus, the total energy of the object is

$$E = \tfrac{1}{2}kA^2 = (0.5)(41.7 \text{ N/m})(0.12 \text{ m})^2 = \underline{0.300 \text{ J}}$$

8. <u>20.1 Hz</u>

9. The period of a pendulum 1.8 m long is

$$T = 2\pi\sqrt{\frac{L}{g}} = (2\pi)\left(\frac{1.8 \text{ m}}{9.81 \text{ m/s}^2}\right)^{1/2} = 2.69 \text{ s}$$

so its frequency is

$$f = \frac{1}{T} = \frac{1}{2.69 \text{ s}} \; 0.372 \text{ Hz}$$

At the bottom of the pendulum's arc, its speed is

$$v = v_{max} = 2\pi f A = (2\pi)(0.372 \text{ s}^{-1})(0.06 \text{ m}) = 0.140 \text{ m/s}$$

This is the speed at which the 1.4-kg bob hits the 2-kg bob. If the collision is elastic, we can find the velocities v_1 and v_2 afterward. Just before the collision, the 2.0-kg pendulum is at rest, so their relative velocity is equal to the velocity of the 1.4-kg pendulum, which is 0.140 m/s. We remember that after an elastic collision, the relative velocity of the masses is reversed, so

$$v_1 - v_2 = -0.140 \text{ m/s} \tag{1}$$

Momentum is conserved, so

$$p_{after} = p_{before}$$

$$m_1 v_1 + m_2 v_2 = m_1 v$$

$$(1.4 \text{ kg})v_1 + (2.0 \text{ kg})v_2 = (1.4 \text{ kg})(0.140 \text{ m/s})$$

Simplifying this equation by dividing out the 1.4 kg gives

$$v_1 + 1.4286v_2 = 0.140 \text{ m/s} \qquad\qquad (2)$$

To solve Equations (1) and (2), we add them to get

$$2v_1 + 0.4286v_2 = 0$$

$$v_2 = \frac{-2v_1}{0.4286} = -4.67v_1$$

Substituting this into either Equation (1) or (2) and solving gives $v_1 = -0.025$ m/s and $v_2 = 0.115$ m/s. Now, these velocities correspond to the maximum speeds of each of the pendulums, so we can find the amplitudes of the first from $v_1 = 2\pi f A$. Thus,

$$A_1 = \frac{v_1}{2\pi f} = \frac{0.025 \text{ m/s}}{(2\pi)(0.372 \text{ Hz})} = 1.07 \times 10^{-2} \text{ m} = \underline{1.07 \text{ cm}}$$

In just the same way, $A_2 = \underline{4.92 \text{ cm}}$.

10. $\underline{0.223 \text{ Hz}}$

11.   (a) The formula for the period of a simple pendulum is

$$T = 2\pi \sqrt{\frac{L}{g}}$$

so

$$L = \frac{gT^2}{4\pi^2} = \frac{(9.81 \text{ m/s}^2)(2.00 \text{ s})^2}{4\pi^2} = \underline{0.994 \text{ m}}$$

(b) The maximum value of the kinetic energy is equal to the total energy:

$$E_{k,\max} = \frac{1}{2}mv_{\max}^2$$

$$v_{\max}^2 = \frac{2E_{k,\max}}{m}$$

The energy of the pendulum is 0.20 J and its mass is 2 kg, so

$$v_{\max}^2 = \frac{2(0.20 \text{ J})}{2 \text{ kg}} = 0.20 \text{ m}^2/\text{s}^2$$

which gives

$$v_{\max} = 0.447 \text{ m/s}$$

But we know that the period is 2.00 s, so the frequency is $f = 1/T = 1/2.00 \text{ s} = 0.50$ Hz, and

$$v_{max} = 2\pi f A$$

Thus, the amplitude is

$$A = \frac{v_{max}}{2\pi f} = \frac{0.447 \text{ m/s}}{2\pi(0.50 \text{ s}^{-1})} = \underline{0.142 \text{ m}}$$

12.   (a) $\underline{126}$ (b) $\underline{1.4 \text{ Hz}}$

13.   (a) The total energy of the swinging child is

$$E = E_{k,max} = \tfrac{1}{2}mv^2_{max} = \tfrac{1}{2}(38 \text{ kg})(1.82 \text{ m/s})^2 = \underline{62.9 \text{ J}}$$

     (b) The fraction of the energy dissipated every cycle is

$$\frac{|\Delta E|}{E} = \frac{2\pi}{Q} = \frac{2\pi}{25} = 0.251$$

so

$$|\Delta E| = (0.251)E = (0.251)(62.9 \text{ J}) = 15.8 \text{ J}$$

is the energy that must be supplied every 3.75 s. The corresponding average power is

$$P = \frac{W}{t} = \frac{15.8 \text{ J}}{3.75 \text{ s}} = \underline{4.22 \text{ W}}$$

CHAPTER 16

Mechanical Waves: Sound

I. Key Ideas

Wave Motion Wave motion is the transport of energy and momentum from one point to another without the bulk transport of matter. Mechanical waves require a physical medium to transport them, whereas electromagnetic waves can propagate in a vacuum. In mechanical waves, a disturbance of a material medium is propagated through the medium because of its physical properties.

Waves on a String If one end of a stretched string is given a twitch, a wave pulse travels down the string. Energy and momentum are carried along the string, but the material of the string itself is only twitching from side to side. The speed at which the wave moves depends on the tension in the string and its mass per unit length. At the end of the string, part or all of the wave pulse is reflected. The way in which it is reflected depends on how the string ends. If it is fastened rigidly or is tied to another string on which the wave speed is slower, the reflected wave is *inverted* relative to the incident wave.

Transverse and Longitudinal Waves Waves on a string are an example of *transverse waves* because the direction of the disturbance is at right angles to the direction in which the wave propagates. There are also *longitudinal waves*, in which the dis-placement of the medium is parallel to the direction of propagation. Sound waves, for example, are longitudinal patterns of compression and rarefaction. Surface waves on water combine transverse and longitudinal motions.

Interference Waves of all sorts obey the principle of *superposition*, which states that the wave disturbances at a point due to two or more different sources simply add. Each wave propagates as if the others weren't there. Two waves may cancel or augment each other, producing destructive or constructive *interference*, respectively.

Wave Speed The speed at which a wave propagates is dependent on the properties of the *medium* through which it moves, not on the motion of the source. For a given kind of mechanical wave, the speed can be derived from Newton's second law. Typically, it's the square root of the ratio of an elastic property of the medium to an inertial property.

Harmonic Waves A sinusoidal disturbance in a medium, as of a string whose end is wiggled back and forth in simple harmonic motion, is a *harmonic wave*. The wave shape at an instant is sinusoidal. The length of one cycle in space is the *wavelength* of the wave.

Sound Waves Harmonic sound waves are sinusoidal disturbances in the density and pressure of a substance such as the air. The human ear is sensitive to sound waves in the frequency range from 20 Hz to 20 kHz, but many important applications use sound waves of much higher frequencies.

The Doppler Effect If the source or the receiver of a sound wave is in motion with respect to the medium (such as air) that propagates the sound, the frequency received will not in general be that of the source. This is the *doppler effect*. If the source and the receiver are approaching one another, the received frequency will be higher than that of the source and vice versa. The change in frequency depends slightly on whether it is the source or the receiver that moves relative to the medium. For electromagnetic waves, such as light, no medium is required to transmit the waves and the doppler effect must be calculated somewhat differently; our formulas for the doppler effect are only approximately correct for electromagnetic waves. If the source moves faster than the wave speed, waves from the source "pile up" behind it as a shock wave.

Intensity Waves transport momentum and energy. The *intensity* of a wave is the power it delivers across a unit area. For a wave that propagates uniformly in all directions from a point source, the intensity varies inversely as the square of the distance from the source. The intensity of a wave can be expressed as its velocity multiplied by the average energy per unit volume of the wave disturbance. The intensity of a wave is proportional to the square of its *amplitude*. The ear can detect a very wide range of sound intensities, so intensity level is often expressed using a logarithmic scale—the *decibel scale*.

II. Key Equations

Key Equations

Wave speed

$$v = \sqrt{F/\mu} \qquad v = \sqrt{B/\rho_0} \qquad v = \sqrt{\frac{\gamma RT}{M}}$$

Frequency and wavelength

$$v = f\lambda$$

Doppler effect (moving source)

$$f' = \frac{1}{(1 \mp u_s/v)}f_0$$

Doppler effect (moving receiver)

$$f' = (1 \pm u_r/v)f_0$$

In each doppler-effect formula, the top sign refers to the case of the source and receiver approaching one another.

Shock-wave angle

$$\sin\theta = \frac{v}{u}$$

Intensity

$$I = \frac{P}{4\pi r^2} \qquad I = \eta v$$

Decibel intensity level

$$\beta = 10\log_{10}(I/I_0)$$

III. Possible Pitfalls

The term "amplitude" means just the same for a sinusoidal mechanical wave as it does for simple harmonic motion—the maximum displacement of a bit of the medium from its undisturbed, equilibrium position. It's not the distance from crest to valley.

We tend to think of waves in terms of transverse waves on a string, but remember that various quantities other than displacement can propagate as waves. Sound is a pressure wave, in that a pattern of excess pressure propagates through the air, as well as a displacement wave. In both the sound wave and the wave on a string, there is a pattern of velocity in the medium that is propagating as well as a pattern of displacement.

It is really, really easy to make sign errors when doing doppler-effect problems. Here is a case where simply memorizing raw formulas will do you absolutely no good. Remember that, in the frequency formulas given, the *wave speed* and the *source* and *receiver speeds* are used, so these are always positive. Whenever the source and receiver are getting closer together, the observed frequency is always *higher* than that of the source; it is always *lower* when the source and receiver are getting further apart. If both source and observer are moving relative to the medium in which sound waves are propagating, I suggest doing the problem in steps.

Mechanical waves propagating through a material medium disturb the medium; that is, they displace it. Be careful not to confuse the velocity of a bit of the medium due to the wave disturbance with the velocity of the wave. The two velocities aren't directly related. The latter has to do with the rate at which the *pattern of disturbance* is moving through the medium. The pattern is the wave. In the case of a longitudinal wave, the two velocities are sometimes even in opposite directions.

All mechanical waves that move through a material medium move in a manner that is *determined* by the medium and its properties. The wave source stimulates the medium, and something that detects the wave is stimulated by the medium; but the properties of the source and the detector are not what determines the motion of the wave.

The equation $v = f\lambda$ is really a relation (or a matter of definition) between frequency and wavelength. The wave speed v is determined by the properties of the medium that transports the waves and may or may not depend on the wavelength. The speed of sound in air, for example, is nearly independent of wavelength, while the speed of ripples on a pond depends strongly on it.

The equation $v = f\lambda$ comes simply from the definitions of frequency and wavelength. It is a fundamental relationship that applies to *any* sort of periodic wave. Maybe because the equation is so simple, it is very frequently mixed up—we forget what divides which and so on. However, it is also a very easy formula to check by making sure that the dimensions of the quantities make sense.

IV.   Questions and Answers

Questions

1.   Solids are typically thousands of times denser than air, yet the speed of sound in a solid is usually greater than it is in air. Why is this?

2.   Why does the sound of a car horn drop in frequency as the car passes close by you? Where is the car when you hear the horn's true pitch?

3.   A Slinky is a children's toy that is just a long, loose-coiled spring. It is quite useful as a wave demonstrator. How would you generate longitudinal waves in a Slinky? What about transverse waves?

4.   A transverse sine wave is propagating along a stretched wire. It comes to a junction where the wire is fastened to a different wire that has half the linear density (mass per unit length) of the first. Does the wave propagate faster or slower beyond the junction? Does its frequency change? Its wavelength?

5. A long, heavy rope is hung from the ceiling, its bottom end free. If you waggle the free end back and forth to send waves up the rope, you'll find that waves don't propagate up at a constant speed. Why not? Does the wave speed increase or decrease as the waves propagate upward?

6. I'm fond of a very basic demonstration of the nature of a wave that goes like this: All the students in the class are organized into a single line, and each student is told to imitate whatever he or she sees the person just ahead doing, ignoring all the others. The front student raises his or her hand or whatever, and the gesture travels all along the line. What properties of the "medium" determine the velocity of the "wave"? Does the "wave" transport energy?

7. In Figure 16-1*a*, two wave pulses are approaching one another along a stretched string. A little later, the string looks like Figure 16-1*b*, with the two pulses more or less cancelling (for this one instant). But in this undisplaced string, where is the energy that the two pulses in the upper sketch were clearly carrying?

Figure 16-1

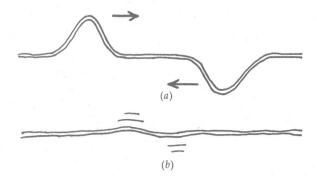

(a)

(b)

8. A tuning fork, which vibrates at a fixed frequency, is being waved back and forth along the direction from it to you. Describe what you hear. What will you hear if the fork is moved from side to side?

9. Consider a wave, say a sound wave, that propagates uniformly in all directions from a point source. How does the amplitude of the wave vary with distance from the source?

10. A nice, pleasant sound level at which to listen to music is 60 dB. Would 120 dB be twice as loud?

Answers

1.   This is because the bulk modulus of the solid is also many thousands of times greater than that of the air. It is the ratio of the bulk modulus to the density that determines the speed of sound in a material.

2.   It drops because of the doppler effect, of course. When the car is approaching you, the frequency you hear is higher than the natural frequency of the horn; when it is receding, you hear a lower frequency. The sound that you hear as the car is just passing you—when, instantaneously, it is moving perpendicular to the direction from you to it—will be the horn's true pitch.

3.   You can generate longitudinal waves in a slinky very easily by tying one end of the spring to the wall and stretching it to a moderate degree. Moving the free end back and forth along the line of the spring will produce easily visible compression waves along the Slinky at a few metres per second (see Figure 16-2a). Transverse waves are generated by moving the free end up and down (see Figure 16-2b).

Figure 16-2

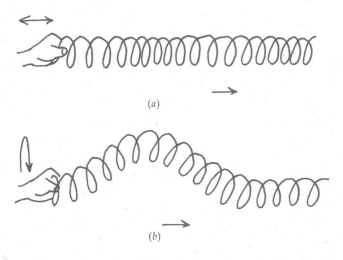

(a)

(b)

4.   The second wire, with the lower linear density, is lighter than the first wire and is under the same tension (since they are fastened end to end). Thus, it will propagate the wave at a higher speed. The frequency is the rate at which a single point on the wire is moving up and down. This clearly has to be the same on both sides of the junction. Thus, on the lighter wire, the wave has the same frequency but is propagating faster, so its wavelength increases.

5.   The reason the wave speed is not constant is that the upper part of the rope is under a higher tension than the lower part because of the weight of the lower part. Since wave speed increases with increasing tension, the wave speed will increase as the waves move upward.

6.   The "wave," the pattern being propagated, is whatever the first person in line is doing that the others imitate in sequence. The medium is the actual line of students; the velocity of propagation is their average separation divided by their average reaction time. Certainly energy is transported; it's the propagation of the wave that sets the last person's arms waving or whatever.

7.   The energy is still right there; it just happens to be all kinetic energy at the instant the sketch is drawn. Although the string is, in the sketch, instantaneously undisplaced, it's still in motion.

8.   You'll hear the pitch rise and fall as the tuning fork is moved toward or away from you, but you'll hear only a steady pitch as it is moved from side to side.

9.   The intensity must decrease as $1/r^2$, where r is the distance from the source, for a wave propagating uniformly in all directions in three dimensions. Thus, the amplitude, which is proportional to the square root of the intensity, must decrease as $1/r$.

10. No. In fact, just what "twice as loud" means isn't clear because loudness is a function of perception; but this isn't it. The 60 dB increase means the sound intensity is increased by a factor of one million.

V. Problems and Solutions

Problems

1.   Far out at sea, your ship is traveling at 12 mi/h in the same direction as the waves, and you notice that one swell passes you every 9 s. If the ocean waves are 500 ft apart, what is the velocity of the waves?

How to Solve It

- The frequency multiplied by the wavelength gives you the velocity of the waves.

- If you know the speed of the waves relative to your ship and the ship is traveling in the same direction as the waves, you can calculate the speed of the waves.

- The wave speed is the speed of the waves relative to the ship plus the speed of the ship.

2. The extreme range of human singing voices is roughly from 53 Hz (A flat below bass low C) to 1267 Hz (E flat above soprano high C). Find the range of wavelengths that corresponds to this frequency range.

How to Solve It

- Having no other information, we usually take the speed of sound in air to be 340 m/s, the value at 20°C.

- Frequency times wavelength gives you the wave speed.

3. (*a*) Air has a molar mass of 28.9 g and the ratio of its specific heat at constant pressure to that at constant volume is $\gamma = 1.40$; helium has a molar mass of 4.00 g and $\gamma = 1.67$. At room temperature (20°C), calculate the speed of sound in air and in He. (*b*) Calculate the frequency in each medium of a sound of wavelength 2.2 m.

How to Solve It

- There is a simple formula for the speed of sound waves in a gas in terms of temperature, molar mass, and the ratio γ of specific heats at constant pressure and at constant volume.

- The speed of sound is the frequency multiplied by the wavelength.

4. You stand on the edge of a canyon 420 m wide and clap your hands. An echo comes back to you 2.56 s later. What is the air temperature? (Use $\tau = 1.40$ for air.)

How to Solve It

- Calculate the speed of sound from the data given. (Don't forget the sound has to go across the canyon and back.)

- There is a simple formula for the speed of sound waves in a gas in terms of temperature, molar mass, and the ratio γ of specific heats at constant pressure and at constant volume.

- Solve for the temperature.

5. In a steel wire, the velocity of longitudinal waves—that is, sound waves within the steel—is just 10 times that of transverse waves on the wire when the wire is under a tension of 6200 N. If Young's modulus for steel is 2.0×10^{11} N/m², what is the diameter of the wire?

How to Solve It
- Young's modulus is used in this case rather than the bulk modulus because the sound waves compress the steel along one direction only.

- It may seem that you need the density of the steel to do this problem, but you don't. Write the formulas for the speed of sound (longitudinal) waves in the wire and for transverse waves on it. Write a formula for the ratio of these speeds.

- What is the relation between the linear density μ and the density ρ? If you've done everything right so far, the mass and length of the wire should cancel out so that you can solve for its diameter.

6. A steel piano wire 75 cm long has a mass of 4.5 g. The speed of transverse waves in the wire is 230 m/s. (*a*) What tension is the wire under? (*b*) To reduce the speed of waves on the wire to 172 m/s without increasing its tension, the wire's mass is increased by wrapping fine bronze wire uniformly around it. What mass of bronze wire must be used?

How to Solve It
- Calculate the linear density μ of the wire and, from it, the tension F in the wire.

- To get the lower speed from a wire under the same tension, the linear density must increase. By how much?

- This extra linear density multiplied by the length gives you the extra mass that must be added to the wire.

7. A woman who has perfect pitch, standing next to a railroad track, is amused to notice that the whistle of an approaching train is sounding a true concert A (440 Hz). After the train has passed her and is receding, the pitch has dropped to a true F natural (352 Hz). (*a*) How fast was the train going? (*b*) What was the actual frequency of the whistle?

How to Solve It

- The first frequency is heard when the train is coming straight toward the woman. Write a formula for the received frequency in terms of the unknown source frequency and source speed.

- Repeat this for the second frequency, which is heard when the train is receding from the woman at the same source speed.

- Eliminate the whistle frequency f_0 from these two equations, and you will be left with something that can be solved for the velocity ratio u/v.

- When you have solved for the velocity u of the train, use that in either of the equations to find f_0.

8. My car's horn sounds a tone of frequency 250 Hz. If I am driving directly at you at 25 m/s, blowing my horn, on a hot day (93°F), (*a*) what is the wavelength of the sound that reaches you? (*b*) What frequency do you hear?

How to Solve It

- The only thing out of the ordinary is that you have to recalculate the speed of sound at the temperature given.

- The wavelength is decreased by the factor $(1 - u_s/v)$ because the sound source is coming at you.

9. A car moves at a speed of 40 km/h directly toward a stationary wall. Its horn emits a signal at a frequency of 300 Hz. (*a*) If the velocity of sound is 340 m/s, find the frequency and wavelength of the sound waves that hit the wall. (*b*) The waves reflect off the wall and are received by the driver of the car. What frequency does she hear?

How to Solve It

- It's simplest to do this in problem steps. What is the frequency of the sound "received" by the wall?

- Sound of this same frequency (f_1, say) is reflected back toward the oncoming car.

- The driver of the oncoming car is now a moving observer receiving waves that originate with frequency f_1. What frequency does she hear?

10. A car travels at 100 km/h, blowing its horn at a frequency of 250 Hz. Taking the speed of sound in air as 340 m/s, what frequency is heard by a receiver who is proceeding at 60 km/h in the same direction (*a*) directly ahead of the car and (*b*) directly behind it?

How to Solve It

• In this problem, both the source and the receiver are moving. You can do each part of the problem in two steps, as in Problem 9.

• Or you can find the formula in the text that has both a moving source and a moving receiver in it. If you don't make a sign mistake, this is more direct.

• Remember, when the source and the receiver are approaching each other, the received frequency is higher than the source frequency; when they are separating, it's lower.

11. Suppose each of the loudspeakers of my stereo system is delivering 1 watt of power in the form of audible sound into my den. (*a*) If I'm sitting 2.5 m away from the speakers, what is the intensity of the sound I hear directly from the speakers (that is, neglecting the sound reflected from the walls of the room)? (*b*) To what decibel level does this correspond?

How to Solve It

• Calculate the direct sound intensity from the source from the power output of the source and the receiver's distance from it.

• Note that, in reality, unless the walls were specially constructed to be nonreflecting, the sound level would be higher than this due to reflections from the walls.

• Convert the intensity you calculated in part (*a*) to a decibel level.

12. Assume that a barking dog can put out (on the average) 1 mW of sound power. Assume that the sound propagates uniformly in all directions. What is the sound intensity level (in dB) 6 m away from a pen containing eight barking dogs?

How to Solve It

• Intensity is power per unit area.

• Don't forget that there is more than one dog.

• Calculate the intensity and convert it to a decibel level.

Solutions

1. Since the frequency observed from on board the ship is one wave every 9 s, the velocity of the waves relative to the ship is given by

$$v' = f\lambda = \left(\frac{1}{9 \text{ s}}\right)(500 \text{ ft}) = 55.6 \text{ ft/s}$$

The actual wave speed relative to the sea is

$$v = v' + u$$

where

$$u = (12 \text{ mi/h})\frac{(5280 \text{ ft/mi})}{(3600 \text{ s/h})} = 17.6 \text{ ft/s}$$

is the speed of the boat. Thus,

$$v = 55.6 \text{ ft/s} + (17.6 \text{ ft/s}) = \underline{73.2 \text{ ft/s}}$$

2.   <u>6.4 m to 26.8 cm</u>

3.   (a) At a temperature of 20°C = 293 K, the speed of sound in a gas is

$$v = \sqrt{\frac{\gamma RT}{M}}$$

so

$$v_{\text{air}} = \sqrt{\frac{(1.40)(8.31 \text{ J/K})(293 \text{ K})}{0.0289 \text{ kg}}} = \underline{343.4 \text{ m/s}}$$

For helium,

$$v_{\text{He}} = \sqrt{\frac{(1.67)(8.31 \text{ J/K})(293 \text{ K})}{0.0040 \text{ kg}}} = \underline{1008 \text{ m/s}}$$

(b) The frequency of a wave of wavelength λ is

$$f = \frac{v}{\lambda}$$

so for $\lambda = 2.20$ m

$$f_{\text{air}} = \frac{343.6 \text{ m/s}}{2.20 \text{ m}} = \underline{156 \text{ Hz}}$$

and

$$f_{\text{He}} = \frac{1008 \text{ m/s}}{2.20 \text{ m}} = \underline{458 \text{ Hz}}$$

Around 150 Hz is a typical frequency for a male speaking voice. If you've ever heard someone speak after inhaling a lungful of helium, you know that the voice sounds unnaturally high and squeaky. This problem illustrates why.

4. A chilly $-5°C$ (268 K)

5. The velocity of sound waves in a fluid is

$$v = \sqrt{B/\rho}$$

By analogy, for the velocity of linear compression waves in a solid, we use Young's modulus:

$$v_{long} = \sqrt{Y/\rho}$$

The speed of transverse waves along the wire is

$$v_{trans} = \sqrt{F/\mu}$$

The ratio of these speeds is

$$\frac{v_{long}}{v_{trans}} = \sqrt{\frac{Y/\rho}{F/\mu}} = \sqrt{\frac{Y\mu}{F\rho}}$$

Now, if m is the mass of the wire, L is its length, and A is its cross-sectional area,

$$\mu = \frac{m}{L} \quad \text{and} \quad \rho = \frac{m}{V} = \frac{m}{AL}$$

so

$$\frac{\mu}{\rho} = A$$

Thus,

$$\frac{v_{long}}{v_{trans}} = \sqrt{\frac{Y\mu}{F\rho}} = \sqrt{\frac{YA}{F}}$$

We know this ratio is 10, so

$$\frac{YA}{F} = 100$$

and

$$A = \frac{100F}{Y} = \frac{(100)(6200 \text{ N})}{2 \times 10^{11} \text{ N/m}^2} = 3.10 \times 10^{-6} \text{ m}^2$$

This area is $\pi r^2 = \frac{1}{4}\pi d^2$, where d is the diameter of the wire. Solving for d gives 1.99×10^{-3} m.

6.   (a) 317 N   (b) 3.55 g

7.   (a) When the train is coming toward the woman at speed u, the frequency she hears is

$$f_1' = \frac{f_0}{1 - u/v} = 440 \text{ Hz}$$

where v is the speed of sound. When the train is moving away,

$$f_2' = \frac{f_0}{1 + u/v} = 352 \text{ Hz}$$

The ratio of these two expressions is

$$\frac{f_1'}{f_2'} = \frac{\dfrac{f_0}{1 - u/v}}{\dfrac{f_0}{1 + u/v}} = \frac{1 + u/v}{1 - u/v} = \frac{440}{352} = 1.25$$

We want the speed of the train, so we solve this for the ratio u/v:

$$1 + u/v = (1.25)(1 - u/v)$$

$$(2.25)(u/v) = 0.25$$

$$u/v = 0.111$$

Thus,

$$u = 0.111v = (0.111)(340 \text{ m/s}) = 37.8 \text{ m/s}$$

(b) Substituting the value of v into the first of the equations in part (a), we find the actual frequency of the whistle is

$$f_0 = f_1'(1 - u/v) = (440 \text{ Hz})[1 - (37.8 \text{ m/s})/(340 \text{ m/s})]$$

$$= (440 \text{ Hz})(1 - 0.111) = 391 \text{ Hz}$$

8.   (a) 1.31 m   (b) 269.1 Hz

9.   (a) The speed of the car is

$$u_c = (40 \text{ km/h}) \frac{1000 \text{ m/km}}{3600 \text{ s/hr}} = 11.11 \text{ m/s}$$

so the wave frequency "that the wall receives" is

$$f_1 = \frac{f_0}{1 - u_s/v} = \frac{f_0}{1 - u_c/v} = \frac{300 \text{ Hz}}{1 - \dfrac{11.11 \text{ m/s}}{340 \text{ m/s}}} = 310.1 \text{ Hz}$$

and the wavelength of the sound coming to the wall is

$$\lambda = \frac{v}{f_1} = \frac{340 \text{ m/s}}{310.1 \text{ Hz}} = \underline{1.096 \text{ m}}$$

(b) The wall reflects sound of frequency 310.1 Hz, and the driver of the car receives the reflected sound with a frequency

$$f_2' = (1 + u_c/v)f_1 = \left(1 + \frac{11.11 \text{ m/s}}{340 \text{ m/s}}\right)(310.1 \text{ Hz}) = \underline{320.3 \text{ Hz}}$$

Notice that in each step I chose the sign that made the received frequency higher than the source frequency because the source and the receiver are approaching one another.

10.   (a) $\underline{258.9 \text{ Hz}}$ (b) $\underline{242.4 \text{ Hz}}$

11.   (a) The intensity at a distance r from the two loudspeakers is

$$I = \frac{P}{4\pi r^2} = \frac{(2)(1 \text{ W})}{(4\pi)(2.5 \text{ m})^2} = \underline{2.55 \times 10^{-2} \text{ W/m}^2}$$

(b) The corresponding decibel level is

$$\beta = \log_{10}(I/I_0) = \log_{10}\left(\frac{2.5 \times 10^{-2} \text{ W/m}^2}{10^{-12} \text{ W/m}^2}\right) = \underline{104 \text{ dB}}$$

This is an extraordinarily loud noise, and reflected sound would make it even louder. One watt can chase you out of the room! You don't buy amplifiers that will put out lots of watts because you need that much acoustical power; rather, it's because good-quality loudspeakers have low efficiency.

12. $\underline{72 \text{ dB}}$

CHAPTER 17

Interference, Diffraction, and Standing Waves

I. Key Ideas

Interference Waves at a point in a medium produced by different sources simply add; this is the *principle of superposition*. The waves may thus either reinforce or cancel one another, depending on whether the displacements they produce have the same or opposite signs. This phenomenon is called *interference*.

Constructive and Destructive Interference The interference of two harmonic waves of the same frequency and wavelength depends on the phase difference between them. If both waves "crest" together at a particular point, the two waves will reinforce each other. At points where the two waves are 180° out of phase, conversely, they will tend to cancel each other. These cases are *constructive* and *destructive interference*, respectively. If the two sources are in phase, constructive interference will occur at points that are equidistant from the sources or that differ in distance by a whole number of wavelengths.

Coherence Two sources need not be in phase to display interference, but they must have a constant phase difference. Such sources are said to be *coherent*. It is easy to make two coherent sound sources, but any two separate light sources are incoherent. Waves from incoherent sources do not show interference.

Diffraction *Diffraction* is the bending or spreading of a wave that occurs when part of a wavefront is limited by an aperture or obstacle. A wave traveling through an aperture that is much smaller than a wavelength spreads out uniformly as though from a point source. On the other hand, if the wavelength is very small compared to the aperture size (this is ordinarily the case for light), the wave propagates approximately in straight lines. This is called the *ray approximation*. Patterns of diffraction can be calculated using Huygens' method, which treats each point on a wavefront as a point source of further waves. Because of diffraction, one cannot use a wave to observe details smaller than about a wavelength.

Beats When two waves of nearly equal frequency are superposed, they move regularly in and out of phase with one another, producing a resultant wave whose intensity varies regularly in time. These variations are called *beats*; the *beat frequency* is the difference between the frequencies of the two waves.

Standing Waves Whenever waves are confined so that they propagate within some finite region of space, there are reflections at the boundaries. At a "fixed" end or boundary, where the wave displacement must go to zero, the reflected wave is reversed; at a "free" end, it is not. Successive reflected waves interfere with one another. At (and only at) certain frequencies, all the reflected waves will interfere constructively to produce a stationary vibration pattern of large amplitude called a *standing wave*. This is an example of resonance—the vibrating system produces a large response when it is driven at one of its "natural" frequencies.

Standing Waves on a String Fixed at Both Ends If a string is fixed at both ends so that it cannot vibrate there, standing waves will "fit" on it only at certain frequencies. The frequencies are those for which the length of the string is a whole number of half wavelengths since the ends of the string must be points of zero displacement, called *nodes*, of the sine wave. The lowest such frequency—the *fundamental* frequency—is that at which the string is one-half wavelength long. Other frequencies at which the string will support standing waves, called the *overtones* or *harmonics*, are integer multiples of the fundamental frequency.

Standing Waves on a String Fixed at One End If a stretched string is fixed at one end but is free to vibrate at the other, the standing wave frequencies are those at which there is a node at one end and an antinode at the other. At the fundamental frequency, therefore, the length of the string is one-quarter wavelength rather than a half wavelength, and only the odd harmonics are present.

Standing Sound Waves Standing sound waves in air are produced by the vibration of an air column of fixed length. In a pipe with both ends closed, for example, reflections of the sound wave at each end produce standing waves if the frequency is right for there to be a displacement node (a pressure antinode) at each end. The harmonic series is the same as that for a string with both ends fixed. Likewise, a pipe open at only one end will have a pressure node at the open end and a pressure antinode at the other, producing frequencies corresponding to those of a stretched string with one end "free." The frequencies are slightly different, however, because the effective position of the pressure node is a little *outside* the open end.

Harmonic Analysis Musical instruments typically produce a mixture of a fundamental frequency and its overtones. The fundamental frequency determines the pitch of the sound; the mixture of harmonic frequencies in the sound determines its *tone quality*. The corresponding waveforms, although periodic, look very different from sine waves, but they can be analyzed as a superposition of sine waves of the various harmonic frequencies. Likewise, sounds of arbitrary tone quality can be produced by superposing sine waves of different frequencies.

II. Key Equations

Key Equations

Displacement in harmonic waves

$$y_1 = A \cos (2\pi f t)$$

$$y_2 = A \cos (2\pi f t + \delta)$$

Beat frequency

$$f_{beat} = f_1 - f_2 = \Delta f$$

Standing waves on a string fixed at both ends or for a pipe closed at both ends

$$\lambda_n = 2\frac{L}{n} \qquad f_n = \frac{nv}{2L} = nf_1$$

(for $n = 1, 2, 3, 4, \ldots$)

Standing waves on a string fixed at one end or for a pipe open at one end

$$\lambda_n = 4\frac{L}{n} \qquad f_n = \frac{nv}{4L} = nf_1$$

(for $n = 1, 3, 5, 7, \ldots$)

III. Possible Pitfalls

Destructive interference means that the two interfering waves are 180° out of phase with one another and so will tend to cancel. The cancellation is total, however, only if the two waves have equal amplitudes.

An air column open at one or both ends has a fixed effective length, and the harmonic frequencies are in integer ratios, but the effective length is not exactly the length of the pipe. The displacement antinode is a little *outside* the open end.

The fundamental frequency of a standing-wave system is the *lowest* frequency at which it can resonate. Correspondingly, the fundamental has the *longest* wavelength of any of the resonance frequencies; the overtones have shorter wavelengths.

Two wave sources need not be vibrating in phase in order for interference to occur, but they must be *coherent*—that is, they must have a definite, fixed phase relation to each other.

Light does not actually propagate in straight lines past an obstruction or through an aperture, but because its wavelength is so short, it appears to do so in ordinary circumstances.

The frequencies of a vibrating string are calculated as if there were nodes at both ends. Clearly this isn't exactly right in the usual experiment, in which one of the ends is being driven. However, the driven end is very nearly a node if the amplitude of the standing wave is large compared to that of the driven end, but this is an approximation.

A vibrating string, such as a guitar string or a piano string, can act as a source of sound waves of the same frequency in air. But notice that the wavelength of the sound waves and the wavelength of the standing waves on the string don't bear any relation to one another.

IV. Questions and Answers

Questions

1.   Imagine that you're listening to orchestral music being played on a stereo that is in another room down the hall. Will diffraction have any effect on the sound you hear?

2.   About how accurately do you think you could tune two guitar strings to the same frequency using beats?

3.   Most orchestral instruments are based on the resonance frequencies of either a stretched string or air in a pipe. What happens to these frequencies as the temperature increases?

4.   The fundamental vibration frequency of an air column in a pipe closed at both ends is 240 Hz; the fundamental frequency is 117.5 Hz if the pipe is open at one end. Why is the ratio of the frequencies approximately 2:1? Why is it not exactly 2:1? What would you expect to be the fundamental frequency of the pipe if it is open at both ends?

5.   What happens to the frequency of a clarinet if it is sounded in an atmosphere of pure helium?

6.   Which of the properties that determine the fundamental frequency of a vibrating system (a string, a pipe, or the like) is normally varied in playing a musical instrument at different frequencies?

7.   How can the phenomenon of beats be used to tune musical instruments?

8.   Why is it that we can hear, but not see, around corners.

9.   When we play chords on the piano, several notes (frequencies) are sounded at once. Why don't we hear beats?

10. It's hard to set up a really convincing demonstration of two-source interference using sound waves in a lecture room. You can set up two loudspeakers vibrating in phase, and as you move from one to another part of the room, you can hear variations in the loudness of the sound, but they aren't terribly striking. Why not?

11. One end of a stretched string is vibrated to produce a standing wave on the string. What is it that determines the amplitude of the standing wave?

Answers

1.   The diffraction of sound waves is the main reason you can hear "around corners" from the music room to the hall and then into your room. The smallest aperture involved is likely to be the width of a doorway, around 1 m. Sound waves with wavelengths much less than 1 m (corresponding to frequencies of about 340 Hz) will be diffracted less effectively. Thus, a smaller part of the source intensity at high frequencies gets to your ear. You will hear high notes, and particularly the high harmonics that give musical sounds their tone quality, less effectively, and the music will sound "flat" or "muffled" to you.

2.   The frequency difference between the two strings will be heard as a beat frequency. The quality of the sound from a plucked string does not remain the same for more than a second or so, so you lose perception of beats if they are more than a few seconds apart; even so, you should be able to match the frequencies to within 0.5 Hz.

3.   The wind-instrument frequencies increase with increasing temperature because the speed of sound in air increases. Since strings will expand a little as the temperature goes up, the tension in a string decreases a little, so the wave speed and fundamental frequency decrease. Thus, as they warm up, wind instruments go sharp and string instruments go flat! This is why an orchestra warms up first and then tunes.

4. The ratio is about 2:1 because the length of the pipe is a half wavelength when both ends are closed and a quarter wavelength when one end is open. It's not exactly 2:1 because the displacement antinode is not exactly at the open end but a little outside, so the effective air column is a little longer than the actual pipe. With both ends open, the fundamental is around 240 Hz again. What matters is not whether the ends are open or closed, but whether the two ends are the same or different. (In fact it would be a little less than 240 Hz—perhaps 232 Hz—because the ends are open.)

5. All the frequencies that the clarinet produces are higher by the ratio of the speed of sound in helium to that in air, which is about 2.5 times. Thus, all its tones sound more than an octave higher. Benny Goodman could really have winged those high ones—but not for long, of course.

6. Most often, it is the length of the system that is varied. You finger the violin string to change its vibrating length, and you open or close holes or insert extra lengths of pipe on a wind instrument to "end" the air column at different places. Length modification is secondary, however, in playing some brass instruments; what you do there is select different ones of the harmonic frequencies of a pipe of fixed length—yes, by how you hold your mouth.

7. The phenomenon of beats is a clearly audible indicator of a small frequency difference. When you sound at once two strings that are supposed to have the same frequency, for instance, any difference in frequency produces beats that you can hear. Then, you adjust the tension in one string until no beats are heard.

8. Because the obstacles, apertures, and such that surround us in everyday circumstances—doorways, for instance—tend to be not very large compared to the wavelengths of audible sound, but all are very large compared to the wavelengths of visible light. Thus, sound waves are significantly diffracted, but light waves are not.

9. We do hear them, in a way. The beat or difference frequencies in this case are tens to hundreds of hertz, and we can't hear them as intensity variations. Due to the way the ear responds to sound waves, however, "tones" at the difference frequencies are present in the combination of sounds we hear and, in some circumstances, are easily distinguished.

10. Because, in almost any room, a large part of the sound you are hearing comes not directly from the sources but has been reflected (one or more times) from the walls. It is only the part coming directly from the sources in which you perceive strong interference maxima and minima, so these are usually not very noticeable.

11. The amplitude of the standing wave grows until the power being dissipated by nonconservative processes in the string is equal to the power being delivered to the string by the signal source.

V. Problems and Solutions

Problems

Unless otherwise instructed, take the speed of sound in air to be 340 m/s.

1. (*a*) A man is sitting in a room directly between two loudspeakers 5 m apart that are vibrating in phase. He is 1.8 m from the nearer speaker. If the lowest frequency at which he observes maximum destructive interference is 122 Hz, what is the speed of sound in air? (*b*) If, instead, he listens from point *B* (see Figure 17-1), what is the lowest frequency at which he will observe destructive interference?

Figure 17-1

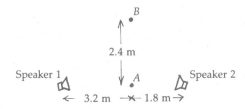

How to Solve It
- When the man observes maximum destructive interference, the waves reaching him are 1/2-cycle out of phase (or 3/2 or 5/2 or so on, but we want the lowest frequency).

- The phase difference in each case arises from a difference in path length from each source to the listener, which must therefore be half a wavelength.

- The velocity of a harmonic wave is its frequency multiplied by its wavelength.

2. Two loudspeakers radiate sound waves in phase at a frequency of 100 Hz. A listener is at a distance of 8.5 m from one speaker and 13.6 m from the other. Either speaker alone would produce a sound intensity level of 75 dB at the position of the listener. (*a*) What is the sound intensity level at the listener? (*b*) What is the intensity level at the listener if the same two speakers are 180° out of phase?

How to Solve It

- In part (*a*), the phase difference is due to the difference in path length.

- If the listener and loudspeakers haven't moved, changing the speakers so that they are out of phase by 180° just changes the phase difference of the two waves at the position of the listener by half a cycle.

- Remember that when two interfering harmonic waves are in phase, their amplitudes simply add.

3. A loudspeaker that is producing sound with a wavelength much larger than the speaker acts as a point source; that is, the sound radiates more or less uniformly in all directions. If the wavelength is very small compared to the size of the speaker, on the other hand, the sound travels straight ahead from the speaker in approximately a straight line. If each speaker in a portable stereo is 18 cm in diameter, for what frequency is the sound wavelength (*a*) ten times and (*b*) one-tenth of the loudspeaker diameter?

How to Solve It

- The factor of 10 in each case is just to put an arbitrary numerical value on "much larger than" or "very small compared to."

- The speed of a harmonic wave is its frequency multiplied by its wavelength.

4. Two loudspeakers 3 m apart and a few metres away from you are producing sound of the same frequency. At about what frequency can you tell by the sound that there are two sources rather than one?

How to Solve It
- A wave cannot be used to observe or locate details much smaller than one wavelength.

- At what frequency do sound waves have a wavelength of 3 m?

5. Two identical strings on your piano are tuned to concert A (440 Hz). Each is under a tension of 1300 N. Over the course of time, one string loosens to the point that, when you strike the two strings, you hear beats every 1.1 s. By how much has the tension in the loose string decreased?

How to Solve It
- From the beats, determine how much the fundamental frequency of the loosened string differs from 440 Hz.

- The frequency has decreased because the transverse wave speed on the string has decreased.

- Knowing how much the wave speed on the string has decreased, you can find how much the tension in it has decreased.

6. When a violin string is played simultaneously with a tuning fork of frequency 264.0 Hz, beats occurring once every 0.65 s are heard in the sound. Tightening the string slightly causes the beats to disappear. What was the initial frequency of the violin string?

How to Solve It
- When you tighten a string, you increase its fundamental frequency.

- Thus, the original frequency of the string was less than 264.0 Hz, and tightening the string has increased it to 264.0 Hz.

- The initial beat frequency was the difference between the initial frequencies of the tuning fork and the violin.

7. A string is stretched between fixed supports 0.70 m apart, and the tension is adjusted until the fundamental frequency of the string is 264 Hz (middle C). If the string has a mass of 4.2 g, what is the tension?

How to Solve It
- The speed of a harmonic wave is its frequency multiplied by its wavelength.

- When the string is vibrating at its fundamental frequency, its length is half the wavelength of the standing wave.

- The speed of a wave on a string is determined by its linear density (mass per unit length) and the tension it is under.

8. The lowest note on the standard piano is A (27.5 Hz). On my piano (a small console) the "string" producing this note is 85 cm long. Its mass is 20 g. What must be the tension in the string?

How to Solve It

- Piano strings are fastened at both ends, so the wavelength of the fundamental note is twice the length of the string.

- The speed of a wave on a string is determined by its linear density (mass per unit length) and the tension it is under.

9. Three successive resonance frequencies in an organ pipe are 273, 364, and 455 Hz. How long is the pipe? Ignore end corrections.

How to Solve It

- Open organ pipes and closed organ pipes (which are open at one end) differ in the ratio of successive resonance frequencies. How do they differ?

- From the given frequencies, determine which kind of pipe you have here.

- In either case, the fundamental frequency, which is determined by the length of the pipe, can be found from the difference of successive standing-wave frequencies.

10. A 5-g steel wire 0.8 m long with one end fastened to a very light string so that it is free to move transversely is under a tension of 840 N. Find the frequencies of the fundamental and the next two harmonics.

How to Solve It

- Because one end of the wire is free, the fundamental wavelength is four times the length of the wire.

- The mass and tension are given; find the velocity of standing waves on the wire.

- Remember that for the wire with one end free, only the odd harmonics occur.

11. The fundamental frequency of a violin string 30 cm long is sounded next to the open end of a closed organ pipe 41 cm long. The strongest standing sound wave in the pipe occurs when the string is vibrating at the pipe's fundamental frequency. This happens when the tension in the string is 220 N. What is the mass of the string?

How to Solve It
- What is the fundamental frequency for a closed organ pipe 0.41 m long? Remember, a closed organ pipe is open at one end.

- If a string 30 cm long has this frequency for a fundamental, find the wave speed on the string.

- From the wave speed and the tension the string is under, find its mass.

12. The maximum range of human hearing is about 20 Hz to 20 kHz. What is the longest organ pipe that could have a fundamental frequency within this range?

How to Solve It
- The longest wavelength corresponds to the lowest frequency.

- Thus, we want an organ pipe whose fundamental frequency is 20 Hz.

- Would the pipe have to be longer if it were open at both ends or closed at one end?

Solutions

1.   (a) The path length difference of the two sound waves is

$$\Delta s = s_2 - s_1 = 3.2 \text{ m} - 1.8 \text{ m} = 1.4 \text{ m}$$

When the man first observes complete destructive interference, the path length difference must be equal to one-half wavelength. Thus,

$$\lambda = 2 \, \Delta s = (2)(1.4 \text{ m}) = 2.8 \text{ m}$$

The speed of sound is therefore

$$v = f\lambda = (122 \text{ Hz})(2.8 \text{ m}) = \underline{342 \text{ m/s}}$$

(b) In the second case, the path length difference must (again) be one-half wavelength. The distance of point B from speaker 1 (see Figure 17-2) is

$$d = \sqrt{(3.2 \text{ m})^2 + (2.4 \text{ m})^2} = 4.0 \text{ m}$$

Figure 17-2

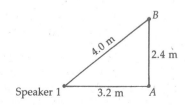

Speaker 1   3.2 m   A
B
4.0 m
2.4 m

and the distance from speaker 2, found in the same way, is 3.0 m. Thus, in this case, the wavelength of the sound is 2.0 m. The speed of the sound hasn't changed, so the lowest frequency for destructive interference is given by

$$f = \frac{v}{\lambda} = \frac{342 \text{ m/s}}{2.0 \text{ m}} = \underline{171 \text{ Hz}}$$

2.   (*a*) The intensity is <u>zero</u>; this can't be expressed on the decibel scale.

(*b*) <u>81 dB</u> (which is four times the intensity level of one speaker alone.)

3.   For a harmonic wave

$$v = f\lambda$$

so

$$f = \frac{v}{\lambda}$$

(*a*) For a wavelength ten times the diameter of the loudspeaker,

$$f = \frac{v}{10D} = \frac{340 \text{ m/s}}{(10)(0.18 \text{ m})} = \underline{189 \text{ Hz}}$$

(*b*) For a wavelength one-tenth the diameter of the loudspeaker,

$$f = \frac{v}{0.1D} = \frac{340 \text{ m/s}}{(0.1)(0.18 \text{ m})} = \underline{18.9 \text{ kHz}}$$

4.   Somewhere around <u>110 Hz</u>.

5.   The frequency of the loosened string is

$$f' = f_0 - f_{\text{beat}} = 440 \text{ Hz} - \frac{1}{1.1 \text{ s}} = 439.09 \text{ Hz}$$

Because $f = v/2L$ and the length L is fixed, the wave velocity on the string and the fundamental frequency are directly proportional:

$$\frac{f'}{f_0} = \frac{v'}{v_0} = \sqrt{\frac{F'/\mu}{F/\mu}} = \sqrt{\frac{F'}{F}}$$

so

$$F' = \left(\frac{f'}{f_0}\right)^2 F = \left(\frac{439.09 \text{ Hz}}{440 \text{ Hz}}\right)^2 (1300 \text{ N})$$

$$= (0.9959)(1300 \text{ N}) = 1294.6 \text{ N}$$

That is, the tension has decreased by 1300 N − 1294.6 N = $\underline{5.4 \text{ N}}$.

6. $\underline{262.5 \text{ Hz}}$

7.   The wavelength of the fundamental standing wave on a string fixed at both ends is

$$\lambda_1 = 2L = (2)(0.70 \text{ m}) = 1.40 \text{ m}$$

Thus, the wave speed is

$$v = f\lambda_1 = (264 \text{ Hz})(1.40 \text{ m}) = 369.6 \text{ m/s}$$

The linear density (mass per unit length) of this string is

$$\mu = \frac{m}{L} = \frac{0.0042 \text{ kg}}{0.70 \text{ m}} = 0.0060 \text{ kg/m}$$

The velocity of wave on a stretched string is

$$v = \sqrt{F/\mu}$$

so

$$F = \mu v^2 = (0.0060 \text{ kg/m})(369.6 \text{ m/s})^2 = \underline{820 \text{ N}}$$

8. $\underline{51.4 \text{ N}}$

9.   The first thing we need to find out is which kind of pipe we have here. Successive resonance frequencies of an open organ pipe are in the ratios of successive integers; those of a closed organ pipe are in the ratios of successive odd integers. Here, the ratios of successive frequencies are

$$\frac{364 \text{ Hz}}{273 \text{ Hz}} = 1.333 = \frac{4}{3} \quad \text{and} \quad \frac{455 \text{ Hz}}{273 \text{ Hz}} = 1.667 = \frac{5}{3}$$

that is, the frequencies are in the ratios 3:4:5, so the pipe is open at both ends. The difference between successive frequencies is the fundamental frequency

$$f_1 = 455 \text{ Hz} - 364 \text{ Hz} = 364 \text{ Hz} - 273 \text{ Hz} = 91.0 \text{ Hz}$$

The corresponding wavelength is

$$\lambda_1 = \frac{v}{f_1} = \frac{340 \text{ m/s}}{91.0 \text{ Hz}} = 3.74 \text{ m}$$

For a pipe open at both ends, the fundamental wavelength is twice the length of the pipe (neglecting end corrections), so this pipe is 1.87 m long.

10. 114.6 Hz, 343.7 Hz, and 572.8 Hz

11. For a closed organ pipe 0.41 m long, the fundamental wavelength is

$$\lambda_1 = 4L = (4)(0.41 \text{ m}) = 1.64 \text{ m}$$

If the speed of sound is 340 m/s, the fundamental frequency is

$$f_1 = v/\lambda_1 = \frac{340 \text{ m/s}}{1.64 \text{ m}} = 207 \text{ Hz}$$

This is the fundamental frequency of the pipe and, therefore, of the stretched string that excites it. The wave speed on the string is therefore

$$v = f_1\lambda_1 = (f_1)(2L) = (207 \text{ Hz})(2)(0.30 \text{ m}) = 124 \text{ m/s}$$

The wave velocity on a stretched string is

$$v = \sqrt{F/\mu}$$

where F is the tension in the string and μ its linear density (mass per unit length). Thus,

$$v^2 = \frac{F}{\mu} = \frac{F}{m/L} = \frac{FL}{m}$$

and

$$m = \frac{FL}{v^2} = \frac{(220 \text{ N})(0.30 \text{ m})}{(124 \text{ m/s})^2} = 4.29 \times 10^{-3} \text{ kg}$$

12. It is an open pipe 8.5 m long.

CHAPTER 18

Electric Fields and Forces

I. Key Ideas

Electric Charge Electric charge is a fundamental property of matter. Some of its manifestations—for instance, that rubbing some substances together causes them to attract small objects—have been known for thousands of years. The modern understanding of electric charge was developed in the eighteenth and nineteenth centuries and was formulated theoretically by Maxwell. Electric charge can be positive or negative. Charge is *conserved*; that is, it is not created or destroyed in natural processes but is only transferred from one object to another. And charge is quantized, that is, amounts of charge can only be multiples of a fundamental amount of charge e.

Electrostatic Force The force between point electric charges is given by *Coulomb's law*; it is directly proportional to each charge and inversely proportional to the square of the distance between them. This is very like the gravitational force between point masses, but the electrostatic force is very much stronger and can be either attractive or repulsive. Charges of opposite sign attract each other, whereas charges of like sign repel. If there are more than two charges, the force on one due to all the others is just the vector sum of the individual forces due to each.

The Electric Field It's convenient to describe the forces exerted by electric charges in terms of an electric field. The field describes a "condition in space" at a point such that, if a hypothetical "test charge" were placed at that point, the force on it would be equal to the value of the electric field multiplied by the charge. Thus, the field is clearly a vector quantity. The electric field due to many point charges is just the (vector!) sum of the fields due to each one. Beyond its usefulness as a calculational device, the electric field allows us to avoid the theoretical difficulty of "action at a distance."

Lines of Force It is often useful to represent the electric field by a *lines-of-force diagram*. Lines of force, or *electric field lines*, are curves drawn such that, at any point, the tangent to the curve is in the direction of the field. Thus, lines of force point straight away in all directions from a positive point charge or converge radially on a negative point charge. Because of the inverse-square dependence of the electric field on distance, if the number of lines terminating at a charge is proportional to the charge, the density of the lines (the number of lines passing through a unit area) at some distance from the charge is proportional to the magnitude of the field at that distance.

Spherical Symmetry A lines-of-force diagram shows immediately that the field outside a uniform spherical shell of charge is the same as if all the charge were a point charge at the center and that the field inside the shell is zero.

Gauss' Law We define the *electric flux* through an area as the component of the electric field perpendicular to the area multiplied by the area. It is represented by the number of lines passing through the area on a lines-of-force diagram. Thus, the *net* flux out through any *closed* surface is proportional to the net charge inside it. This is *Gauss' law*; it expresses a very general property of the electric field. In a few, very special cases of symmetric charge distribution, Gauss' law can be used to find the electric field.

Electric Dipole An *electric dipole* is a pair of equal and opposite point charges separated by a distance. Its *dipole moment* is the product of either charge and the distance between them. In a uniform electric field, there is no net force on a dipole (although there is a torque); in a nonuniform field, there is a force in the direction in which the field is stronger.

Intermolecular Forces Atoms and molecules are electrically neutral structures of positive and negative charge. In some molecules—water is a very important example—the centers of positive and negative charge do not coincide, so the molecule possesses a permanent intrinsic dipole moment. Such a molecule is called *polar*. In an external electric field, the positive and negative charges of even nonpolar molecules are displaced, and an electric dipole is induced in them. Electrical interaction between dipoles is the source of forces between molecules.

II. Numbers and Key Equations

Numbers

Coulomb's constant

$$k = 8.99 \times 10^9 \text{ N·m}^2/\text{C}^2$$

Fundamental unit of electric charge

$$e = 1.60 \times 10^{-19} \text{ C}$$

Key Equations

Coulomb's law

$$F = k \frac{q_1 q_2}{r^2}$$

Electric field

$$\mathbf{E} = \frac{\mathbf{F}}{q_0}$$

Electric field of a point charge

$$E = \frac{kg}{r^2}$$

Electric field of a spherical "shell"

$$E = \frac{kQ}{r^2} \quad \text{(if } r > R)$$

$$E = 0 \quad \text{(if } r < R)$$

Electric flux

$$\Delta \phi = E_n \, \Delta A$$

Gauss' law

$$\phi_{\text{net}} = \Sigma E \, \Delta A = 4\pi k Q_{\text{in}}$$

Electric dipole moment

$$\mathbf{p} = q\mathbf{L}$$

III. Possible Pitfalls

We sometimes speak of the fundamental charge e as "the charge of an electron," but e is a positive quantity. The charge of an electron, properly, is $-e$.

If you keep track of signs properly, Coulomb's law will tell you the direction—that is, whether the force is one of attraction or one of repulsion—as well as its magnitude. Even if signs get mixed up, just remember that like charges always repel one another and charges of opposite sign attract.

The electric fields due to several charges simply add. The field due to charges q_1 and q_2 at some point is just the field due to q_1 added to the field due to q_2. They add as *vectors*, remember.

You have to be forever careful about signs. The force on a *positive* charge is in the direction of the electric field, but the force on a negative charge is in the direction *opposite* to that of E.

Gauss'-law calculations of E are very simple except when they're impossible. You can only use

Gauss' law in some highly symmetric cases; in essence, you must be able to identify a surface on which the field is everywhere either zero or a constant.

An electric dipole consists of *equal* charges of opposite sign, not just any pair of positive and negative charges.

IV.   Questions and Answers

Questions

1. In Figure 18-1, $\mathbf{F}_1$ is the force on q due to q_1, and $\mathbf{F}_2$ is the force on q due to another charge q_2. The total force on charge q is $\mathbf{F} = \mathbf{F}_1 + \mathbf{F}_2$. (*a*) If q is a positive charge, what is the sign of q_1? (*b*) Which of q_1 and q_2 is larger in magnitude?

Figure 18-1

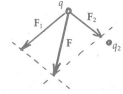

2. You have an isolated conductor with a large electric charge on it. If you bring uncharged little bits of paper or some other light material near it, you will observe that no matter what the sign of the charge on the conductor, the bits are attracted to it. If they touch the conductor, however, they are likely to be repelled by it thereafter. Can you explain this behavior?

3. At point P in Figure 18-2, E is found to be zero. (*a*) What can you say about the signs of q_1 and q_2? (*b*) What can you say about their magnitudes?

Figure 18-2

4. A positively charged glass rod attracts a light object suspended by a thread. Does it follow that the object is negatively charged? If, instead, the rod repels it, does it follow that the suspended object is positively charged?

5. What is the advantage of thinking of the force on a charge at a point P as being exerted by an "electric field" at P rather than by other charges some distance away? Is the convenience of the field as a calculational device worth inventing a new physical quantity? Or is there more to the field concept than this?

6. Would the physical universe be different if the proton were negatively charged and the electron were positively charged?

7. Would the physical universe be any different if the proton's charge were very slightly larger in magnitude than the electron's?

8. When an object is given a positive charge, does its mass increase or decrease?

Answers

1. (*a*) Charge q_1 repels the positive q and so is itself positive. (*b*) From the drawing, q_1 is farther away from q than q_2 is yet exerts a greater force on it than q_2 does. Hence, $q_1 > q_2$.

2. The electric field of the large conductor polarizes (that is, induces an electric dipole moment in) the bits of paper along the field direction. The field of a finite charge distribution is always nonuniform, so there is a force on the bits of paper. Because the end of the induced dipole that has a charge opposite that of the large conductor is always closer to the conductor, the force exerted on the dipole is always attractive. When the bits of material actually touch the conductor, some charge is transferred to them; thereafter, they have charges of the same sign as that on the large conductor and so are repelled by it.

3. (*a*) The contributions of the two charges to the electric field cancel, so their contributions must be of opposite sign. Since both the charges are in the same direction from P, the charges themselves must be opposite in sign. (*b*) The contributions to the field from q_1 and q_2 must be of the same size if they are to cancel. Since q_2 is farther away from P, it must be larger.

4. In the first case, the object may not be charged at all (see Question 2). If the suspended object is repelled, however, it can only have a positive charge since the "induced-dipole" force will always be attractive. This is probably the reason that electrical attraction was known to the ancients, whereas electrical repulsion was not.

5. Philosophically, most of us find it more acceptable to think of a force that acts here being exerted by something here. But, if the philosophical problems don't interest you, and if you never do anything with the electric field other than electrostatics, the electric field is just a calculational device that you may or may not think worthwhile. In situations in which charges move or change, however, the electric field must be considered a real, separate element of the system. It propagates the information about the change through space at a fixed speed, transports energy, and so forth.

6. No; which we call positive and which negative is wholly arbitrary.

7. Yes, it would make and enormous difference. Every object
would have a net positive charge, and electrostatic forces between
objects (as opposed to gravitation and contact forces) would domi-
nate macroscopic physics.

8. Its mass decreases slightly since the way we give objects a positive charge is to remove electrons from them.

V.   Problems and Solutions

Problems

1. Two point charges are separated by a distance of 20 cm. The
numerical value of one charge is twice that of the other. If each
charge exerts a force of magnitude 45 N on the other, find the
charges.

How to Solve It
- Write Coulomb's law for the force between charges of magnitude q
  and $2q$.

- Solve for q.

2. The mass of an electron is 9.11×10^{-31} kg. How far apart
would two electrons have to be in order for the electric force exerted by each on the other to be equal to its weight?

How to Solve It
- Coulomb's law gives you the force each electron exerts on the
  other if they are a distance r apart.

- The Coulomb force is to be equal to the electron's weight; from this
  you can solve for r.

3. A spherical birthday-party balloon 25 cm in diameter contains
helium at room temperature (say 20°C) and at a pressure of
1.3 atm. If one electron could be stripped from every helium
atom in the balloon and removed to a satellite orbiting the earth
22,000 miles up, with what force would the balloon and the
satellite attract each other?

How to Solve It

- From the data given, find the number of helium atoms in the balloon and the charge of this number of electrons.

- When the electrons are removed, the helium is left with an equal positive charge.

- Use Coulomb's law to find the force between these two charges when they are 22,000 miles apart.

4. A charge $q_1 = 0.6$ μC is at the origin, and a second charge $q_2 = 0.8$ μC is on the x axis at $x = 5.0$ cm. (*a*) Find the force (magnitude and direction) that each charge exerts on the other. (*b*) How would your answer change if q_2 were -0.8 μC?

How to Solve It

- The magnitude of each force is found from Coulomb's law—you know both charges and the distance between them.

- The direction of each force depends on the signs of the two charges.

5. When a test charge of $+5$ nC is placed at a certain point, the force that acts on it is 0.08 N, directed northeast. (*a*) If the test charge were -2 nC instead, what force would act on it? (*b*) What is the electric field at the point in question?

How to Solve It

- The force on the test charge at a certain point is its charge multiplied by the electric field there.

- If the field stays the same, the force is directly proportional to the charge.

- The value of the field can be found from the given force.

6. Near the surface of the earth, there is an electric field, downward, of magnitude approximately 100 N/C. What charge (magnitude and sign) would have to be placed on a penny of mass 3 g to cause it to rise into the air with an upward acceleration of 0.19 m/s^2?

How to Solve It

- Draw a free-body diagram of the penny, and from it determine what the force on the penny must be.

- From the force and the known electric field, determine the charge that the penny would have to have.

7. Two charges are placed on the x axis, $+5$ μC at the origin and -10 μC at $x = 10$ cm. (*a*) Find the electric field on the x axis at $x = 6$ cm. (*b*) At what point(s) on the x axis is the electric field zero?

How to Solve It
- Calculate the electric field due to each point charge at $x = 10$ cm; the total field is the vector sum of the two.

- In order for the field to be zero, the individual fields due to the two charges must have opposite signs and the same magnitude. This tells you (how?) that the point of zero field must be at $x < 0$.

- Find the exact position by writing a formula for the total field and solving for x.

8. In the situation sketched in Figure 18-3, the electric field at the origin is found to be zero. If q_1 is 10^{-7} C, what is q_2?

How to Solve It
- Calculate the field due to q_1 at the origin.

- The field due to q_2 at the origin has the same magnitude; use this to find the magnitude of q_2.

- Find the sign of q_2 from the fact that the field contributions must cancel.

Figure 18-3

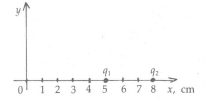

9. Consider a very large plane sheet that carries a uniformly distributed charge Q. (*a*) Use Gauss' law to find the electric field due to the sheet. (*b*) What field would be created by two equal but oppositely charged parallel plane sheets? The simplest way to express your answers is in terms of the surface charge density σ (the charge per unit area) on the sheet.

How to Solve It
- The key to applying Gauss' law this way is to use the symmetry of the situation. Suppose the plane sheet is large enough so that you can consider it to be infinite. What must be the direction of the field? How do you suppose the field compares on one side of the sheet and the other?

- To use Gauss' law to find the magnitude of the field, we must find a surface to apply it to on which the field is the same everywhere (or else zero). Draw a cylindrical surface with ends parallel to the sheet. Does this meet the requirement?

- The flux through the sides of such a surface is zero since E is parallel to the sides. You just need to calculate the flux out through the ends.

10. Use Gauss' law to find the electric field just outside the surface of a sphere carrying a uniform surface charge density σ (charge per unit area).

How to Solve It
- Remember that to use Gauss' law to calculate E, you must exploit the symmetry of the situation.

- In this situation, the field must be directed radially and must depend only on the distance from the origin.

- Calculate the flux through an imaginary spherical surface concentric with the sphere and just outside it.

11. An electric dipole on the x axis consists of charges of ± 0.05 μC at $x = \pm 0.5$ cm, respectively. (a) Find the electric field (magnitude and direction) at $y = 5$ cm on the y axis. (b) Repeat the calculation for a dipole on the x axis that is twice as long but has the same electric dipole moment.

How to Solve It
- Calculate the electric field of each of the two point charges at $y = 5$ cm. Remember that the electric field is a vector.

- Add the two point-charge fields to get the field due to the dipole.

- In part (b), if the dipole moment is the same, the charge must be half as much.

12. In the Bohr model, the hydrogen atom consists of an electron in a circular orbit around the nucleus of radius $a_0 = 5.29 \times 10^{-11}$ m. What is the electric dipole moment of two point charges $\pm e$ separated by a distance a_0?

How to Solve It
- The value of the dipole moment is simply the charge on either end times their separation.

- As the electron orbits the atom, the direction of the dipole moment is continuously changing.

13. Figure 18-4 shows the electric field diagram for two point charges. The magnitude of q_1 is 40 μC. (a) What is the sign of q_1? (b) What is the value of q_2? (c) In the diagram, how many lines of force terminate at infinity?

How to Solve It
- Lines of force begin on positive charges and end on negative charges.

Figure 18-4

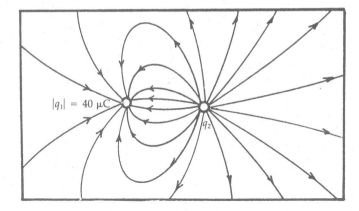

$|q_1| = 40 \ \mu C$

q_2

- The number of lines beginning or ending at a charge is proportional to the magnitude of the charge.

14. An electron with an initial speed of 5.0×10^5 m/s enters a region in which there is an electric field directed along its direction of motion. If the electron travels 5.0 cm in the field before being stopped, what is the magnitude of the electric field?

How to Solve It
- Remember that the force on an electron, which is negatively charged, is in the direction opposite that of the electric field.

- From the given speed and distance, calculate the acceleration of the charge.

- Find the force on the electron and, from that, the magnitude of the electric field. The mass of the electron was given in Problem 2.

Solutions

1. Call one charge q; then the other has charge $2q$. (As this problem is stated, we don't care about directions, so we can ignore signs.) The force that each exerts on the other is given by Coulomb's law:

$$F = k \frac{q_1 q_2}{r^2} = k \frac{(q)(2q)}{r^2} = \frac{2kq^2}{r^2}$$

so

$$q^2 = \frac{Fr^2}{2k} = \frac{(45 \ \text{N})(0.20 \ \text{m})^2}{(2)(8.99 \times 10^9 \ \text{N·m}^2/\text{C}^2)} = 1.00 \times 10^{-10} \ \text{C}^2$$

and

$$q = 1.00 \times 10^{-5} \text{ C}$$

The other charge is $2q = 2.00 \times 10^{-5}$ C.

2. 5.08 m

3. The ideal-gas law can be used to find the number of moles of helium

$$PV = nRT \qquad \text{or} \qquad n = \frac{PV}{RT}$$

The volume of the balloon is

$$V = \frac{4}{3}\pi r^3 = \frac{4}{3}(\pi)(0.125 \text{ m})^3 = 8.18 \times 10^{-3} \text{ m}^3 = 8.18 \text{ L}$$

and the temperature $T = 20 + 273 = 293$ K, so

$$n = \frac{(1.3 \text{ atm})(8.18 \text{ L})}{(0.0821 \text{ L·atm/mol·K})(293 \text{ K})} = 0.442 \text{ mol}$$

The number of helium atoms is n times Avogadro's number:

$$nN_A = (0.442 \text{ mol})(6.02 \times 10^{23} \text{ atoms/mol})$$

$$= 2.66 \times 10^{23} \text{ atoms}$$

If one electron were removed from each atom, the charge removed would be

$$(2.66 \times 10^{23} \text{ electrons})(-1.60 \times 10^{-19} \text{ C/electron})$$

$$= -4.26 \times 10^4 \text{ C}$$

If two charges of this magnitude are $(22{,}000 \text{ mi})(1609 \text{ m/mi}) = 3.54 \times 10^7$ m apart, the force they exert on one another is

$$F = k\frac{q_1 q_2}{r^2}$$

$$= (8.99 \times 10^9 \text{ N·m}^2/\text{C}^2)\frac{(4.26 \times 10^4 \text{ C})(-4.26 \times 10^4 \text{ C})}{(3.54 \times 10^7 \text{ m})^2}$$

$$= -1.30 \times 10^4 \text{ N}$$

This is about 1.5 tons; the electrostatic force is strong.

4. (a) 1.73 N; they repel each other.   (b) The force is the same, but they attract.

5.   (a) Since $F = qE$, the force in a given field is proportional to the charge q. Thus, if the force on 5 nC is 0.08 N northeast, that on -2 nC is

$$\frac{(-2 \text{ nC})}{(5 \text{ nC})} (0.08 \text{ N northeast}) = -0.032 \text{ N northeast}$$

$$= \underline{0.032 \text{ N southwest}}$$

(b) The field is

$$E = \frac{F}{q} = \frac{0.08 \text{ N}}{5 \times 10^{-9} \text{ C}} = \underline{1.60 \times 10^7 \text{ N/C northeast}}$$

6. $\underline{-3.00 \times 10^{-4} \text{ C}}$

7.   (a) The electric field is just the vector sum of the fields of the two point charges alone. The field at $x = 6$ cm due to the charge at the origin is

$$E_1 = k \frac{q}{r^2} = (8.99 \times 10^9 \text{ N·m}^2/\text{C}^2) \frac{5 \times 10^{-6} \text{ C}}{(0.06 \text{ m})^2}$$

$$= 1.25 \times 10^7 \text{ N/C}$$

This field is directed away from the positive charge, that is, toward $+x$. In the same way, the field due to the charge at $x = 10$ cm is

$$E_2 = (8.99 \times 10^9 \text{ N·m}^2/\text{C}^2) \frac{10^{-5} \text{ C}}{(0.04 \text{ m})^2} = 5.62 \times 10^7 \text{ N/C}$$

This field is directed toward the 10-μC charge, so it is also directed toward $+x$. Thus,

$$E = E_1 + E_2 = (1.25 \times 10^7 \text{ N/C}) + (5.62 \times 10^7 \text{ N/C})$$

$$= \underline{6.87 \times 10^7 \text{ N/C}}$$

(b) From the solution to part (a), we can see that the zero field point can't be between the two charges because there the two field contributions add. Furthermore, it can't be to the right of the charges because, to cancel, the two contributions must be of equal magnitudes, so the point must be farther from the larger than from the smaller charge. If x is the point where the field is zero, then $x < 0$.

$$E = E_1 + E_2 = k \frac{q_1}{(x)^2} + k \frac{q_2}{(x - 10 \text{ cm})^2} = 0$$

so

$$\frac{q_1}{(x)^2} = -\frac{q_2}{(x - 10 \text{ cm})^2}$$

Rearranging and substituting in, we obtain

$$\frac{(x - 10 \text{ cm})^2}{x^2} = -\frac{q_2}{q_1} = -\frac{-10 \text{ }\mu\text{C}}{5 \text{ }\mu\text{C}} = 2$$

So

$$x - 10 \text{ cm} = \sqrt{2}x$$

and

$$x = \underline{-24.1 \text{ cm}}$$

8. $\underline{-2.56 \times 10^{-7} \text{ C}}$

Figure 18-5

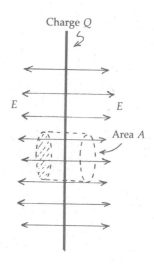

Charge Q

9.   (a) If the plane sheet is essentially infinite, we're always just as close to one edge as the other; thus, by symmetry, the field must be perpendicular to the plane sheet. Likewise, the left side of the sheet looks just the same as the right, so the field must be the same on both sides. Since lines of force start from a positive charge, we get the diagram shown in Figure 18-5. Consider the cylindrical surface shown in the diagram. There is no flux into the sides of the cylinder because they are parallel to the field. Since the field points out of both ends of the cylinder, the net flux is

$$\phi_{net} = EA + EA = 2EA = 4\pi k Q_{in}$$

In this equation, Q_{in} is the charge inside the surface. If the uniform charge density on the plane is σ, then $Q_{in} = \sigma A$, and

$$2EA = 4\pi k \sigma A$$

or

$$\underline{E = 2\pi k \sigma}$$

(b) The field due to two such plane sheets will be just the sum of the individual fields; the only difference is that field lines due to the negatively charged sheet go to that sheet. Thus, the fields cancel outside the two sheets and add in between, as illustrated in Figure 18-6. Thus,

$$\underline{E = 0} \text{ (outside)} \qquad \text{and} \qquad \underline{E = 4\pi k \sigma} \text{ (inside)}$$

Figure 18-6

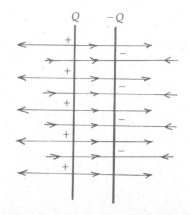

10. $\underline{E = 4\pi k \sigma}$. It's interesting that this result is the same as for the field between parallel sheets with equal and opposite charges.

11.   (*a*) The field of the dipole is just the sum of the two point-charge fields. From Figure 18-7, the magnitude of the field of each charge is

$$E_1 = \frac{kq}{r^2}$$

where *r* can be found using the pythagorean theorem. Thus,

$$E_1 = \frac{(8.99 \times 10^9 \text{ N·m}^2/\text{C}^2)(5 \times 10^{-8} \text{ C})}{\sqrt{(0.005 \text{ m})^2 + (0.05 \text{ m})^2}} = 1.78 \times 10^5 \text{ N/C}$$

To find the field of the dipole, the two point-charge fields must be added as vectors. From the diagram, we can see that the *y* components cancel and the *x* components add. To calculate the *x* components, we need the angle θ. From the diagram,

$$\tan \theta = \frac{0.5 \text{ cm}}{5 \text{ cm}} = 0.100$$

so

$$\theta = 5.71°$$

Therefore,

$$E_{1x} = E_1 \sin \theta = (1.78 \times 10^5 \text{ N/C}) \sin (5.71°)$$

$$= 1.77 \times 10^4 \text{ N/C}$$

This is the contribution of each charge, so the field of the dipole is 3.54 × 10⁴ N/C, parallel to the dipole moment in the minus *x* direction.

   (*b*) If we repeat the calculations in just the same way for a dipole twice as long with charges half as large, we get almost the same thing: $E = 3.39 \times 10^4$ N/C.

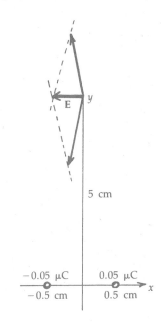

Figure 18-7

12. 8.47×10^{-30} C·m

13.   (*a*) Electric field lines begin on positive charges and end on negative charges. Thus, the charge on the left is <u>negative</u>:

$$q_1 = -40 \ \mu\text{C}$$

   (*b*) The number of lines that originate or terminate at a charge is proportional to the magnitude of the charge. Twelve lines end at the charge on the left (q_1), and 18 lines start from the other. Thus, the magnitude of q_2 is

$$|q_2| = \frac{(18)}{(12)} (40 \ \mu\text{C}) = 60 \ \mu\text{C}$$

Lines start on positive charges and end on negative ones, so

$$q_2 = \underline{+60\ \mu C}$$

(c) From very far away, this pair of charges looks like a net charge of $+20\ \mu C$, so <u>6 lines</u> go off to infinity.

14. <u>14.2 N/C</u>

CHAPTER 19

Electrostatics

I. Key Ideas

Electric Potential Electrostatics is the study of the interactions and effects of electric charges at rest. Charges exert forces on one another, and thus, if they are displaced, they do work on and exchange energy with one another. The work that must be done on a test charge q_0 to displace it in an electric field due to other charges (which is to say, the negative of the work done on the test charge *by* the field) is the increase in its electrostatic potential energy. As this work is always proportional to the test charge q_0, what is determined by the field is the change in *potential energy per unit charge*. This is defined the change in the electric potential.

Potential of a Point Charge Only differences of potential are defined, so the potential can be chosen to be zero at any convenient point. Conventionally, we take the potential to be zero infinitely far away from a charge distribution. With this choice, the potential of a point charge decreases inversely with the distance from the charge. The potential is then the work that would have to be done to bring a unit positive charge in from infinity to the point in question. The potential at any point due to a distribution of charges is the sum of the potentials at that point due to each charge.

Electric Field and Potential In a *uniform* electric field, the potential difference between any two points is just the field times the component of the distance between the points parallel to the field direction. On a lines-of-force diagram, the E field lines point in the direction of decreasing electric potential.

Conductors There is a tremendous variation among materials (around 20 orders of magnitude) in their ability to *conduct* electricity; that is, in the extent to which charge can move about freely through them. Some materials are *conductors*: these allow very free movement of electric charge because their structures are such that they contain electrons that are essentially free, that is, not bound to individual atoms.

Conductors in an Electric Field If an external electric field is applied to a conductor, the free charge in the conductor moves (in practice, nearly instantaneously) in such a way that the resultant field inside the conductor is zero. Any net charge on the conductor must be on its outside surface. (It is because of this property that isolated conductors can be charged by induction.) The free surface charge produces an electric field that is *perpendicular* to the surface of the conductor.

Equipotential Surfaces Surfaces drawn perpendicular to lines of the electric field are always at the same potential. We call these *equipotential surfaces*; to move a charge on such a surface requires no work. The entire body of a conductor in electrostatic equilibrium is an equipotential.

Charge Sharing Conductors not in contact with one another are not in general at the same potential. The potential difference between them depends on their shapes, separation, and charges. If they are brought into contact, the charge on them will rearrange itself so that electrostatic equilibrium is reestablished. In equilibrium, both conductors together are an equipotential, and the field inside both of them is zero. The transfer of charge from one conductor to another is called *charge sharing*. If one conductor makes contact *inside* another, all the charge on them flows to the outer surface of the outer conductor. How much charge can be placed on a conductor in this way is limited only by *dielectric breakdown* in the air or whatever insulating material surrounds the conductor. Because the surface of a conductor is an equipotential, charge tends to concentrate on sharp curves and corners.

Capacitance A *capacitor* consists of two conductors that carry equal and opposite charges. Its capacitance is the ratio of the charge on either conductor to the potential difference between them. The capacitance is determined by the geometry of the conductors. A typical capacitor might be a pair of parallel conducting plates; its capacitance would depend on the area and separation of the plates.

Dielectrics An insulating material—a *dielectric*—between the electrodes of a capacitor changes its capacitance. The electric field induces a surface charge on the dielectric surfaces (it "polarizes the dielectric"). The field created due to this surface charge tends to cancel the field applied to the dielectric. The net electric field inside the capacitor is thus reduced, and the capacitance is correspondingly increased.

Combinations of Capacitors When capacitors are connected in parallel (see Figure 19-1a), the same potential difference appears across each; thus, the charges stored, and so the capacitances, just add. In a series connection (see Figure 19-1b), a larger potential difference is needed to put the same charge on each capacitor, so the capacitance of the combination is less than that of either one.

Figure 19-1

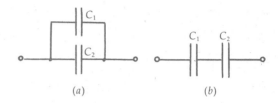

(a)   (b)

Energy in the Electric Field Work is required to charge a capacitor; the charged capacitor stores this work as electrostatic potential energy. Another way to say this is that energy is stored in the electric field set up in the capacitor. This result is general—energy is stored in any electric field. The *energy density* (the energy per unit volume) in the capacitor depends only on the electric field in it.

II.   Numbers and Key Equations

Numbers

$$1 \text{ volt (V)} = 1 \text{ J/C}$$

$$1 \text{ N/C} = 1 \text{ V/m}$$

$$1 \text{ farad (F)} = 1 \text{ C/V}$$

$$1 \text{ electronvolt (eV)} = 1.60 \times 10^{-19} \text{ C·V}$$

$$= 1.60 \times 10^{-19} \text{ J}$$

Permittivity of free space

$$\epsilon_0 = 8.85 \times 10^{-12} \text{ C}^2/\text{N·m}^2$$

$$= 8.85 \times 10^{-12} \text{ F/m}$$

Key Equations

Work done against an electric field

$$\Delta W = F_x \, \Delta x = -q_0 E_x \, \Delta x$$

Potential difference

$$\Delta V = \frac{\Delta U}{q_0} = -E_x \, \Delta x$$

Point charge

$$E = \frac{kq}{r^2} \qquad V = \frac{kq}{r} \qquad U = q_0 V = \frac{kqq_0}{r}$$

Field at the surface of a conductor

$$E = 4\pi k\sigma$$

Charge density on the surface of a sphere

$$\sigma = \frac{V}{4\pi kr}$$

Capacitance

$$C = \frac{Q}{V}$$

Parallel-plate capacitor

$$V = Ed \qquad C = \frac{A}{4\pi kd} = \epsilon_0 \frac{A}{d}$$

Dielectric in a capacitor

$$E' = \frac{E_0}{K} \qquad C = KC_0$$

Capacitors in parallel

$$C_{\text{eff}} = C_1 + C_2$$

Capacitors in series

$$\frac{1}{C_{\text{eff}}} = \frac{1}{C_1} + \frac{1}{C_2}$$

Energy stored in a capacitor

$$U = \tfrac{1}{2}QV = \tfrac{1}{2}CV^2 = \tfrac{1}{2}\frac{Q^2}{C}$$

Energy density in an electric field

$$\eta = \frac{\text{energy}}{\text{volume}} = \tfrac{1}{2}\epsilon_0 E^2$$

III.   Possible Pitfalls

Remember that only differences in potential energy, and thus in electrical potential, are defined. The position at which the potential is zero is arbitrary, although, as a convention, we usually take the potential of a finite charge distribution to be zero at infinite distance.

Potential is a scalar quantity; it has no direction. Don't mix it up with the electric field, which is a vector; they're different quantities.

The electrostatic force is always in the direction of decreasing potential energy; the electric field is always in the direction of decreasing potential. These directions aren't necessarily the same since the electrostatic force on a negative charge is in a direction opposite that of the electric field.

The force on a point charge is in the direction of the electric field only if the charge is positive; in particular, electrons go the other way.

The capacitance of any two conductors depends only (if there is no dielectric) on the geometry of the conductors, that is, on their size, shape, and position. In any case, it does not depend on the charge on the conductors or on the potential difference between them.

We almost always deal with differences in potential, not absolute values, and so we often write just "V" rather than "ΔV" for potential difference. Make sure you know which you're talking about. It also doesn't help that "V" is used to stand for "volt"; results like "$V = 3$ V" can be confusing.

Adding a dielectric to a capacitor *weakens* the field in the capacitor because of the opposing polarization electric field. This means that a given charge on the plates corresponds to a smaller potential difference, so the capacitance is *increased*. It's easy to get thinking backwards about this.

IV.   Questions and Answers

Questions

1.   At point P in Figure 19-2, the electric potential is zero. (As usual, we take the potential to be zero at infinite distance.) What can you say about the two charges? Are there any other points of zero potential on the line of the two charges?

Figure 19-2

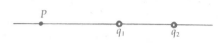

2.   Lines of force for the electric field of a system of two point charges are shown in Figure 19-3. Draw on the diagram what you think a few equipotential surfaces for this system would look like.

Figure 19-3

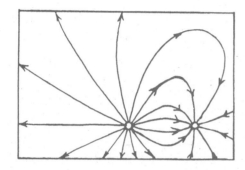

3.   How could you use "charge sharing" to produce charges q_1 and q_2 on two conductors such that $q_1 = 3q_2$? Could you make $q_1 = -3q_2$?

Figure 19-4

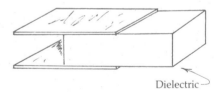

Dielectric

4.   A slab of dielectric material is half in, half out of a parallel-plate capacitor, as shown in Figure 19-4. There is a force on the slab. Does the force tend to draw the slab in or spit it out? Why?

5.   Figure 19-5 shows a square metal plate that has a net positive charge; it is hung from a fine insulating thread, say. Sketch what you think the equipotential surfaces and lines of force in the plane of the drawing would look like.

Figure 19-5

6.   The bound surface charge on a dielectric in a capacitor is always less, even though it may be only slightly less, than the free charge on the plates. Why must this be so?

7.   Explain physically why you would expect the potential energy of a pair of charges of the same sign to be positive and that of a pair of charges having opposite signs to be negative.

8.   If the electric potential throughout some region of space is zero, does it necessarily follow that the electric field is zero? If the electric field throughout a region is zero, does it necessarily follow that the electric potential is zero?

9. If the voltage across a capacitor is doubled, by how much does the energy stored change?

10. A charged insulated rod is used to charge a conductor by induction. Can it be used again and again to charge other conductors?

11. A point charge is moved a small distance in a region of space where there is an electric field. How can it be moved so that its electric potential is unchanged? How can it be moved so that its electric potential is increased?

Answers

1. Since the charges make opposite contributions to the potential, their signs must be opposite; since q_1, although closer, contributes to the potential in the same amount, q_2 must have greater magnitude. There will be one more point on the line between the two charges at which the potential is zero. (You know this because the potential far away in either direction must have the same sign, that of q_2. Thus, the potential function must go through zero at two points on the axis.)

2. Each equipotential surface must be everywhere perpendicular to the lines of force, so they must look something like what is sketched in Figure 19-6.

Figure 19-6

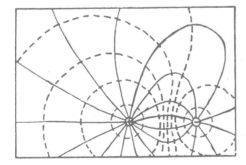

3. Put a charge q on a spherical conductor. Touch it with a second, uncharged spherical conductor whose radius is one third that of the first. When you separate the conductors, the smaller one takes away a charge of $q/4$, leaving a charge of $3/4q$ on the larger conductor. To get opposite signs, induce equal and opposite charges on two identical spherical conductors that are initially uncharged; then proceed as before.

4. The force tends to draw the slab in because the charge on each plate attracts the (opposite) charge induced on the nearer surface of the dielectric. Another way to say this is that inserting the dielectric reduces the stored energy in the capacitor and thereby decreases the potential energy of the system, and force always acts in the direction of decreasing potential energy.

5. This is a case in which it's easier to draw the equipotential surfaces. Since the square conductor itself is an equipotential, the equipotential surfaces near it in its own plane must be squares surrounding the conductor. The farther away you go, the more nearly the square conductor approximates a point charge, so the equipotentials must become more and more circular. Once you've drawn the equipotentials, you know that the lines of force must be everywhere perpendicular to them. Sufficiently far away, they must go uniformly out in all directions, as from a point charge. These considerations give you something like what's sketched in Figure 19-7. Notice that the lines of force tend to concentrate at the corners of the square.

Figure 19-7

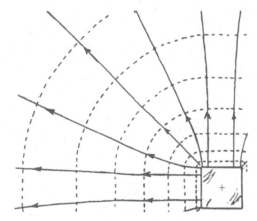

6. If this were not so, the electric fields from equal but opposite free and polarization charge densities would cancel completely, and then there would be no net field in the dielectric to polarize it!

7. Like charges repel one another, so to put them in proximity I have to do some work on them. This work is recoverable, so it goes into increased potential energy of the charges. If they have opposite signs, they attract each other, and as they come together, I must do negative work on them to restrain them; their potential energy thus decreases as they approach one another.

8. If the electric potential throughout some region of space is zero, it is constant, so there is no electric field. When the electric field throughout some region of space is zero, no work is done on a charge moving in that region, so the electric potential is constant but not necessarily zero. The answers are thus, respectively, yes and no.

9. Since the stored energy is proportional to V^2, it increases by a factor of four when V is doubled.

10. Yes, it can be used over and over until the charge leaks off the rod into the air or something. The charge on the insulated rod induces the charge on the conductor but is not affected by the charge on the conductor since the two never touch.

11. If the charge is moved perpendicular to the electric field, no work is done on it, so its potential energy is unchanged. To increase its potential energy, the charge must be moved against the direction of the force on it. Thus, you would move a positive charge against the field direction or a negative charge along it.

V. Problems and Solutions

Problems

1. Three point charges, each $-0.5\ \mu C$, are located at the vertices of an equilateral triangle 0.5 m on a side, as in Figure 19-8. (*a*) What is the electric potential energy of charge q_1 due to the other two? (*b*) What is the electric potential at this point?

Figure 19-8

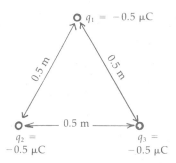

How to Solve It
- Calculate the potential energy of q_1 due to each of the other two charges.

- Remember that potential energy is a scalar quantity; when you have contributions due to more than one charge, they just add.

- The electric potential is potential energy per unit charge.

2. A point charge $q_0 = 0.5\ \mu C$ is at the origin. (*a*) Find the potential at $x = 0.8$ m (taking it to be zero at infinite distance). (*b*) A second particle with charge $q = 1\ \mu C$ and mass 0.08 g is placed at $x = 0.8$ m. What is its potential energy? (*c*) If charge q is released from rest, what will its speed be when it reaches $x = 2$ m?

How to Solve It
- Calculate the potential directly from the charge and distance.

- The electric potential is the potential energy per unit charge, so you can calculate the potential energy of charge q directly.

- Recalculate the potential energy at $x = 2$ m. The decrease in potential energy will show up as an increase in kinetic energy, as the electrostatic force is conservative.

• Form its kinetic energy at $x = 2$ m, find the speed of charge q.

3. Two equal point charges, each -0.02 μC, are placed on the x axis as shown in Figure 19-9. Find the electric potential (a) at point A ($x = 5$ cm) and (b) at point B ($y = 5$ cm).

Figure 19-9

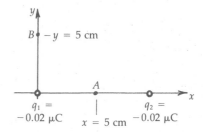

How to Solve It
• The potential at either point due to each of the point charges can be calculated directly from the charge and distance.

• The total potential is just the sum of the two in each case.

4. Two point charges are placed on the x axis: $+0.5$ μC at $x = 0$ and -0.2 μC at $x = 10$ cm. At what point(s) on the x axis is the electrical potential equal to zero?

How to Solve It
• For the total potential to be zero, the potentials due to each charge alone must have the same magnitude (and opposite signs).

• In order for the potentials due to the individual charges to be the same, the charge of larger magnitude must be farther away.

5. Polyethylene has a dielectric constant of 2.3 and a dielectric breakdown strength of about 5×10^7 V/m; air has a dielectric constant of 1.0 and a dielectric strength of about 3×10^6 V/m. If you have two plane conductors of area 0.5 m², what is the largest capacitance you can make that will withstand 600 V (a) with air for a dielectric? (b) With polyethylene?

How to Solve It
• There is a formula for the capacitance of a parallel-plate capacitor. Write this in such a way that the capacitance depends explicitly on the dielectric strength E_{max} of the material between the plates.

• The largest capacitance will correspond to the closest plate spacing.

• However, the plates cannot be so close together that E_{max} times d is less than 600 V!

6. A conducting sphere of radius 40 cm is isolated in air. How much charge can be placed on it before dielectric breakdown occurs in the air around it? The dielectric strength of air is around 3×10^6 V/m.

How to Solve It
- The field at the surface of the sphere is the same as if all the charge on the sphere were located at its center.

- Thus, outside the sphere, the electric field decreases as $1/r^2$, where r is the distance from the center. So if breakdown occurs anywhere, it will be right at the surface of the sphere.

- Just find what this charge must be if the electric field is 3×10^6 V/m at the surface.

7. A parallel-plate capacitor is made by sandwiching 0.1-mm sheets of paper (dielectric constant 3.7) between three sheets of aluminum foil and connecting the outer aluminum sheets together (see Figure 19-10). A "sandwich" with an area of 10 m^2 is fabricated this way. (To be practical it would then have to be rolled or folded up so as to fit in a small package.) What is the capacitance of this capacitor?

Figure 19-10

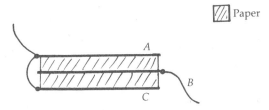

How to Solve It
- The key to this problem is to regard the "sandwich" as two capacitors. If you imagine the sheet of aluminum foil in the middle to be two plates connected together, you can see this. Since each side of one capacitor is connected to one side of the other, the two are in parallel.

- For each of the two capacitors, you can calculate the capacitance from the given dimensions and dielectric constant.

- The capacitances of capacitors in parallel simply add.

8. A parallel-plate capacitor has a capacitance of 5.0 μF and a plate separation of 0.2 mm. There is just air between the plates. (*a*) What is the area of the plates? (*b*) What is the largest charge that can be stored on the capacitor before breakdown occurs? The dielectric strength of air is around 3×10^6 V/m.

How to Solve It

- There is a formula for the capacitance of a parallel-plates capacitor; solve it for the plate area.

- The dielectric strength of air multiplied by the plate separation gives you the breakdown voltage.

- The capacitance multiplied by the potential difference is the charge stored on the plates.

9. You have a bucketful of identical capacitors, each with a capacitance of 1.0 μF and a voltage rating of 250 V. By connecting them in series and in parallel you are to come up with a combination that has a capacitance of 0.75 μF and a voltage rating of 1000 V. How many capacitors do you need?

How to Solve It

- Since you want the combination to withstand a potential difference greater than one capacitor will, you need several in series.

- Calculate the capacitance of a string of enough capacitors in series to withstand 1000 V.

- Connect enough such strings in parallel to give you an effective capacitance of 0.75 μF.

10. Three capacitors have capacitances 10 μF, 15 μF, and 30 μF. What is the effective capacitance if the three are connected (*a*) in parallel and (*b*) in series?

How to Solve It

- Capacitances in parallel simply add.

- For capacitors in series, it is $1/C$ that just adds. Don't forget, though, that the answer is $1/C_{eff}$—you have to invert it again to get C_{eff}.

Figure 19-11

11. Three 0.18-μF capacitors are connected in parallel across a 12-V battery, as shown in Figure 19-11. The battery is then disconnected. Next, one capacitor is carefully disconnected so that it doesn't lose any charge and is reconnected "upside down"—that is, with its positively-charged side at *b* and its negatively-charged side at *a*. (*a*) What is the potential difference across the capacitors now? (*b*) By how much has the stored energy of the capacitors changed in the process?

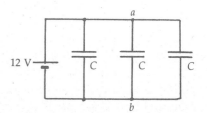

How to Solve It

- When the battery is disconnected, there is no connection between the positive and negative sides of any of the capacitors.

- After reversing one capacitor you have—again—three identical capacitors in parallel. How much charge is on each side of the group of capacitors now?

- Knowing the charge, find the potential drop across each capacitor.

- Since there is now a lower potential difference across the "same" configuration of capacitors, the stored energy must have decreased.

12. What is the effective capacitance of the network of three capacitors shown in Figure 19-12?

Figure 19-12

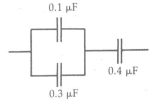

0.1 μF

0.3 μF

0.4 μF

How to Solve It

- The capacitances of the two capacitors that are connected in parallel just add.

- Now what you have is the effective capacitance of that pair in series with the 0.4-μF capacitor; calculate the effective capacitance of that combination.

13. (*a*) You want to store 10^{-5} C of charge on a capacitor, and you have available only 100 V with which to charge it. What must be the value of the capacitance? (*b*) You want to store 10^{-3} J of energy on a capacitor, and you have available only 100 V with which to charge it. What must be the value of the capacitance?

How to Solve It

- In part (*a*), calculate the capacitance from the known charge and voltage drop.

- In part (*b*), write a formula for the energy stored in the capacitor in terms of the capacitance and the voltage drop across it; then solve this for C.

14. A parallel-plate capacitor has a plate separation of 1.5 mm and is charged to 600 V. If an electron leaves the negative plate, starting from rest, how fast is it going when it hits the positive plate?

How to Solve It

- While it is moving from one plate to the other, the electron is acted on only by the force exerted on it by the electric field.

- This is a conservative force, so the decrease in the potential energy of the electron as it crosses the gap is equal to the kinetic energy it gains.

- From the mass and kinetic energy of the electron when it reaches the positive plate, calculate its speed.

15. A capacitor made of parallel plates 2.0 mm apart is charged to 500 V. (*a*) What is the electric field between the plates? (*b*) What is the electrostatic energy density in the space between the plates? (*c*) If the plates are squares 0.6 m on a side, what is the total energy stored in the capacitor? (*d*) What is the capacitance of the capacitor?

How to Solve It
- From the voltage drop and separation, calculate the electric field in the region between the plates.

- The electrostatic energy density between the plates of the capacitor depends only on the electric field.

- Multiply the energy density by the volume of the region between the plates to get the total energy stored.

- Knowing the potential difference and the stored energy, you can find the capacitance; what you get should agree with what you would get directly from $C = \epsilon_0 A/d$.

Solutions

1. (*a*) Each of the other charges is -0.5 μC at a distance of 0.5 m from q_1, so the potential energy of q_1 due to either of the other charges is the same. For q_2,

$$U_2 = k \frac{q_1 q_2}{r}$$

$$= (8.99 \times 10^9 \text{ N·m}^2/\text{C}^2) \frac{(-5 \times 10^{-7} \text{ C})(-5 \times 10^{-7} \text{ C})}{0.5 \text{ m}}$$

$$= 4.495 \times 10^{-3} \text{ J}$$

So that due to q_2 and q_3 together is

$$U = (2)(4.495 \times 10^{-3} \text{ J}) = \underline{8.99 \times 10^{-3} \text{ J}}$$

(b) The electric potential is the potential energy per unit charge:

$$V = \frac{U}{q_1} = \frac{8.99 \times 10^{-3} \text{ J}}{-5.0 \times 10^{-7} \text{ C}} = -1.80 \times 10^4 \text{ J/C}$$

$$= \underline{-1.80 \times 10^4 \text{ V}}$$

2.   (a) 5620 V   (b) 5.62×10^{-3} J   (c) 9.18 m/s

3.   (a) The potential due to each point charge is found from

$$V = \frac{kq}{r}$$

Here q_1 and q_2 are equal, and each charge is a distance 5 cm from point A, so

$$V_{1A} = V_{2A} = \frac{(8.99 \times 10^9 \text{ N·m}^2/\text{C}^2)(-2 \times 10^{-8} \text{ C})}{(5 \times 10^{-2} \text{ m})} = -3596 \text{ V}$$

The potential at A is simply the sum of the two:

$$V_A = V_{1A} + V_{2A} = (2)(-3596 \text{ V})$$

$$= \underline{-7192 \text{ V}}$$

(b) Point B is also 5 cm from one of the charges, but it is a distance $\sqrt{(5 \text{ cm})^2 + (10 \text{ cm})^2} = 11.18$ cm from the other (see Figure 19-13). Thus,

$$V_{2B} = \frac{(8.99 \times 10^9 \text{ N·m}^2/\text{C}^2)(-2 \times 10^{-8} \text{ C})}{0.1118 \text{ m}} = -1608 \text{ V}$$

and

$$V_B = V_{1B} + V_{2B} = (-3596 \text{ V}) + (-1608 \text{ V}) = -\underline{5204 \text{ V}}$$

Figure 19-13

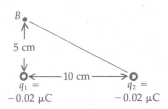

4. $x = \underline{7.14 \text{ cm}}$; $x = \underline{16.67 \text{ cm}}$

5.   (a) The capacitance of a parallel-plate capacitor containing a material of dielectric constant K is

$$C = K\epsilon_0 \frac{A}{d}$$

but

$$E = \frac{V}{d} = \frac{600 \text{ V}}{d}$$

so

$$d = \frac{600 \text{ V}}{E}$$

The electric field E has its maximum (breakdown) value when the plate spacing d is minimum and thus when the capacitance is maximum. Thus, for air

$$d_{min} = \frac{600 \text{ V}}{3 \times 10^6 \text{ V/m}} = 2 \times 10^{-4} \text{ m}$$

and

$$C_{max} = K\epsilon_0 \frac{A}{(2 \times 10^{-4} \text{ m})}$$

$$= (1)(8.85 \times 10^{-12} \text{ F/m}) \frac{0.5^2}{2 \times 10^{-4} \text{ m}} = \underline{2.2 \times 10^{-8} \text{ F}}$$

(b) Likewise, with polyethylene ($K = 2.3$ and $E_{max} = 5 \times 10^7$ V/m) as the dielectric, we get $C_{max} = \underline{8.5 \times 10^{-7} \text{ F}}$.

6. $\underline{5.34 \times 10^{-5} \text{ C}}$

7. Physically, the "sandwich" arrangement is equivalent to two identical capacitors, each having a plate spacing of 0.1 mm, an area of 10 m^2, and paper as a dielectric, connected in parallel (see Figure 19-14).

Figure 19-14

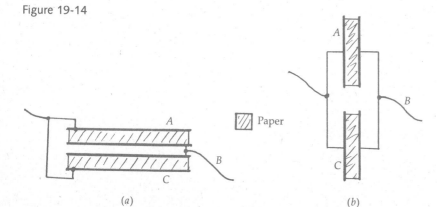

(a)   (b)

The capacitance of each of these is

$$C = K\epsilon_0 \frac{A}{d} = (3.7)(8.85 \times 10^{-12} \text{ F/m}) \frac{10 \text{ m}^2}{10^{-4} \text{ m}}$$

$$= 3.27 \times 10^{-6} \text{ F} = 3.27 \text{ μF}$$

The capacitance of the two in parallel is thus 6.54 μF.

8.   (a) 113 m²   (b) 0.0030 C

9.   Each capacitor can withstand 250 V, and the combination is to withstand 1000 V. Since the potential drops across series capacitors add, this means that we want strings of four capacitors in series. The capacitance of a string of four 1.0-μF capacitors in series is given by

$$\frac{1}{C_{\text{eff}}} = \frac{1}{C} + \frac{1}{C} + \frac{1}{C} + \frac{1}{C} = \frac{4}{C}$$

so

$$C_{\text{eff}} = \frac{1.0 \text{ μF}}{4} = 0.25 \text{ μF}$$

If we put three 0.25-μF capacitors in parallel, their effective capacitance is $(3)(0.25 \text{ μF}) = 0.75 \text{ μF}$. Thus, we would connect three such strings, each consisting of four capacitors in series, in parallel, as sketched in Figure 19-15. We would thus need 12 capacitors.

Figure 19-15

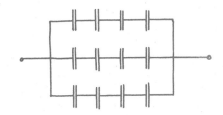

10.   (a) 5 μF   (b) 55 μF

11.   (a) The charge on each of the three capacitors to begin with was

$$Q_1 = CV = (0.18 \text{ μF})(12 \text{ V}) = 2.16 \text{ μC}$$

with the positive charge on the top plate and the negative charge on the bottom. The moment you reconnect the inverted capacitor, it has a charge $+Q_1$ on the bottom plate and $-Q_1$ on the top (see Figure 19-16). The total charge of $Q_1 + (-Q_1) + Q_1 = Q_1$ immediately redistributes itself (equally, by symmetry) on the three capacitors, so each one now carries a charge $Q_1/3$. The potential difference across them, then, is

$$V = \frac{1}{3}\frac{Q_1}{C} = \frac{1}{3}\left(\frac{2.16 \text{ μC}}{0.18 \text{ μF}}\right) = 4.0 \text{ V}$$

Figure 19-16

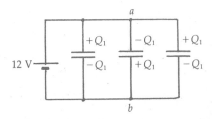

(b) The energy initially stored on each of the three capacitors was

$$U_1 = \tfrac{1}{2}CV^2 = \tfrac{1}{2}(0.18\ \mu F)(12\ V)^2 = 1.30 \times 10^{-5}\ J$$

It has decreased to

$$U = \tfrac{1}{2}(0.18\ \mu F)(4\ V)^2 = 1.44 \times 10^{-6}\ J$$

or $\tfrac{1}{9}$ of its original value.

12. 0.2 μF

13. (a) The capacitance must be

$$C = \frac{Q}{V} = \frac{10^{-5}\ C}{100\ V} = 10^{-7}\ F$$

(b) The stored energy is given by

$$U = \tfrac{1}{2}CV^2$$

and therefore

$$C = \frac{2U}{V^2} = \frac{(2)(10^{-3}\ J)}{(100\ V)^2} = 2 \times 10^{-7}\ F$$

14. 1.45×10^7 m/s

15. (a) The electric field is

$$E = \frac{V}{d} = \frac{500\ V}{0.002\ m} = 2.5 \times 10^5\ V/m = 2.5 \times 10^5\ N/C$$

(b) The energy density in the field is

$$\eta = \tfrac{1}{2}\epsilon_0 E^2$$

$$= \tfrac{1}{2}(8.85 \times 10^{-12}\ C^2/N{\cdot}m^2)(2.5 \times 10^5\ N/C)^2 = 0.277\ J/m^3$$

(c) If the plates are squares 0.6 m on a side, then the volume of the region of space between the plates is

$$Vol = Ad = (0.6\ m)^2(2 \times 10^{-3}\ m) = 7.20 \times 10^{-4}\ m^3$$

The stored energy is thus

$$U = (\eta)(\text{Vol}) = (0.277 \text{ J/m}^3)(7.2 \times 10^{-4} \text{ m}^3) = \underline{1.99 \times 10^{-4} \text{ J}}$$

(d) This stored energy is

$$U = \tfrac{1}{2}CV^2$$

so

$$C = \frac{2U}{V^2} = \frac{(2)(1.99 \times 10^{-4} \text{ J})}{(500 \text{ V})^2} = \underline{1.59 \times 10^{-9} \text{ F}}$$

CHAPTER 20

Electric Current and Circuits

I. Key Ideas

Current Electric current is charge in motion. The amount of current is the charge per unit time that flows past some point. The direction of the current is conventionally taken as that of the flow of hypothetical positive charges, even though in most cases the current actually consists of negatively-charged electrons moving in the opposite direction. This distinction is only occasionally important. Current flows in the direction of the electric field in a conductor.

Drift Velocity The actual motion of electrons in a metal is complex. In the absence of an electric field, the free electrons are in random motion at high velocities due to thermal agitation. In this condition, the average velocity of the electrons is zero. When an electric field is applied, it accelerates the electrons only a little (until they collide with ions or other electrons) but all in the *same* direction. The effect of the field is thus to superimpose on the random thermal motion an average "drift" opposite to the field direction. Typically, this *drift velocity* is very small—perhaps 10^{-4} m/s.

Ohm's Law Electric current in a conductor is caused by an electric field. (Note that the conductor here is not in electrostatic equilibrium.) There is thus a difference in electric potential along the direction of current flow. Empirically, in most materials, the current is directly proportional to the potential difference; this is *Ohm's law*. Materials that obey Ohm's law are called *ohmic* materials.

Resistance The *resistance* between two points in a material is the ratio of the potential difference between the points to the current flow in the material. If the material is ohmic, the resistance is constant. The resistance of an object depends on its dimensions and on the *resistivity* of the material. (Note the analogy to the definition of thermal resistance.) The reciprocal of the resistivity is the *conductivity*.

Resistivity and Temperature The resistivity of a material depends on temperature. Under ordinary conditions, the resistivity of a metal increases nearly linearly with temperature. For many metals, the resistivity below a certain critical temperature is exactly zero. In such a *superconductor*, a current flow can be sustained indefinitely even in the absence of any electric field.

Power Dissipation When a current flows in an ordinary conductor, electrical energy is continually dissipated as heat. Microscopically, what happens is that when the electrons collide with positive ions in the conductor, the direction of drift is "forgotten," so the kinetic energy of their drift becomes part of the random thermal energy of the conductor. This dissipation of electrical energy in a resistive material is *Joule heating*. The power dissipated is the product of the potential difference across the resistance multiplied by the current in it.

Sources of emf   To keep a steady current flowing in a conductor, a source of energy is needed. A device, such as a battery or generator, that converts chemical, mechanical, or some other form of energy into electrical energy is called a *source of electromotive force* or *emf*. An ideal source of emf would be one that maintains a constant potential difference across its terminals regardless of the current drawn from it. Real sources of emf have an effective internal resistance, so the terminal voltage decreases with increasing current.

Combinations of Resistors   Resistances connected in series so that the same current flows through each of them, one after another, simply add. For resistances connected in parallel, the equivalent resistance is found by adding the inverses of the resistances and inverting the result. Note that these rules are just opposite those for combinations of capacitors.

Kirchhoff's Rules   Kirchhoff's rules are applications of the conservation of energy and charge to electric circuits; they are general and suffice to solve any circuit: (1) The net change in potential around any closed path in the circuit must be zero. (2) The total of all currents entering any junction point less the total current leaving it must be zero.

RC **Circuits**   In circuits containing both a resistance and a capacitance, the electric potential cannot change instantaneously. The characteristic time scale of charging or discharging a capacitance C through a resistor R is the *time constant RC*. In the discharge curve shown in Figure 20-1, RC is the time required for the charge on the capacitor to decrease to 0.37 of its initial value.

Meters   Instruments designed to measure voltage or current are based on some device that responds (mechanically or digitally) to a current through it. An ideal voltmeter would have infinite resistance; an ideal ammeter, zero resistance. Real meters have finite internal resistance. By putting the proper external resistances in series or parallel with a galvanometer, voltmeters, ammeters, or ohmmeters to read various ranges can be made.

II.   Numbers and Key Equations

Numbers

$$1 \text{ ampere (A)} = 1 \text{ C/s}$$
$$1 \text{ ohm } (\Omega) = 1 \text{ V/A}$$

Key Equations

Electric current

$$I = \frac{\Delta Q}{\Delta t}$$

Resistance

$$R = \frac{V}{I}$$

Ohm's law

$$V = IR \qquad (R \text{ constant})$$

Resistance of a conductor

$$R = \rho \frac{L}{A}$$

Conductivity and resistivity

$$\sigma = \frac{1}{\rho}$$

Figure 20-1

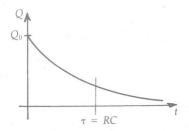

Variation in resistivity with temperature

$$\rho = \rho_{20}[1 + \alpha(t - 20\ C°)]$$

Power supplied by a source of emf

$$P = \mathcal{E}I$$

Power dissipation in a resistor

$$P = IV = I^2R = \frac{V^2}{R}$$

Terminal voltage

$$V_a - V_b = \mathcal{E} - IR$$

Resistors in series

$$R_{eq} = R_1 + R_2 + \cdots$$

Resistors in parallel

$$\frac{1}{R_{eq}} = \frac{1}{R_1} + \frac{1}{R_2} + \cdots$$

Capacitor discharge

$$\frac{\Delta Q}{\Delta t} = -\frac{Q}{RC}$$

Time constant for an RC circuit

$$\tau = RC$$

Circuit symbols

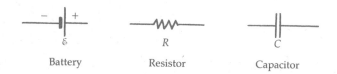

| Battery | Resistor | Capacitor |

III. Possible Pitfalls

Outside a source of emf, current flows from the positive to the negative terminal; inside, it flows back from the negative to the positive terminal. Think of a source of emf as a "charge pump," with the positive terminal as its "high-pressure" outlet.

When you calculate the power dissipated in a resistor using $P = I^2R$, make sure you are using the current that's flowing in *that* resistor.

Remember that the rules for the equivalent resistance of resistors in series and parallel are exactly *opposite* the rules for combining capacitors. In these formulas, R behaves like $1/C$, and vice versa.

The rules for the equivalent resistance of series and parallel resistors apply only for exactly those combinations. It's easy to look at two resistances in a circuit, for instance, and think they're in series when actually another branch forks off between them.

When a capacitor is charged through a resistor, not all of the energy drawn from the source of emf goes into charging the capacitor; some of it is dissipated in the resistance. (As it happens, the energy goes half and half.)

The *critical temperature* below which a metal becomes superconducting has nothing at all to do with the critical temperature of a gas!

Resistance depends both on geometry and on a material property, the resistivity.

When you apply Kirchhoff's voltage rule around a loop, be careful to remember that the potential *drops* as you go past a resistor in the direction of the current, but it increases if you are going "upstream." Likewise, the voltage increases as you go through a seat of emf from the negative to the positive terminal, and vice versa. It's terribly easy to make sign errors. It will help if you ask yourself if you are moving to higher or lower potential as you pass each element.

Likewise, be careful with the signs (directions) of the currents when applying Kirchhoff's current rule. Pick some sign convention and stick to it. For instance, you might call all currents flowing into a junction positive and all those flowing out of it negative; then the sum of all the currents is zero.

You have to assign a direction to the current in every branch of a circuit to apply Kirchhoff's rules. If you guess wrong on the direction of one or more currents, *it doesn't matter at all*. All that will happen is that when you solve the problem, that current will come out negative.

When you calculate the effective resistance of a parallel combination by adding $1/R$'s, don't forget that you have to invert the result again to get the final answer.

Voltmeters are connected in parallel with the potential difference being measured; if they are not to affect the circuit, you want them to draw as little current as possible and so to have a very high resistance. Ammeters are connected *into* the current being measured, so they should have as little resistance as possible.

V.   Questions and Answers

Questions

1. We distinguish the direction of a current in a circuit. Ought we to consider it a vector quantity?

2. We justified a number of electrostatic phenomena with the argument that there can be no electric field in a conductor. Now we say (Ohm's law) that the current in a conductor is proportional to the potential difference and thus to the electric field in the conductor. Is there a contradiction here?

3. The average drift speed of the electrons in a wire carrying a steady current is a constant. But the electric field in the wire is doing work on the electrons. Where is the energy going?

4. When 120 V is applied to the filament of a 75-watt light bulb, the current drawn is 0.63 A. When a potential difference of 3 V is applied to the same filament, the current is 0.086 A. Is the filament an ohmic material?

5. Under ordinary conditions, the drift speed of electrons in a metal is around 10^{-4} m/s or less. Then how is it that, when you flip the wall switch, it doesn't take an hour or so for a light bulb several metres away to come on?

6. Does the time required to charge a capacitor through a given resistance depend on the applied emf? Does it depend on the total amount of charge to be placed on the capacitor?

7. To change the charge on a capacitor in an *RC* circuit takes a time of the order of *RC*. Another way to say this is that the voltage across a capacitance in an *RC* circuit can't be changed instantaneously. Give a simple physical explanation of why this result is to be expected.

8. Today, ordinary strings of Christmas-tree lights contain eight or so bulbs connected in parallel across a 110-V line. Forty years ago, most strings contained eight bulbs connected in series across the line. What would happen if you put one of the old-style bulbs into a modern Christmas-tree light set? (The sockets are made differently to prevent you from doing this.)

9. Consider the circuit in Figure 20-2. Can you simplify it by replacing resistors connected in series or parallel with equivalent single resistances?

10. Commercial voltmeters will often be specified as "20,000 ohms per volt full scale" or "1000 ohms per volt." What does this mean? Which of these two is the "better" voltmeter?

11. Two wires, *A* and *B*, have the same physical dimensions but are made of different materials. If *A* has twice the resistance of *B*, how do their resistivities compare?

12. For a given source of emf, which remains constant, will more heat develop in a large external resistance connected across it or a small one. What about a real source of emf, which will have some nonzero internal resistance?

13. Would you always want to use the most sensitive possible galvanometer as a basis for making ammeters and voltmeters with various scales?

Figure 20-2

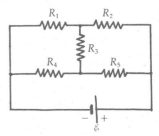

Answers

1. No, the only "direction" that current has is one way or the other. To call such an essentially one-dimensional quantity a vector would have no meaning.

2. There is no contradiction. The electric field in a conductor must be zero if the conductor is in electrostatic equilibrium. A conductor in which charges are moving is clearly not in equilibrium.

3. The energy is dissipated as heat in the resistance of the wire. When an electron collides with something, it "forgets which way it was going" since the random thermal motions of electrons and ions in the wire are so much larger than the drift speed. Thus, when it collides, the extra kinetic energy the electron has gained from the field becomes a contribution to the random, disorganized thermal motions in the material. As far as being accelerated by the field is concerned, the electron starts over again after each collision.

4. Just calculating the resistance of the filament gives 190 Ω when

120 V is applied and 35 Ω when 3 V is applied. Thus, the behavior of the filament "from the outside" is nonohmic. In fact, the filament is made from tungsten, an ordinary metal that is ohmic. What's happening is that, because the filament is in a vacuum, increasing the current greatly increases its temperature. With 3 V across it, the filament is at around 700 K; applying 120 V raises its temperature to nearly 2800 K. The resistance of tungsten, like that of most conductors, increases with increasing temperature.

5.   The electric field, which is what sets the electrons moving, is set up along the whole length of the wire nearly instantaneously, so current flow in the light bulb begins almost at once.

6.   This depends on what you mean. The time to charge it to a given fraction of its steady-state voltage and charge depends only on the time constant RC. The time to charge it partially to a given potential—to 10 V, say—will be shorter if a larger emf is applied.

7.   Because energy is stored in the electric field set up in the capacitor, to charge its plates requires that work be done. The resistor in series with the battery limits the rate at which the battery can supply the needed energy, so the charging requires a finite time.

8.   The bulbs in the old series string were designed to operate with around 110 V/8 $\approx$ 14 V across each bulb. The modern parallel connection puts the full 110 volts across each one. In a modern string, the current in our old-style bulb would be much more than its filament was designed for, and it would burn out at once, possibly spectacularly.

9.   No, there are no series or parallel resistor combinations in this circuit.

10. A voltmeter specified as "1000 ohms per volt" for example, is one that draws a current of

$$\frac{1 \text{ V}}{1000 \ \Omega} = 0.0001 \text{ A} = 1 \text{ mA}$$

when its meter movement indicates full scale. A good voltmeter should draw as little current as possible, so 20,000 Ω/V (which requires a more sensitive meter movement) is better.

11. Since resistance is directly proportional to resistivity, A has twice the resistivity of B.

12. If the terminal voltage remains constant—that is, if the internal resistance of the source of emf is negligible—then more power will be delivered to and more heat will be developed in the smaller resistor. If the source's internal resistance is not zero, the situation is more complicated, because the terminal voltage will decrease as current is drawn from the source.

13. In any given application, the more sensitive meter will disturb the circuit being measured less, but it may also be easier to destroy if you do something dumb when you wire it into the circuit.

V. Problems and Solutions

Problems

1.   A piece of 14-gauge copper wire (0.163 cm in diameter) is 14 m long. (*a*) What is its resistance? (*b*) If a potential difference of 1 V is applied across the wire, what current flows in it? (*c*) What is the electric field in the wire? The resistivity of copper is 1.7×10^{-8} $\Omega \cdot$m.

How to Solve It
- Calculate the resistance of the wire.

- Ohm's law can be used to find the current in the wire.

- The electric field in the wire is the voltage drop per unit length.

2.   A power transmission line is made of copper 1.8 cm in diameter. If the resistivity of copper is 1.7×10^{-8} $\Omega \cdot$m, find the resistance of a mile of this line.

How to Solve It
- Convert the length and radius of the line to metres.

- Calculate the cross-sectional area of the line.

- Calculate the resistance directly from the formula relating it to resistivity, length, and area.

3.   An electric heater consists of a single resistor connected across a 110-V line. It is used to heat 200 g of water in a cup (to make instant coffee) from 20°C to 90°C in 2.7 min. Assuming that 90 percent of the energy drawn from the power source goes into heating the water, what is the resistance of the heater?

How to Solve It

● Find the heat in calories required to raise the temperature of the water by the given amount and convert it to joules.

● This is 90 percent of the heat input from the heater in 2.7 min. Use this information to find the heater power in watts.

● From the supply voltage and the power input, calculate the resistance of the heater.

4. For a silver wire 0.1 inch in diameter and 100 feet long carrying a current of 25 A, find (*a*) the resistance, (*b*) the potential difference between the ends of the wire, (*c*) the electric field in it, and (*d*) the rate at which heat is being generated in the wire. The resistivity of silver is 1.6×10^{-8} $\Omega \cdot$m.

How to Solve It

● Calculate the resistance of the wire. Watch out for units—you need to convert everything to SI units before plugging valves into a formula.

● Use Ohm's law to get the potential difference across the wire and, from this, the electric field.

● The voltage drop across a resistor multiplied by the current flowing through it gives the power dissipated.

5. A battery has a terminal voltage of 11.6 V when it is delivering 30 A to the starter motor of a car. Under different conditions, it delivers 80 A and its terminal voltage is 10.7 V. Find the internal resistance and the emf of the battery.

How to Solve It

● Write an expression for the terminal voltage of the battery in terms of its emf and internal resistance and the current being drawn from it.

● Substitute in the numbers given for each of the two cases.

● The two equations you get can be solved simultaneously for the internal resistance and the emf of the battery.

6. A potential difference of 3.6 V is applied between points *a* and *b* in the circuit shown in Figure 20-3. Find the current in each of the resistors and the total current drawn from the power source.

Figure 20-3

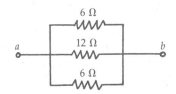

How to Solve It

● Since the resistors are in parallel, the same potential difference appears across each one.

● Use Ohm's law to get the current through each resistor.

● The total current for resistors in parallel is just the sum of the three individual currents.

7. A potential difference of 7.5 V is applied between points *a* and *c* in the circuit shown in Figure 20-4. Find the difference in potential between points *b* and *c*.

Figure 20-4

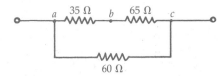

How to Solve It

● Find the effective resistance of the series combination of two resistors.

● Use Ohm's law to find the current through these two resistors.

● Use Ohm's law again to find the voltage drop across the 65-Ω resistor, which is between points *b* and *c*.

● Notice that you don't need to know the value of the third (60-Ω) resistance.

8. In the circuit of Figure 20-5, a potential difference of 5 V is applied between points *a* and *b*. Find (*a*) the equivalent total resistance, (*b*) the current in each resistor, and (*c*) the power being dissipated in each resistor.

Figure 20-5

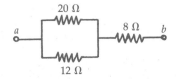

How to Solve It

● First, find the equivalent resistance of the two parallel resistors; then find the equivalent resistance of this in series with the 8-Ω resistor.

● Ohm's law then tells you the total current that flows from *a* to *b*, and this is the current in the 8-Ω resistor.

● This current divides between the two parallel resistors such that the voltage drop across each of them is the same.

● The power dissipated in a resistor carrying current I is I^2R.

9. In the circuit shown in Figure 20-6, the internal resistance of the batteries is negligible. Find (*a*) the current in each resistor and (*b*) the potential difference between points *a* and *b*.

Figure 20-6

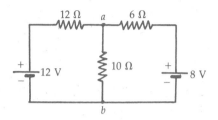

How to Solve It

● There are three different branches in the circuit, so you need three equations to solve for the currents in them.

● Kirchhoff's voltage rule gives you two independent equations, one for each of the two loops. Kirchhoff's current rule, applied at either junction, gives you a third.

● Solve the three equations simultaneously for the three branch currents.

● Apply Ohm's law to the 10-Ω resistor to get the potential difference between *a* and *b*.

10. An old car battery has an emf of 12.0 V and an internal resistance of 0.2 Ω. Assume the starter motor across it is a load of 0.6 Ω (see Figure 20-7). (*a*) Find the current delivered to the load. (*b*) A new battery with an emf of 12.4 V and an internal resistance of 0.01 Ω is connected in parallel with the old battery and the starter motor to give it a boost. Now what current is being delivered to the starter motor? (*c*) What power is being delivered to it? (*d*) What power is being drawn from each battery?

Figure 20-7

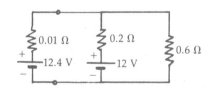

How to Solve It

● The old battery is driving the series combination of its internal resistance and the starter motor. Find the current drawn from it from Ohm's law, and then use the current to find the terminal voltage.

● When the new battery is connected in parallel with the old one, the resulting circuit is identical, except for the numbers, with that in Problem 9; solve it in the same way.

● The power being drawn from each battery is the terminal voltage multiplied by the current drawn from the positive terminal. Be careful with signs!

11. In the circuit shown in Figure 20-8 on page 280, find (*a*) the current in each resistor, (*b*) the power supplied by each source of emf, and (*c*) the power dissipated in each resistor.

Figure 20-8

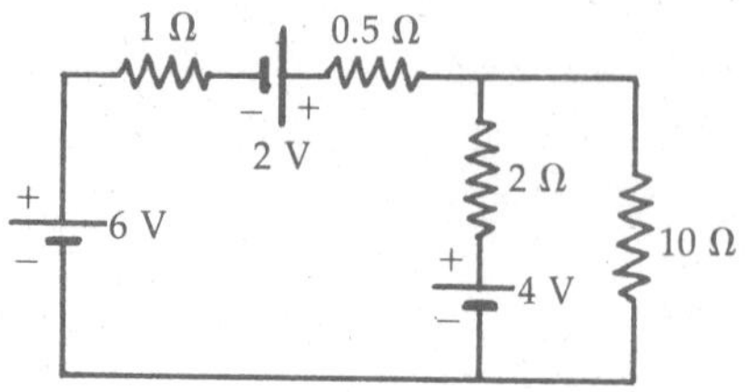

1 Ω   0.5 Ω
2 V
6 V
2 Ω
4 V
10 Ω

How to Solve It
● There are three different branches in the circuit, so you need three equations to solve for the currents in them.

● Kirchhoff's voltage rule gives you two independent equations, one from each of the two loops, and Kirchhoff's current rule, applied at either junction, provides a third.

● Solve the three equations simultaneously for the three branch currents.

● Knowing the currents, you can calculate the power supplied by each battery and that dissipated in each resistance. (By the conservation of energy, the total power supplied ought to equal the total power dissipated.)

12. Find the current in each of the three branches of the circuit sketched in Figure 20-9.

How to Solve It
● This is almost the same circuit as in Problems 9 and 11, except for the numbers; solve it for the three branch currents in just the same way.

● Be careful with signs!

Figure 20-9

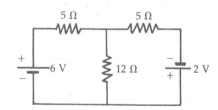

5 Ω   5 Ω
6 V   12 Ω   2 V

13. A 12-V battery with an internal resistance of 0.6 Ω is used to charge an 0.2-μF capacitor through a 5-Ω resistor. Find (a) the initial current drawn from the battery, (b) the final charge on the capacitor, and (c) the time constant of the circuit.

How to Solve It
● Note that the equivalent resistance of this circuit is the series combination of the external resistor and the internal resistance of the battery.

● Initially, the capacitor is uncharged, so at first the 12 V appears across the resistance in the circuit, and Ohm's law gives the initial current.

- When the capacitor is fully charged, all the 12-V potential drop appears across it.

- The resistance of the *RC* circuit times its capacitance gives you the time constant.

14. A 10-µF capacitor carries an initial charge of 80 µC. (*a*) If a resistance of 25 Ω is connected across it, what is the initial current that flows in the resistor? (*b*) What is the time constant of the circuit?

How to Solve It
- Calculate the voltage across the capacitor when it is fully charged.

- When the resistance is connected across this capacitor, this same voltage initially appears across the resistor; get the resulting current from Ohm's law. Of course, the current decreases as the capacitor discharges.

- Resistance times capacitance gives you the time constant.

15. In a certain galvanometer, full-scale deflection corresponds to a current of 0.2 mA. The galvanometer's internal resistance is 50 Ω. (*a*) What resistance must be placed in parallel with it to make an ammeter that reads 3 mA full scale? (*b*) What resistance must be placed in series with it to make a voltmeter that reads 1 V full scale?

How to Solve It
- In part (*a*), the combination of the galvanometer resistance and the parallel resistor must pass a current of 3 mA when current in the galvanometer branch itself is 0.2 mA.

- In part (*b*), the combination of the galvanometer resistance and the external series resistor must be carrying a current of 0.2 mA for a full-scale reading when the potential drop across it is 1 V.

16. The meter *M* in Figure 20-10 is calibrated with readings from 0.0 to 1.0, but no units are given. When resistor *R* is 10 Ω, the meter reads 0.70; when *R* is 20 Ω, the meter reads 0.42. Find the internal resistance of the meter and the current required for full-scale deflection.

Figure 20-10

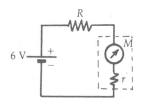

How to Solve It
- If I_{max} is the current corresponding to full-scale deflection, then the series combination of 10 Ω and the meter resistance draws a current of $0.70I_{max}$ from the 6-V battery. Express this as an equation.

- The series combination of 20 Ω and the meter resistance draws a current of $0.42I_{max}$. Express this as a second equation.

- Eliminate I_{max} from these two equations to get the internal resistance r of the meter.

- Substitute the value of r into one of the two equations to get I_{max}.

Solutions

1. (a) The resistance of the wire is given by

$$R = \rho \frac{L}{A} = (1.7 \times 10^{-8}\ \Omega\cdot m)\ \frac{14\ m}{\frac{\pi}{4}\ (1.63 \times 10^{-3}\ m)^2}$$

$$= \underline{0.114\ \Omega}$$

(b) By Ohm's law, the current is

$$I = \frac{V}{R} = \frac{1\ V}{0.114\ \Omega} = \underline{8.77\ A}$$

(c) The electric field in the wire is

$$E = \frac{V}{L} = \frac{1\ V}{14\ m} = \underline{0.071\ V/m}$$

Notice that only a very small field is required to drive a substantial current through the wire.

2. $\underline{0.108\ \Omega}$

3. To raise the temperature of 200 g of water by 70 C° (assuming there are negligible heat losses) requires heat equal to

$$Q = mc\ \Delta T = (200\ g)(1.00\ cal/g\cdot C°)(70\ C°)$$

$$= 1.4 \times 10^4\ cal$$

$$= (1.4 \times 10^4\ cal)(4.184\ J/cal) = 5.86 \times 10^4\ J$$

This is only 90 percent of the energy drawn from the power source, so the energy delivered to the heater is

$$\frac{5.86 \times 10^4\ J}{0.90} = 6.51 \times 10^4\ J$$

If this energy is delivered in 2.7 min (60 s/min) = 162 s, the power being delivered is

$$P = \frac{E}{t} = \frac{6.51 \times 10^4 \text{ J}}{162 \text{ s}} = 402 \text{ W}$$

The power P dissipated in a resistor is V^2/R, so

$$R = \frac{V^2}{P} = \frac{(110 \text{ V})^2}{402 \text{ W}} = \underline{30.1 \text{ }\Omega}$$

4.   (a) $\underline{0.0962 \text{ }\Omega}$ (b) $\underline{2.41 \text{ V}}$ (c) $\underline{0.0789 \text{ N/C}}$ (d) $\underline{60.2 \text{ W}}$

5.   In the circuit sketched in Figure 20-11, $\mathcal{E}$ is the emf, and r is the internal resistance of the storage battery, and R is the external load (the starter motor). From the circuit, the terminal voltage is

$$V_T = \mathcal{E} - Ir$$

so in the first case

$$11.6 \text{ V} = \mathcal{E} - (30 \text{ A})(r)$$

and in the second

$$10.7 \text{ V} = \mathcal{E} - (80 \text{ A})(r)$$

Subtracting the two equations gives $0.9 \text{ V} = (50 \text{ A})(r)$ so

$$r = \underline{0.018 \text{ }\Omega}$$

Substituting this value back into the first equation gives

$$11.6 \text{ V} = \mathcal{E} - (30 \text{ A})(0.018 \text{ }\Omega)$$

and

$$\mathcal{E} = \underline{12.1 \text{ V}}$$

Figure 20-11

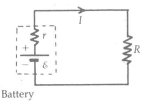

Battery

6.   Reading from the top, the resistors carry $\underline{0.6 \text{ A}, 0.3 \text{ A}, \text{ and}}$ $\underline{0.6 \text{ A}}$, so the total current is $\underline{1.5 \text{ A}}$.

7.   The 35-Ω and 65-Ω resistors connected in series have an equivalent resistance of $35 \text{ }\Omega + 65 \text{ }\Omega = 100 \text{ }\Omega$. Since there is 7.5 V across these two resistors, the current in them is

$$I = \frac{V}{R} = \frac{7.5 \text{ V}}{100 \text{ }\Omega} = 0.075 \text{ A}$$

The potential difference across the 65-Ω resistor is

$$V = IR = (0.075 \text{ A})(65 \text{ } \Omega) = \underline{4.88 \text{ V}}$$

8.   (a) $\underline{15.5 \text{ } \Omega}$ (b) $\underline{0.323 \text{ A}}$ in the 8-Ω resistor, $\underline{0.121 \text{ A}}$ in the 20-Ω resistor, and $\underline{0.202 \text{ A}}$ in the 12-Ω resistor   (c) $\underline{0.835 \text{ W}}$, $\underline{0.293 \text{ W}}$, $\underline{0.490 \text{ W}}$, respectively

9.   The first step in solving a Kirchhoff's-rules problem is to label each current-carrying branch and guess a direction for the current. This has been done in Figure 20-12. We can get two equations using Kirchhoff's voltage rule, which says that the net potential drop around any closed loop in the circuit must be zero. Starting from b and going clockwise around the left-hand loop gives

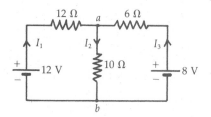

Figure 20-12

$$+12 \text{ V} - (12 \text{ } \Omega)(I_1) - (10 \text{ } \Omega)(I_2) = 0 \qquad (1)$$

and going clockwise around the right-hand loop gives

$$+(10 \text{ } \Omega)(I_2) + (6 \text{ } \Omega)(I_3) - 8 \text{ V} = 0 \qquad (2)$$

Simplifying these two equations a little bit, we get

$$1.2I_1 + I_2 = 1.2 \text{ A} \qquad (1a)$$

$$I_2 + 0.6I_3 = 0.8 \text{ A} \qquad (2a)$$

Applying Kirchhoff's current rule at point a gives $I_1 + I_3 = I_2$, so

$$I_3 = I_2 - I_1 \qquad (3)$$

We use this equation to substitute for I_3 in Equation (2a):

$$I_2 + (0.6)(I_2 - I_1) = 0.8 \text{ A}$$

$$1.6I_2 - 0.6I_1 = 0.8 \text{ A}$$

$$I_2 - 0.375I_1 = 0.5 \text{ A} \qquad (2b)$$

If we subtract Equation (2b) from Equation (1a), the I_2's cancel out:

$$1.575I_1 = 0.7 \text{ A}$$

so

$$I_1 = \underline{0.444 \text{ A}}$$

Substituting this back into Equation (2b), we get

$$I_2 - (0.375)(0.444 \text{ A}) = 0.5 \text{ A}$$

and

$$I_2 = \underline{0.667 \text{ A}}$$

Finally, from Equation (3)

$$I_3 = I_2 - I_1 = 0.667 \text{ A} - 0.444 \text{ A} = \underline{0.223 \text{ A}}$$

(b) The potential difference between a and b is

$$V_{ab} = (10 \ \Omega)I_2 = (10 \ \Omega)(0.667 \text{ A}) = \underline{6.67 \text{ V}}$$

10. (a) $\underline{15.0 \text{ V}}$ (b) Of the 21.25 A from the new battery, $\underline{20.31 \text{ A}}$ goes to the starter motor and $\underline{0.94 \text{ A}}$ is shoved backwards through the old battery, charging it.   (c) $\underline{247 \text{ W}}$ (d) $\underline{259 \text{ W}}$ from the new battery; $\underline{\text{none}}$ from the old, since 11.4 W is being delivered to it.

11. (a) The circuit is redrawn in Figure 20-13 with the branch currents labeled. Two Kirchhoff voltage (loop) equations and one current (junction) equation can be written. These are

Figure 20-13

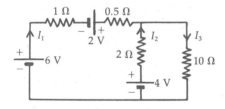

$$6 \text{ V} - (I_1)(1 \ \Omega) + 2 \text{ V} - (I_1)(0.5 \ \Omega) + (I_2)(2 \ \Omega) - 4 \text{ V} = 0 \qquad (1)$$

$$4 \text{ V} - (2 \ \Omega)(I_2) - (10 \ \Omega)(I_3) = 0 \qquad (2)$$

$$I_1 + I_2 - I_3 = 0 \qquad (3)$$

Simplifying Equations (1) and (2), we get

$$3I_1 - 4I_2 = 8 \text{ A} \qquad (1a)$$

$$I_2 + 5I_3 = 2 \text{ A} \qquad (2a)$$

We can eliminate I_3 between Equations (2a) and (3):

$$I_2 + 5(I_1 + I_2) = 5I_1 + 6I_2 = 2 \text{ A} \qquad (4)$$

Solving Equations (1a) and (4), we get $I_1 = \underline{1.474 \text{ A}}$ and $I_2 = -\underline{0.895 \text{ A}}$; thus, from Equation (3), $I_3 = \underline{0.579 \text{ A}}$. The minus sign means that current I_2 is actually in the opposite direction from that indicated in Figure 20-8.

(b) The power supplied by the 6-V battery is

$$P = VI = (6 \text{ V})(1.474 \text{ A}) = \underline{8.84 \text{ W}}$$

The 2-V battery supplies $(2 \text{ V})(1.474 \text{ A}) = \underline{2.95 \text{ W}}$, and the 4-V battery supplies $(4 \text{ V})(-0.895 \text{ A}) = \underline{-3.58 \text{ W}}$. What the minus sign means is that current is being driven backward through the 4-V battery. If it isn't a storage battery, this power is being dissipated as heat.

(c) The power dissipated in the 1-Ω resistor is

$$P = I^2R = (1.474 \text{ A})^2(1 \text{ Ω}) = \underline{2.17 \text{ W}}$$

In the 0.5-Ω resistor, its $(1.474 \text{ A})^2(0.5 \text{ Ω}) = \underline{1.09 \text{ W}}$; in the 2-Ω resistor, its $(-0.895 \text{ A})^2(2 \text{ Ω}) = \underline{1.60 \text{ W}}$; and in the 10-Ω resistor, it's $(0.579 \text{ A}) (10^2 \text{Ω}) = \underline{3.35 \text{ W}}$. Notice that the total power supplied and the total dissipated do agree—8.21 W.

12. From left to right, $\underline{0.869 \text{ A}}$, $\underline{0.138 \text{ A}}$, and $\underline{0.731 \text{ A}}$.

13. The circuit is sketched in Figure 20-14. Notice that the internal resistance of the voltage source is in series with the external 5-Ω resistor, so effectively, the capacitor is being charged through a resistance of 5.6 Ω.

Figure 20-14

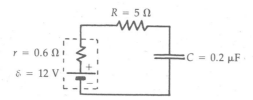

(a) The capacitor is initially uncharged, so the instant the circuit is closed, all the emf of the source appears across the resistance, and

$$I = \frac{V}{R} = \frac{12 \text{ V}}{5.6 \text{ Ω}} = \underline{2.14 \text{ A}}$$

(b) The capacitor charges until the entire potential difference of 12 V appears across it and no more current flows. Thus, after a long time, the charge on the capacitor is

$$Q = CV = (2 \times 10^{-7} \text{ F})(12 \text{ V}) = 2.4 \times 10^{-6} \text{ C} = \underline{2.4 \text{ μC}}$$

(c) The time constant of the circuit is

$$RC = (5.6 \text{ Ω})(2.0 \times 10^{-7} \text{ F}) = \underline{1.12 \times 10^{-6} \text{ s}}$$

14.   (a) $\underline{0.320\ \text{A}}$ (b) $\underline{2.5 \times 10^{-4}\ \text{s}}$

15.   (a) The potential difference across the galvanometer when it is reading full scale is

$$V = IR = (2 \times 10^{-4}\ \text{A})(50\ \Omega) = 0.01\ \text{V}$$

Thus, we want the shunted meter—the ammeter—to be drawing 3 mA when there is a voltage of 0.01 V across it, so the combination of the galvanometer and the parallel resistor (see Figure 20-15) must have an equivalent resistance of

Figure 20-15

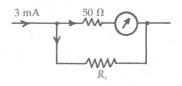

$$R_{\text{eq}} = \frac{V}{I} = \frac{0.01\ \text{V}}{3 \times 10^{-3}\ \text{A}} = 3.33\ \Omega$$

The galvanometer's internal resistance $r = 50\ \Omega$, so

$$\frac{1}{R_{\text{eq}}} = \frac{1}{r} + \frac{1}{R_s}$$

$$\frac{1}{3.33\ \Omega} = \frac{1}{50\ \Omega} + \frac{1}{R_s}$$

From which

$$\frac{1}{R_s} = 0.280\ \Omega^{-1} \quad \text{and} \quad R_s = \underline{3.57\ \Omega}$$

(b) To make a voltmeter, a series resistor is used so that when there is 1 V across the series circuit, 0.2 mA (the full-scale deflection current of the galvanometer) flows in it (see Figure 20-16). The equivalent resistance is $r + R$, so

Figure 20-16

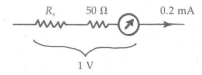

$$R_{\text{eq}}I = V$$

$$(50\ \Omega + R)(2 \times 10^{-4}\ \text{A}) = 1\ \text{V}$$

$$R = 50\ \Omega = 5000\ \Omega$$

$$R = \underline{4950\ \Omega}$$

16. $r = \underline{5\ \Omega}$, $I_{\text{max}} = \underline{0.571\ \text{A}}$

CHAPTER 21

The Magnetic Field

I. Key Ideas

Magnetism Magnetism is a fundamental property of matter; naturally occurring magnetic force has been known since ancient times. Every magnet has two points called *poles* where the magnetic force is strongest. The poles are designated "north" and "south" because a magnet suspended freely will orient itself in approximately a north-south direction; the earth itself is a magnet. Like poles of two magnets repel each other and opposite poles attract with a force that is inversely proportional to the square of the distance between them. Magnetic poles exist only in equal and opposite pairs, however; a single pole cannot be isolated. The source of the magnetic field is not, in fact, a "magnetic charge" but electric charge in motion, that is, electric current.

The Magnetic Field That there is a magnetic field at a point can be demonstrated by suspending a compass needle there; the needle will align itself with the magnetic field, with its north pole in the direction of the field. The *magnetic field* **B** is defined in terms of its effect on a moving point charge. The magnetic force on a charge is proportional to the charge, its speed, the magnetic field, and the sine of the angle between the field direction and that of the motion of the charge. Thus, there is no force on a charge moving along the field direction. The direction of the magnetic force is perpendicular to *both* the velocity of the charge and the direction of the field, according to the right-hand rule (see Figure 21-1).

Figure 21-1

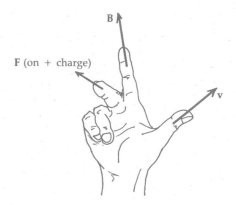

Force on a Current The magnetic force on a current-carrying wire in a magnetic field is the sum of the forces on all the moving charge carriers in the wire. The force on a current "element" is perpendicular to both the magnetic field and the current direction.

Force on a Magnet A permanent magnet consists of equal and opposite magnetic poles; it is the equivalent of equal and opposite magnetic point "charges" separated by a distance d: a dipole. (A north pole is considered positive.) In a *uniform* magnetic field, there is zero net force on the magnet, but there is a *torque* on it that tends to align the dipole along the field direction. The torque is given by the magnetic (dipole) moment times the component of the magnetic field perpendicular to it.

Current Loop A small current loop in a magnetic field acts much like a little bar magnet: there is a torque but no net force on it in a uniform magnetic field that tends to align the magnetic moment of the loop with the field. The magnetic moment of the loop is the current in it multiplied by its area. The torque on a current loop is basic to the functioning of a galvanometer.

Magnetic Field Map Just as we did for the electric field, we can map a magnetic field by drawing lines that, at every point, are in the direction of **B**. Such a diagram has many of the same properties as an electric field diagram, although the lines of the magnetic field do *not* give the direction of the magnetic force on a charge. Magnetic field lines leave the north (+) pole of a magnet and enter the south (−) pole.

Magnetic Force on a Moving Point Charge No work is done on a point charge moving in a magnetic field because the force is perpendicular to the direction of motion; thus, the magnetic force changes only the direction of the velocity. A particle moving perpendicular to a magnetic field tends to move in a circle around the field direction. If the velocity of the particle also has a component along the field, this component is unaffected, and the particle's path will be a helix wrapped around the field direction. Perpendicular magnetic and electric fields can exert zero net force on a particle with a particular velocity determined by the magnitudes of the fields; such "crossed fields" function as a *velocity selector*. Observations of the trajectories of electrons in electric and magnetic fields made possible the first measurements of electron properties.

Cyclotron Frequency The frequency of the circular motion of a charged particle in a magnetic field

is independent of the particle's speed or the radius of its orbit. A *cyclotron* is a particle accelerator that is based on this constant frequency; it uses an alternating high voltage to accelerate particles repeatedly as they circulate in a uniform magnetic field.

The Sources of Magnetism Isolated magnetic poles are not known to exist, although there is no reason why they should not. The source of magnetism is electric current; microscopic circulating currents are the source of natural magnetism. Thus, the lines of **B** are closed curves without beginning or end. The magnetic field at a point due to an "element of current" is given by the *Biot-Savart law*. The field is perpendicular both to the direction of the current element and to the position vector from the current element to the field point, as indicated in Figure 21-2. The Biot-Savart law is an inverse-square force law analogous to Coulomb's law for electric charge. The field due to an extended circuit is found by summing the contributions of all the current elements; almost always calculus is needed to do this.

Figure 21-2

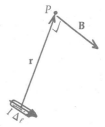

Magnetic Force Between Parallel Wires Parallel currents exert magnetic forces on each other; currents in the same direction attract each other; currents in opposite directions repel. It is on this phenomenon that the definition of the ampere—from which other electrical and magnetic units are derived—is based.

Ampere's Law An equation stating a general property of the magnetic field, somewhat like

Gauss' law for the electric field, *Ampere's law* relates the tangential components of the magnetic field summed around a closed path to the current through the path. In a few special cases where there is very high symmetry, Ampere's law can be used to calculate the field.

Magnetic Field of a Solenoid The magnetic field due to a current loop is like that of a small magnet. A *solenoid* is a closely wound helical coil; it can be treated as many coaxial current loops placed side by side. The magnetic field outside a solenoid is the same as that of a bar magnet of the same shape. The field inside a long solenoid is essentially uniform, except near the ends; the field outside is very much weaker. The field of an actual bar magnet can be considered to be due to an *amperian current* on its surface, which is equivalent to the resultant of the aligned atomic current loops within the magnet.

Magnetism in Matter In some materials, an external magnetic field tends to align atomic magnetic moments, which increases the external field; this is *paramagnetism*. In some materials, below a critical "Curie" temperature, a large-scale alignment occurs that remains after the external field is removed; this is *ferromagnetism*. For materials without a permanent magnetic moment, an external field induces a small magnetic moment tending to reduce the field; this is *diamagnetism*.

II. Numbers and Key Equations

Numbers

1 tesla (T) = 1 N/A·m

1 gauss (G) = 10^{-4} T

Permeability of free space

$$\mu_0 = 4\pi \times 10^{-7} \text{ T·m/A}$$

$$= 4\pi \times 10^{-7} \text{ N/A}^2$$

Key Equations

Magnetic force on a moving charge

$$F = qvB \sin \theta$$

$$= qv_\perp B = qvB_\perp$$

Magnetic force on a segment of straight, current-carrying wire

$$F = I\ell B \sin \theta = I\ell B_\perp$$

Magnetic force on a current element

$$F = I \Delta\ell B \sin \theta = I \Delta\ell B_\perp$$

Magnetic pole strength

$$q_m = \frac{F}{B}$$

Magnetic moment

$$\mathbf{m} = q_m\boldsymbol{\ell}$$

Torque on a magnetic moment

$$\tau = mB \sin \theta$$

Magnetic moment of a current-carrying coil

$$m = NIA$$

Charged-particle orbit in a magnetic field

$$r = \frac{mv}{qB}$$

Cyclotron frequency

$$f = \frac{qB}{2\pi m}$$

(*m* here is mass, not magnetic moment)

Velocity "selected" in crossed *E* and *B* fields

$$v = \frac{E}{B}$$

Biot-Savart law

$$\Delta B = \frac{\mu_0}{4\pi} \frac{I \, \Delta \ell \sin \theta}{r^2}$$

Magnetic field at the center of a current loop

$$B = \frac{\mu_0 I}{2r}$$

Magnetic field due to a long, straight wire

$$B = \frac{\mu_0 I}{2\pi r}$$

Magnetic force per unit length between parallel currents

$$\frac{F}{\Delta \ell} = \frac{\mu_0}{2\pi} \frac{I_1 I_2}{r}$$

Speed of electromagnetic waves

$$c = (\mu_0 \epsilon_0)^{-1/2}$$

Ampere's law

$$\Sigma B_t \, \Delta \ell = \mu_0 I_{\text{total}}$$

Magnetic field inside a solenoid

$$B = \mu_0 \left(\frac{N}{\ell}\right) I$$

Magnetic field in a material

$$B = B_0 + \chi_m B_0$$

III.   Possible Pitfalls

The electric and magnetic fields behave in ways that are fundamentally different; don't confuse them. The electric field is always in the direction of the force it exerts on a charged particle, but the magnetic field is always *perpendicular* to the direction of the force that it exerts.

Remember that magnetic field lines are in the direction of **B**, *not* that of the force on an electric charge. (There is no way to map the latter as the force depends on the direction of motion of the charge.)

The magnetic force on a moving electric charge is *perpendicular* to both the field direction and the direction of motion of the charge.

The path of a moving charged particle is a circle around the direction of the magnetic field *only* in the special case in which its motion is perpendicular to the field. The general motion of a charge is in a helical path around the field direction as an axis.

The magnetic force acts only on charges that have a component of velocity perpendicular to the magnetic field. A charge at rest or one moving along the field direction experiences no force.

The magnetic force between parallel currents is attractive if the currents are in the same direction, repulsive if they are in opposite directions. This might be opposite from the way you would expect them to behave if you reason from the behavior of point charges. Watch out!

The effect of an external applied magnetic field on matter can be either so as to augment the external magnetic field (paramagnetism) or to oppose it (diamagnetism). As a result, magnetic susceptibility can be either positive or negative. An electric field, by contrast, always polarizes matter such that the polarization field opposes the applied field.

There is no force on a closed current loop in a uniform magnetic field, only a torque. There would be a nonzero net force on the loop in a *nonuniform* field.

IV.   Questions and Answers

Questions

1.   Both electric and magnetic fields exert a force on a moving charge. In a particular case, how could you tell whether it is an electric or a magnetic force that is causing a moving charge to deviate from a straight-line path?

2.   A magnetic field is applied to a current-carrying wire, but no force acts on the wire. How can this be?

3.   A velocity selector consists of crossed electric and magnetic fields, with the **B** field directed straight up. A beam of positively charged particles passing through it from left to right is undeflected by the fields. (*a*) In what direction is the electric field? (*b*) The particle beam is reversed so that it passes from right to left. Is it deflected? If so, in what direction? (*c*) A beam of electrons (negatively charged) of the same velocity is passed through from left to right. Is it deflected? If so, in what direction?

4.   The magnetic susceptibility of a material can be either positive or negative. Why?

5.   A bar magnet is effectively a magnetic dipole, and its behavior in a magnetic field is analogous to that of an electric dipole in an electric field. To continue the analogy, what is the magnetic counterpart of a point charge?

6.   If there were such a thing as an isolated magnetic charge (pole), what would its magnetic field be like? What effect do you suppose it would have on a magnet?

7.   We draw lines of force for the electric field such that the direction of a line at any point is the direction of the force on a point charge at that point. There's an obvious visual convenience to this convention. Why can't lines of the magnetic field be mapped in the same way?

Figure 21-3

8.   The bending of the paths of charged particles in a magnetic field is used to help identify the particles seen in high-energy particle detectors such as cloud and bubble chambers. The tracks typically look like the sketch in Figure 21-3, in which an electron is moving in the plane of the drawing in a uniform magnetic field perpendicular to the page. Why does the electron spiral inward rather than continuing in a circular path?

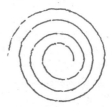

9. Accelerator physicists refer to crossed E and B fields as a velocity selector. In the same sense, the deflection of charged particles in a strong magnetic field perpendicular to their motion can be thought of as a momentum selector. How is this?

10. Two flat, square, current-carrying loops lying in the plane of the page are shown in Figure 21-4. In terms of these, relate the following statements about magnetic forces: Like poles repel, whereas opposite poles attract. Parallel currents attract each other, whereas antiparallel currents repel.

Figure 21-4

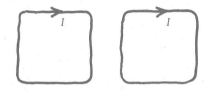

11. Think of a long solenoid of N total turns carrying current I as a helical spring. If the solenoid is stretched along its length without changing N or I, does the field inside it change? How about its magnetic moment?

12. In Figure 21-5, a mass hangs in equilibrium on the end of a spring. If a current is passed through the spring, which way does the mass start to move? If a steady current continues to flow in the spring, will the mass undergo simple harmonic motion thereafter?

Figure 21-5

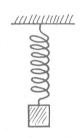

13. Consider the magnetic field due to two long, straight wires carrying parallel currents. Is the field at a point between them in their mutual plane weaker or stronger than the field due to one of the wires alone would be? What if the currents are in opposite directions?

14. The neutron, which has no charge, has a nonzero magnetic moment. Can you explain how this might be?

15. Explain why it is that a magnet will pick up an unmagnetized iron nail but not an otherwise indentical aluminum nail.

Answers

1. The simplest way I can think of is to reverse the direction in which the charge moves. If the force is magnetic, which depends on the velocity of the charge, it will reverse; if it is electric, the force will be in the same direction.

2. The magnetic field is along the direction of the wire.

3. (a) The electric field is perpendicular both to the particle velocity and to the magnetic field. (b) Yes; reversing the direction of the velocity changes the direction of the magnetic force but not that of the electric force. Thus, the two forces no longer cancel. The particles are deflected in the direction of the electric field. (c) No. Changing the sign of the charges reverses both forces, and they still cancel.

4. A positive magnetic susceptibility indicates a paramagnetic material; a negative susceptibility, a diamagnetic material.

5. The magnetic analog of a point charge would be an isolated magnetic pole (a "monopole"), although it seems that there is no such thing.

6. We would expect that its magnetic field would have the properties of the electric field due to a point electric charge. Thus, it would be radially inward or outward, decreasing in magnitude as $1/r^2$. Since the field would be nonuniform, it would attract a magnetic dipole just as the nonuniform field of an electric charge attracts an electric dipole.

7. Lines of the magnetic field are mapped this way, if we mean the force on an isolated magnetic pole; but this isn't very useful as there is apparently no such thing. There is no single direction for the magnetic force on an electric point charge since the force depends on which way the particle is going.

8. As they pass through matter, the charged particles lose energy and slow down. As their speed decreases, so does the radius of curvature of their orbit in the magnetic field.

9. Because the momentum of a charged particle—rather than its kinetic energy, say, or its speed—is the quantity that determines its path in a magnetic field. The radius of curvature of a particle of charge q in a field B is

$$R = \frac{mv}{qB}$$

10. Since the currents in the two loops are in the same sense, the magnetic moments of the two loops are parallel (see Figure 21-6). If we think of these as little bar magnets, they're aligned with north pole next to north, and south pole next to south. Thus, they repel each other. In terms of the currents, the strongest force is between the two currents that are closest together; these are antiparallel and thus repel each other.

11. The number of turns per unit length changes (it decreases) as the coil is stretched, so the magnetic field inside, which is

Figure 21-6

$B = \mu_0(N/\ell)I$, decreases. The magnetic moment, $m = NIA$, remains the same.

12. Attractive magnetic forces between adjoining turns of the spring (think of them as parallel wires) will cause the spring to compress longitudinally. The mass will be lifted up. This force will get stronger as the coils get closer together, so the spring will continue to collapse further. Thus, the motion is certainly not harmonic. This effect can be quite strong, and high-field laboratory solenoids have to be constructed very ruggedly to keep them from "imploding."

13. The magnetic field due to two parallel currents is weaker than the field due to a single current would be because the individual fields of the two wires tend to cancel each other in the region between them. If the currents are in opposite directions, their fields add in the region between them, and the resultant field is stronger than that due to either wire alone.

14. The most one can say is that this is evidence that the neutron has an internal structure of positive and negative charge in motion. Although the charges cancel, their contributions to the magnetic moment apparently do not.

15. This occurs because the magnetic moment induced by the magnet in the (ferromagnetic) iron nail is orders of magnitude larger than that induced in the (paramagnetic) aluminum nail. The force that the nonuniform magnetic field of the magnet exerts on the iron nail is correspondingly much stronger.

V.   Problems and Solutions

Problems

1.   A wire of mass 40 g slides without friction on two horizontal conducting rails spaced 0.8 m apart (see Figure 21-7 on page 296). A steady current of 100 A flows in the circuit formed by the wire

Figure 21-7

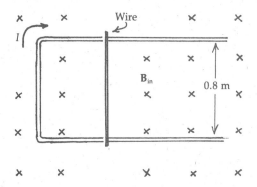

and the rails. A uniform magnetic field of 1.2 T, directed into the plane of the drawing, acts on it. (*a*) In which direction in the diagram will the wire accelerate? (*b*) What is the magnetic force on the wire? (*c*) How long must the rails be if the wire, starting from rest, is to reach a speed of 200 m/s?

How to Solve It
- The direction of the force follows directly from the right-hand rule.

- Find the magnetic force on the current-carrying wire. The rails that form the rest of the circuit are fixed, so the force on them doesn't matter.

- Calculate the acceleration of the moving wire, assuming this magnetic force is the only one that acts on it. The rest is just a kinematics problem.

2.   A straight wire 0.35 m long carrying a current of 1.4 A is in a uniform magnetic field. The wire makes an angle of 53° with the direction of the magnetic field. If the force on the wire is 0.200 N, what is the magnitude of the field?

How to Solve It
- Write down the formula for the force on a current-carrying wire of length ℓ.

- Don't forget the factor $\sin \theta$, which comes in because the wire is not perpendicular to the magnetic field.

- Solve the formula you get for *B*.

3.   There is a uniform magnetic field of magnitude 2.2 T in the positive z direction. Find the force on a particle of charge −1.2 nC if

its velocity is (*a*) 1.0 km/s in the *yz* plane in a direction that makes an angle of 40° with the *z* axis and (*b*) 1.0 km/s in the *xy* plane in a direction that makes an angle of 40° with the *x* axis.

How to Solve It
● In each part, you should draw a diagram to make sure you have the field and velocity directions right.

● The magnetic force on a positive charge is found directly from the right-hand rule, but the force on a negative charge is in the opposite direction.

● When the velocity of a moving charge is perpendicular to the magnetic field, the magnitude of the force is just *qvB*.

4. An electron moves with a velocity of 10^7 m/s in the *xy* plane at an angle of 45° to both the +*x* and +*y* axes. There is a magnetic field of 3.0 T in the +*y* direction. Find the force (magnitude and direction) on the electron.

How to Solve It
● Use the formula for the magnetic force on a moving point charge. Remember that the charge of an electron is negative.

● Don't forget the factor sin θ, which comes in because the electron's velocity is not perpendicular to the magnetic field.

5. A small permanent magnet is suspended as a compass needle that is free to rotate in a horizontal plane. It is 2 cm long and has a cross-sectional area of 0.1 cm². The local magnetic field of the earth is 0.61 G, directed 70° from the horizontal. Its horizontal component is north and south. (*a*) If the maximum torque on the horizontal needle is 6.0×10^{-5} N·m, what is its magnetic moment? (*b*) If this moment were due to a current loop having the same area as the cross-sectional area of the magnet, what would the current in the loop be?

How to Solve It
● The torque is of maximum magnitude when the magnetic moment is perpendicular to the field. This is the case when the compass needle is oriented east-west.

● The magnetic moment of a current loop is *m* = *IA*.

6. A small bar magnet, 3 in long, is placed at an angle of 40° to the direction of a uniform magnetic field of magnitude 1000 G. The

observed torque on the magnet is 0.42 N·m. What is (*a*) the magnetic moment and (*b*) the pole strength of the magnet?

How to Solve It

- Use the torque to calculate the magnetic moment of the magnet.

- The magnetic moment of the magnet is equal to the pole strength
  times its length.

- The magnet is three inches long—be careful to keep your units
  straight.

7.   A cyclotron is to use a magnetic field of 1.1 T to accelerate protons. (*a*) What must be the frequency of the accelerating voltage?
(*b*) If the radius of the magnet poles is 0.30 m, what is the maximum kinetic energy (put it in MeV) to which protons can be accelerated? (*c*) If the alternating voltage applied to the "dees" has a
maximum value of 36 kV, how many orbits must the protons make
to reach their maximum energy?

How to Solve It

- The cyclotron frequency of a given particle depends only on the
  magnetic field.

- Clearly, the radius of the proton's orbit cannot be greater than the
  radius of the magnet poles. Knowing the frequency and orbital radius gives you the velocity and thus the kinetic energy of the protons. (The mass of a proton is 1.673×10^{-27} kg).

- The protons in the cyclotron are "boosted" by the accelerating voltage twice in each orbit.

8.   A beam of 4-MeV protons from an accelerator is deflected by a
bending magnet, as shown in Figure 21-8. The radius of curvature
of the protons is 20 cm. What must the magnetic field be in the
region between the magnet poles?

Figure 21-8

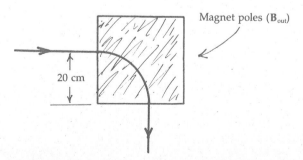

Magnet poles ($\mathbf{B}_{out}$)

20 cm

How to Solve It
- This is a straightforward application of the formula for the radius of curvature of a charged particle orbit in a magnetic field.

- Find the velocity of the protons from the given kinetic energy.

- The magnetic field can be found from the velocity and the radius of curvature.

9. A beam of protons moves in the $+z$ direction in a region of space in which there are crossed electric and magnetic fields. (*a*) If the electric field is 500 V/m and the protons move at a constant speed of 10^5 m/s, what is the magnitude of the magnetic field? (*b*) If the electric field is in the $-y$ direction, what is the direction of the magnetic field?

How to Solve It
- In order for the protons to move at a constant velocity, the magnetic and electric forces that act on a proton must be equal in magnitude.

- Use this fact to calculate the magnetic field.

- The electric and magnetic forces on a proton must be in opposite directions.

10. In Figure 21-9, two long, straight wires parallel to the x axis are at $y = \pm 2.5$ cm. Each wire carries a current of 16 A in the $+x$ direction. Find the magnetic field on the y axis at (*a*) $y = 0$, (*b*) $y = 1$ cm, and (*c*) $y = 4$ cm.

Figure 21-9

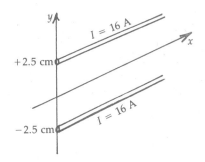

How to Solve It
- At each point, calculate the field due to each wire using the formula for the magnetic field of a long, straight current.

- The field at each point is the sum of the fields of the two wires.

- Adding magnetic fields is vector addition; be careful with directions.

11. Two straight conducting rods 1.0 m long, exactly parallel and separated by 0.85 mm, are connected by an external voltage source and a 17-Ω resistance, as shown in Figure 21-10. The 0.5-Ω rod "floats" above the 2.5-Ω rod, in equilibrium. If the mass of each rod is 25 g, what must the emf $\mathcal{E}_0$ of the voltage source be?

How to Solve It
- If the upper rod is in equilibrium, the magnetic force on it must be equal to its weight, which you can find.

Figure 21-10

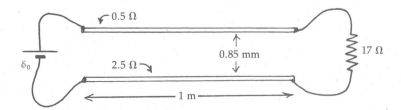

- The two rods are in series, so the same current flows in them both; from the force and the distance between them, you can find what this current is.

- The emf drives this current through the total series resistance of the circuit.

12. Three very long, straight wires are at the corners of a square of side d, as shown in Figure 21-11. The magnitudes of the currents in the three wires are the same, but the two diagonally opposite currents are directed into the page while the other one is directed outward. Find the magnetic field (magnitude and direction) at the fourth corner of the square.

Figure 21-11

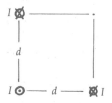

How to Solve It

- Use the formula for the field due to a long straight wire to find the contribution of each wire to the magnetic field at the fourth corner.

- Draw a diagram to show the three field contributions, and add them (as vectors, of course).

13. In Figure 21-12, find the magnetic field (magnitude and direction) at point P. The two curved sections are semicircles with radii of 10 cm and 16 cm, both centered at point P; the current in the loop is 80 A.

Figure 21-12

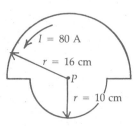

How to Solve It

- The magnetic field at P is the sum of the fields due to the two curved sections and the fields due to the two straight segments.

- The field due to each straight segment is zero from the Biot-Savart law.

- The field due to each semicircle at its center is half the field of a circular loop.

- Be careful with signs when you calculate the total field.

14. The earth's magnetic field at the equator can be taken as 0.7 G directed north. At the center of a flat circular coil of 10 turns of wire, 1.4 m in diameter, the coil's magnetic field exactly cancels the earth's field. (*a*) What must be the current in the coil? (*b*) How should the coil be oriented?

How to Solve It
- What current must flow in the coil in order to produce a magnetic field at its center equal to the earth's field?

- In order to cancel the earth's magnetic field, that of the coil must be directed south.

15. You want to wind a solenoid 3.5 cm in diameter and 16 cm long, in which the magnetic field will be 250 G when a current of 3 A flows in it. What total length of wire do you need?

How to Solve It
- From the given field, current, and length of the solenoid, you can calculate the number of turns that the solenoid must have.

- Use this and the length of a single turn to find the total length of wire you need.

Solutions

1. (*a*) The wire will accelerate in the direction of the magnetic force on it. In Figure 21-7, by the right-hand rule, this is clearly to the right.
   (*b*) The force on a wire of length ℓ, carrying current I, in a perpendicular magnetic field is

$$F = I\ell B_\perp$$

$$= (100 \text{ A})(0.8 \text{ m})(1.2 \text{ T}) = \underline{96 \text{ N}}$$

   (*c*) Thus, the acceleration of the wire is

$$a = \frac{F}{m} = \frac{96 \text{ N}}{0.040 \text{ kg}} = 2400 \text{ m/s}^2$$

It starts from rest and travels a distance s with this acceleration, so

$$v^2 = 2as$$

and the distance it must go to develop a speed of 200 m/s is

$$s = \frac{v^2}{2a} = \frac{(200 \text{ m/s})^2}{(2)(2400 \text{ m/s}^2)} = \underline{8.33 \text{ m}}$$

2. <u>0.511 T</u>

3. (*a*) Figure 21-13*a*, shows **v** and **B**. According to the right-hand rule, the magnetic force on a positive charge would be out of the page, so that on the negative charge here is into the page—that is, in the $-x$ direction. The magnitude of the force is

$$F = qvB \sin \theta$$
$$= (1.2 \times 10^{-9} \text{ C})(10^3 \text{ m/s})(2.2 \text{ T}) \sin 40°$$
$$= \underline{1.70 \times 10^{-6} \text{ N}}$$

(*b*) In Figure 21-13*b*, the field **B** is out of the page in the $+z$ direction. It is thus perpendicular to **v**, so the magnitude of the force is

$$F = qvB_\perp$$
$$= (1.2 \times 10^{-9} \text{ C})(10^3 \text{ m/s})(2.2 \text{ T}) = \underline{2.64 \times 10^{-6} \text{ N}}$$

Its direction is in the xy plane, perpendicular to **v**, as indicated in Figure 21-13*b*.

4. <u>3.40×10^{-12}N in the $-z$ direction</u>

5. (*a*) The horizontal component of **B** is to the north, so when the compass needle is oriented east-west it makes an angle of 90° with the earth's magnetic field. In this case the torque on the needle is just

$$\tau = mB$$

so

$$m = \frac{\tau}{B} = \frac{6.0 \times 10^{-5} \text{ N·m}}{0.61 \times 10^{-4} \text{ T}} = 0.984 \text{ N·m/T} = \underline{0.984 \text{ A·m}^2}$$

(*b*) The magnetic moment of a current loop is

$$m = IA$$

so the current in the loop would be

$$I = \frac{m}{A} = \frac{0.984 \text{ A·m}^2}{10^{-5} \text{ m}^2} = \underline{9.84 \times 10^4 \text{ A}}$$

6. (*a*) <u>6.53 A·m²</u>   (*b*) <u>85.7 A·m</u>

Figure 21-13

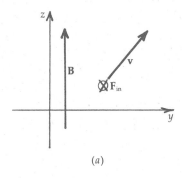

(*a*)

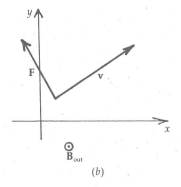

(*b*)

7.   (*a*) The accelerating voltage must have a frequency equal to the cyclotron frequency of a proton, which is

$$f = \frac{qB}{2\pi m} = \frac{(1.60 \times 10^{-19} \text{ C})(1.1 \text{ T})}{(2\pi)(1.673 \times 10^{-27} \text{ kg})} = \underline{1.67 \times 10^7 \text{ Hz}}$$

(*b*) The velocity of protons that are moving at this frequency in a circular orbit of radius 0.30 m is

$$v = 2\pi rf = (2\pi)(0.30 \text{ m})(1.67 \times 10^7 \text{ Hz}) = 3.15 \times 10^7 \text{ m/s}$$

Their kinetic energy is thus

$$E_k = \tfrac{1}{2}mv^2 = \tfrac{1}{2}(1.673 \times 10^{-27} \text{ kg})(3.15 \times 10^7 \text{ m/s})^2$$

$$= 8.30 \times 10^{-13} \text{ J}$$

$$= \frac{(8.30 \times 10^{-13} \text{ J})}{(1.60 \times 10^{-19} \text{ J/eV})} = 5.2 \times 10^6 \text{ eV} = \underline{5.2 \text{ MeV}}$$

(*c*) Each time the orbiting protons cross the gap between the "dees," they are accelerated across a potential difference of 36 kV and thus gain kinetic energy $e \, \Delta V = 36 \text{ keV} = 0.036 \text{ MeV}$. Thus, they are accelerated

$$\frac{5.2 \text{ MeV}}{0.036 \text{ MeV}} = 144 \text{ times}$$

Since this happens twice in each orbit, they must make <u>72 orbits</u>.

8.   <u>1.45 T</u>

9.   (*a*) In order for the net force on the protons to be zero, the electric and magnetic forces must be equal in magnitude. Thus,

$$qE = qvB$$

so

$$B = \frac{E}{V} = \frac{500 \text{ V/m}}{10^5 \text{ m/s}} = 0.005 \text{ T} = \underline{50 \text{ G}}$$

(*b*) The electric field and therefore the force on a proton are in the $-y$ direction; thus, the magnetic force must be in the $+y$ direction in order to cancel the electric force. To produce a force in the $+y$ direction on protons traveling in the $+z$ direction, the magnetic field must be in the <u>$+x$ direction</u>.

10.   (*a*) <u>zero</u>   (*b*) <u>1.22 G in the $-z$ direction</u>   (*c*) <u>2.63 G in the $+z$ direction</u>

11. The magnetic force on the "floating" rod must equal its weight:

$$F = mg = (0.025 \text{ kg})(9.81 \text{ m/s}^2) = 0.245 \text{ N}$$

The force per unit length of the two parallel rods on each other is

$$\frac{F}{\ell} = \frac{\mu_0}{2\pi} \frac{I^2}{r}$$

so

$$I^2 = \frac{2\pi F r}{\mu_0 \ell}$$

$$= \frac{2\pi(0.245 \text{ N})(0.85 \times 10^{-3} \text{ m})}{(4\pi \times 10^{-7} \text{ N/A}^2)(1 \text{ m})} = 1.04 \times 10^3 \text{ A}^2$$

and

$$I = 32.3 \text{ A}$$

is the current in the two rods. Since the two rods and the connecting wires are all in series, the resistance through which this current is flowing is $0.5 \ \Omega + 17 \ \Omega + 2.5 \ \Omega = 20 \ \Omega$. Thus

$$\mathcal{E}_0 = IR = (32.3 \text{ A})(20 \ \Omega) = \underline{646 \text{ V}}$$

12. The resultant field is

$$B = \frac{\mu_0 I}{2\sqrt{2}\pi d} = 0.113 \frac{\mu_0 I}{d}$$

directed downward and to the right.

13. The field at P due to each straight segment is zero because the field due to a current element at a point on the line of the current is zero. The field at P is thus just the sum of the fields due to the two semicircles. Each of these is half the field at the center of a circular loop. The field due to the 10-cm half-loop is

$$B_1 = \frac{1}{2} \frac{\mu_0 I}{2r} = \frac{\mu_0 I}{4r}$$

$$= \frac{(4\pi \times 10^{-7} \text{ T·m/A})(80 \text{ A})}{(4)(0.10 \text{ m})} = 2.51 \times 10^{-4} \text{ T} = 2.51 \text{ G}$$

and that due to the 16-cm section is $B_2 = 1.57$ G. In Figure 21-12, the current in the loop is counterclockwise, so each of these field contributions is directed out of the page; thus, they just add, and

$$B = B_1 + B_2 = 2.51 \text{ G} + 1.57 \text{ G} = \underline{4.08 \text{ G}}$$

14. (a) <u>7.80 A</u>   (b) The coil must be in a vertical plane with its axis directed north and south.

15. The magnetic field inside a long solenoid is

$$B = \mu_0 \left(\frac{N}{\ell}\right) I$$

The required number of turns is thus

$$N = \frac{B\ell}{\mu_0 I} = \frac{(250 \times 10^{-4} \text{ T})(0.16 \text{ m})}{(4\pi \times 10^{-7} \text{ N/A}^2)(3 \text{ A})} = 1060$$

The length of wire that each turn takes is

$$\pi d = (\pi)(3.5 \text{ cm}) = 11.0 \text{ cm} = 0.11 \text{ m}$$

so the total length of wire needed is $(1060)(0.11 \text{ m}) = \underline{117 \text{ m}}$.

CHAPTER 22

Magnetic Induction

I. Key Ideas

Electromagnetic Induction Electric current can be induced in matter by changing magnetic fields; this phenomenon, discovered in the early nineteenth century, is described by Faraday's law. We define *magnetic flux* as the component of the magnetic field perpendicular to the plane of a surface multiplied by the area of the surface; its definition is analogous to that for electric flux. On a magnetic-field diagram, magnetic flux is represented by the total number of lines that pass through the surface. If the magnetic flux through a surface changes in time, for whatever reason, an emf, equal in magnitude to the rate at which the magnetic flux is changing, is induced in a loop that bounds the surface. This is *Faraday's law*.

Lenz's Law *Lenz's law* states that the induced emf is always in such a direction as to oppose whatever change induced it. If this were not so, electromagnetic induction would violate the conservation of energy.

Motional emf A change in magnetic flux may be due to the motion of part of a circuit through the magnetic field; the emf induced in this case is a *motional emf*. This motional emf is induced whether or not there is a complete circuit; if there is not, free charges in the moving conductor are moved so as to create an electric field that produces an equilibrium between electric and magnetic forces.

Eddy Currents When the magnetic flux through a large piece of bulk metal changes, circulating currents called *eddy currents* are set up in the metal. Resistive heating due to these currents dissipates power, so in practice, eddy currents are usually unwanted.

Inductance Since current is the source of the magnetic field, the magnetic flux through a circuit is related to the current in that circuit and in other nearby circuits. A current-carrying circuit thus affects itself magnetically, by induction. The *self-inductance* of a circuit is a quantity that measures the strength of this effect; it is determined by the geometry of the circuit. Inductance is straightforward to define but, in most cases, difficult to calculate. (For a long, tightly-wound solenoid, it is easy to calculate because the magnetic field is uniform inside a solenoid.) If the current in a circuit changes, an emf is induced in the circuit that is proportional to its self-inductance. By Lenz's law, this *back emf* always opposes whatever change induced it. Changing current in a circuit will also induce an emf in another circuit nearby. The *mutual inductance*, which measures the strength of this effect, depends on the geometry of the two circuits.

LR Circuit. In a circuit that contains both an inductance and a resistance, the current cannot change instantaneously because of the inductance. Instead, it changes exponentially over a time characterized by a *time constant* given by L/R. (This is analogous to the time constant of an *RC* circuit; it is the time required for the current, starting from zero, to reach 63 percent of its maximum value.) Thus, work is required to start a current moving through an inductor, and we can say that there is energy stored in the inductor or in its magnetic field.

Generators and Motors In practice, almost all the electrical energy we use is produced by electromagnetic induction in generators. A *generator* is essentially just a coil that is rotated (by a steam engine, say) in a magnetic field. The magnetic flux through the rotating coil changes sinusoidally, inducing a sinusoidal emf in the coil. An electric *motor* is essentially the opposite of a generator. In a motor, an alternating current supplied to the coil in an external magnetic field creates a torque that causes the coil to rotate.

II.   Numbers and Key Equations

Numbers

1 weber (Wb) $= 1$ T·m²

$= 1$ N·m/A $= 1$ V·s

1 henry (H) $= 1$ Wb/A $= 1$ Ω·s

Key Equations

Magnetic flux

$$\phi_m = NB_\perp A = NBA \cos \theta$$

Faraday's law

$$\mathcal{E} = (-)\frac{\Delta \phi_m}{\Delta t}$$

Motional emf

$$\mathcal{E} = B\ell v$$

Magnetic flux and self-inductance

$$\phi_m = LI$$

Self-inductance of a solenoid

$$L = \mu_0 \frac{N^2 A}{\ell}$$

Back emf in an inductor

$$\mathcal{E} = -L \frac{\Delta I}{\Delta t}$$

Mutual inductance of coaxial solenoids

$$M_{12} = \frac{\phi_{m2}}{I_1}$$

$$M_{12} = M_{21} = \mu_0 \frac{N_1 N_2 \pi r_1^2}{\ell}$$

Maximum current in an LR circuit

$$I_{max} = \frac{\mathcal{E}_0}{R}$$

Time constant for an LR circuit

$$\tau = \frac{L}{R}$$

Energy stored in an inductor

$$U = \tfrac{1}{2}LI^2$$

Magnetic energy density

$$\eta_m = \frac{B^2}{2\mu_0}$$

Generator emf

$$\mathcal{E} = \mathcal{E}_{max} \sin \omega t \qquad \mathcal{E}_{max} = NBA\omega$$

Circuit symbol

Inductor

III. Possible Pitfalls

Faraday's law is used only to determine the magnitude of the induced emf. The minus sign in the equation tells you nothing directly; it's there only to remind you of Lenz's law, from which you can infer the sense of the induced emf.

The fact that the emf from a generator varies sinusoidally with time has nothing to do with the shape of the rotating coil or the like; it's due only to the coil being rotated at constant angular speed.

A large magnetic flux is not what causes a large induced emf; it's a flux that is *rapidly changing*. The rate at which the flux is changing determines the induced emf.

Steady magnetic fields do not induce an emf. Only a changing magnetic flux does that. Electromagnetic induction is not important in dc circuits, except perhaps when they are being switched on and off.

Inductance (either self or mutual) is a geometric property. It depends only on the shape and position of circuits and not on the voltages or currents in them.

Lenz's law says that the induced emf is directed so as to oppose the *change* in flux that produced it, not the flux itself. The induced emf may be directed so as to increase or decrease the existing flux, depending on the circumstances.

IV. Questions and Answers

Questions

1. Two conducting loops with a common axis are placed near each other, as sketched in Figure 22-1. If a current I_A is suddenly set up in loop A, as sketched, will there also be a current in loop B? If so, in which direction? What is the direction of the force that loop A exerts on loop B?

2. A given length of wire is wound into a solenoid. How will its self-inductance be changed if it is rewound into another coil of (*a*) twice the length or (*b*) twice the diameter?

3. In a popular demonstration of electromagnetic induction, a metal plate is suspended in midair above a large ac electromagnetic coil, as indicated in Figure 22-2. How does this work? If you try this, one thing you'll notice is that the plate gets quite hot. (In fact, you can cap this demonstration by frying an egg on the plate!) Why is this? Would the trick work if the plate were made of an insulating material?

4. A bar magnet is projected at some initial velocity along the axis of a long, frictionless, horizontal conducting pipe. Describe the subsequent motion of the magnet.

5. A bar magnet is dropped down a long, vertical conducting pipe. Describe its subsequent motion.

Figure 22-1

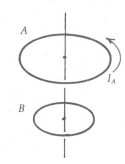

Figure 22-2

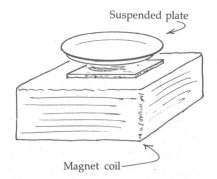

6. Where does the energy that heats your toaster come from?

7. A motor will sometimes burn out when its load is suddenly increased. Why? What is it that "burns out"?

8. As the conducting bar in Figure 22-3 is moved to the right, is an electric field set up in it? If so, in what direction? What causes it?

9. As the conducting bar in Figure 22-3 is moved to the right, does an external force have to act on it in order to keep it moving at a constant speed?

10. What change must you make in the current in an inductor to double the energy stored in it?

11. A simple RL circuit is shown in Figure 22-4. When switch S is opened, a spark jumps between the switch contacts. Why?

Figure 22-3

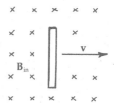

Figure 22-4

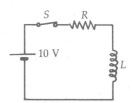

Answers

1. There will be an induced current in loop B while the current in loop A is changing. If the current in A is in the direction shown and is increasing, its flux through B will be increasing. Thus, by Lenz's law, the current induced in B will be in such a direction as to produce an opposite flux through B. Thus, while I_A is increasing, I_B is opposite I_A. Since the currents are antiparallel, the two loops repel while the current in A is increasing. Once I_A has reached a steady value, no current flows in B and there is no force between the loops.

2. The self-inductance of a solenoid is given by $L = \mu_0 N^2 A \ell$. (*a*) If the length ℓ of the coil is doubled, the inductance L is cut in half. (*b*) If the diameter is doubled, you can wind only half as many turns on it with a given length of wire, so N is halved. At the same time, though, A is increased by a factor of 4, so $N^2 A$ is unchanged and so is L.

3. Eddy currents are induced in the plate by the changing magnetic flux produced by the ac electromagnetic coil. According to Lenz's law, these currents are so directed as to be repelled by the field. The result is an upward force that balances the plate's weight. The plate gets hot because the eddy currents dissipate energy (power loss $= I^2 R$) as they flow through the plate. Plainly, then, the demonstration wouldn't work with a nonconducting plate.

4. As the bar magnet moves along the pipe, it induces currents in the walls of the pipe. By Lenz's law, these currents are in such a direction as to oppose the change that induced them, that is, to retard the motion of the magnet. Since, as the question is stated, there are no other forces on the magnet, it must eventually slow down and stop.

5. Currents are induced in the walls of the pipe by the moving magnet, just as in the previous question. But in this case there is also the force of gravity pulling it down. The speed of the magnet will increase, approaching a terminal speed at which the gravitational and magnetic forces are equal.

6. From the electric power line, by which your toaster is ultimately connected, through various lines and transformers and such, to an electromagnetic generator somewhere. The generator gets its energy from whatever energy source is cranking it—most usually a heat engine powered by burning something.

7. When the load is increased, the rotating coil is suddenly slowed, perhaps even stopped momentarily. The emf it induces in the primary coil, which produces the magnetic field, decreases drastically and the current drawn increases greatly. The large power drawn is all dissipated as heat in this primary coil, which is what burns out.

8. As the conducting bar moves through the magnetic field, the free electrons in it are forced to the bottom end of the bar, leaving a positive charge at the top. This separated charge creates an electric field in the bar directed from the positive to the negative charge, that is, down.

9. No; once the conducting bar is in motion and the free charges in it have stabilized, no charge is moving along the bar and thus there is no magnetic force in the direction of the bar's motion.

10. The stored energy in an inductor is proportional to I^2, so the current must be increased by a factor of $\sqrt{2}$ in order to double the stored energy.

11. When the switch is opened, the current drops suddenly toward zero. This induces a very large back emf in the inductor, which appears across the switch gap. The only way current can flow across the gap is by dielectric breakdown of the air in the gap; this is the spark.

V. Problems and Solutions

Problems

1.   A coil of 30 turns of wire, 6 cm in diameter, is placed in a uniform magnetic field directed perpendicular to the plane of the coil. Initially, the value of the magnetic field is 1.0 T. Beginning at time $t = 0$, the field is increased at a uniform rate until it reaches 1.3 T at $t = 10$ s. The field remains constant thereafter. What is the induced emf in the coil at (a) $t < 0$, (b) $t = 5$ s, and (c) $t > 10$ s?

How to Solve It
- Faraday's law relates the emf induced in the coil to the rate at which the flux through it is changing.

- Calculate the magnetic flux through the coil for both values of B.

- What you want to know is the rate at which the flux is changing in each part of the problem.

2.   A 50-turn coil of cross-sectional area 5 cm² has a resistance of 20 Ω. The plane of the coil is perpendicular to a uniform magnetic field of 1.0 T. The coil is suddenly reversed in direction over a period of 0.2 s. What average current flows in the coil during this time?

How to Solve It
- Calculate the flux through the coil when it is perpendicular to B.

- When you flip the coil over, the magnetic flux changes from ϕ_m to $-\phi_m$; the net change is therefore $2\phi_m$.

- When you have found the emf induced in the coil, use Ohm's law to find the current.

3.   A long solenoid 2 cm in diameter and 20 cm long consists of 5000 turns of wire. (a) What is the field inside the solenoid when a current of 5 A flows in it? (b) What is the self-inductance of the solenoid? (c) If the current in the solenoid drops to 0 in 0.1 s, what back emf is generated in it?

How to Solve It
- Just use the appropriate formula from Chapter 21 to get the field inside the solenoid.

- The inductance of the solenoid is the flux through it divided by the current.

● Find the emf from the inductance and the rate of change of the current in the coil.

4. A circular coil of 100 turns and radius 10 cm is in a magnetic field of magnitude 650 G directed perpendicular to the coil. (*a*) What is the magnetic flux through the coil? (*b*) The magnetic field at the coil is increased steadily to 1000 G over a time interval of 0.5 s. What emf is generated in the coil?

How to Solve It
● Since the coil is perpendicular to *B*, the flux through *each* turn of it is just the field multiplied by the area.

● The rate at which the flux changes in time is equal to the induced emf (Faraday's law).

5. A rod 25 cm long moves at 10 m/s in a plane perpendicular to a magnetic field of 1600 G. Its velocity is perpendicular to its length. Find (*a*) the magnitude of the magnetic force on an electron in the rod, (*b*) the magnitude of the electric field created in the rod, and (*c*) the potential difference *V* between the ends of the rod.

How to Solve It
● First, calculate the magnetic force on an electron.

● Since there is no complete circuit, this emf causes negative electrons to be displaced to one end of the rod, leaving a net positive charge at the other.

● In equilibrium, the electric field caused by this separation of charge must exert a force on an electron equal and opposite to that of the magnetic field.

● The electric field multiplied by the length of the rod is the potential difference between its ends.

Figure 22-5

6. A square coil 10 cm on a side, consisting of 30 turns of wire, is between the poles of a large electromagnet that produces a uniform magnetic field of 6 kG. (See Figure 22-5.) The resistance of the coil is 0.82 Ω. Assume that the field drops sharply to zero at the edge of the magnet, as suggested by the figure. If the coil is drawn steadily out of the field in a time of 5 s, (*a*) what current and (*b*) what total charge flow in the coil? Assume that the leading edge of the coil is at the edge of the magnet at time $t = 0$. (*c*) If the coil is drawn out in a time of 1 s, what total charge flows?

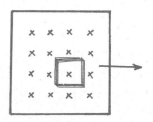

How to Solve It

- While the coil is moving through the uniform field, the motional emf induced in its leading and trailing edges is the same, so there is no net emf around the loop and no current flows.

- As the coil moves out of the field, the motional emf induced in the trailing edge is the net emf in the coil. Use Ohm's law to calculate the current in it.

- The current multiplied by the time it takes for the coil to be drawn out of the field is the net charge that flows during this interval.

7. A solenoid 55 cm long and 2 cm in diameter consists of 800 turns of wire. A small circular coil of 20 turns, 2.2 cm in diameter, is wrapped around the solenoid (see Figure 22-6). Find the mutual inductance of the two coils.

Figure 22-6

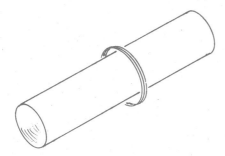

How to Solve It

- Calculate the field inside the solenoid and then use it to find the flux through the small coil.

- Remember that the magnetic field outside the solenoid is effectively zero.

- The mutual inductance is the flux through the small coil divided by the current in the solenoid.

8. A solenoid with a diameter of 3 cm and 400 turns is 15 cm long. It carries a current of 1 A. Find (*a*) the magnetic field inside the solenoid, (*b*) the magnetic flux through it (assuming the field to be uniform), and (*c*) the inductance of the solenoid.

How to Solve It

- Calculate the magnetic field inside the solenoid.

- From this field, the number of turns, and the cross-sectional area of the solenoid, find the flux through it.

- The inductance of the solenoid is the flux per unit current.

9. A solenoid of 1600 turns, cross-sectional area 6 cm², and length 20 cm carries a current of 2.80 A. (*a*) What is the magnetic field inside this solenoid? (*b*) What is its inductance? (*c*) What is the stored energy in the solenoid? (*d*) Divide your answer to (*c*) by the internal volume of the solenoid to get the magnetic energy density (the stored energy per unit volume). (*e*) Calculate the magnetic energy density directly from the magnetic field and compare your answer to what you got in (*d*).

How to Solve It
- Calculate the magnetic field inside the solenoid and its inductance.

- From the inductance and current in the solenoid, calculate the energy stored in the solenoid.

- This divided by the volume of the solenoid should agree with the magnetic energy density calculated directly from the magnetic field.

10. A solenoid is made of 1500 turns of #22 copper wire (diameter 0.644 mm and resistance of 0.0530 Ω/m that is uniformly distributed on a coil form 2 cm in diameter and 18 cm long. Find (*a*) the inductance, (*b*) the resistance, and (*c*) the time constant of the solenoid.

How to Solve It
- You have all the dimensions of the solenoid, so you can calculate its inductance.

- Find the resistance of the coil from the total length of wire in the windings.

- The coil can be considered an inductor and resistor in series; the time constant of such a combination is L/R.

11. A rectangular coil with sides 0.1 m by 0.25 m has 500 turns of wire. It is rotated about its long axis in a magnetic field of 5.8 kG. At what frequency must the coil be rotated to generate a maximum emf of 110 V?

How to Solve It
- Calculate the flux in the coil when it is oriented perpendicular to B.

● The maximum emf is the flux multiplied by the angular velocity of rotation.

● Find the frequency from the angular velocity.

12. An electromagnetic generator consists of a coil of 100 turns of wire with an area of 400 cm² rotating at 60 rev/s in a magnetic field of 0.25 T. What maximum emf is generated in the coil?

How to Solve It
● Calculate the flux in the coil when it is oriented perpendicular to B.

● The maximum emf is the flux multiplied by the angular velocity of rotation.

Solutions

1.   The area of the coil is $\pi r^2 = (\pi)(0.03 \text{ m})^2 = 2.83 \times 10^{-3} \text{ m}^2$. Thus, when $B = 1.0$ T, the flux through the coil is

$$\phi_m = NB_\perp A = (30)(1.0 \text{ T})(2.83 \times 10^{-3} \text{ m})^2 = 0.0849 \text{ Wb}$$

Likewise, when $B = 1.3$ T, $\phi_m = 0.1104$ Wb.
In both (a) and (c) the field is constant; no matter what its value, therefore, the flux is not changing in time and so $\mathcal{E} = 0$.
(b) Between $t = 0$ and $t = 10$ s, the flux changes at a uniform rate. Thus, at any time in that interval, the rate at which ϕ_m is changing is

$$\mathcal{E} = \frac{\Delta\phi_m}{\Delta t} = \frac{0.1104 \text{ Wb} - 0.0849 \text{ Wb}}{10 \text{ s}} = \underline{2.55 \times 10^{-3} \text{ V}}$$

2. $\underline{12.5 \text{ mA}}$

3.   (a) The field inside a long solenoid is

$$B = \mu_0 \left(\frac{N}{\ell}\right) I = (4\pi \times 10^{-7} \text{ N/A}^2) \left(\frac{5000}{0.20 \text{ m}}\right) (5 \text{ A}) = \underline{0.157 \text{ T}}$$

(b) The flux through the solenoid (assuming B is constant from end to end) is

$$\phi_m = NB_\perp A = (5000)(0.157 \text{ T})(\pi)(0.01 \text{ m})^2 = 0.247 \text{ Wb}$$

The flux per unit current is the inductance of the solenoid:

$$L = \frac{\phi_m}{I} = \frac{0.247 \text{ Wb}}{5 \text{ A}} = \underline{0.0493 \text{ H}}$$

(c) The current decreases from 5 A to zero in a time of 0.1 s, so the back emf induced in the solenoid is

$$\mathcal{E} = L\frac{\Delta I}{\Delta t} = (0.0493 \text{ H})\frac{0 - 5 \text{ A}}{0.1 \text{ s}} = \underline{-2.47 \text{ V}}$$

The negative sign means only that the emf induced is in a sense so as to oppose the change that induced it, in this case, so as to try to maintain the current in the coil.

4. (a) $\underline{0.204 \text{ Wb}}$ (b) $\underline{0.220 \text{ V}}$

5. (a) The magnetic force on a point charge q moving at velocity v through a perpendicular magnetic field B is

$$F = qvB_\perp$$

$$= (1.60 \times 10^{-19} \text{ C})(10 \text{ m/s})(0.160 \text{ T}) = \underline{2.56 \times 10^{-19} \text{ N}}$$

The direction of the force on an electron is as indicated in Figure 22-7. (Don't forget that the charge of an electron is negative.)

(b) In equilibrium, the electric field must produce a force that is equal and opposite to the magnetic force, so

$$E = \frac{F}{q} = \frac{2.56 \times 10^{-19} \text{ N}}{(1.60 \times 10^{-19} \text{ C})} = 1.60 \text{ N/C} = \underline{1.60 \text{ V/m}}$$

(c) The potential difference between the ends of the rod is

$$V = E\ell = (1.60 \text{ V/m})(0.25 \text{ m}) = \underline{0.40 \text{ V}}$$

Figure 22-7

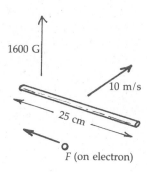

6. (a) $\underline{43.9 \text{ mA}}$ (b) $\underline{0.220 \text{ C}}$ (c) $\underline{0.220 \text{ C}}$ The total charge that flows is independent of how you draw the coil out.

7. The field inside the solenoid is

$$B_s = \mu_0\left(\frac{N}{\ell}\right)I_s = (4\pi \times 10^{-7} \text{ N/A}^2)\left(\frac{800}{0.55 \text{ m}}\right)(I_s)$$

$$= (1.83 \times 10^{-3} \text{ T/A})(I_s)$$

and the field outside a long solenoid is effectively zero. The flux through any single loop surrounding the solenoid is the field inside the solenoid multiplied by the cross-sectional area of the solenoid:

$$\phi_{m1} = BA = (1.83 \times 10^{-3} \text{ T/A})(I_s)\pi(0.01 \text{ m})^2$$

$$= (5.75 \times 10^{-7} \text{ Wb/A})(I_s)$$

The small coil has 20 such turns, so the total flux through it is

$$\phi_{mc} = N\phi_{m1} = (20)(5.75 \times 10^{-7} \text{ Wb/A})(I_s)$$
$$= (1.15 \times 10^{-5} \text{ Wb/A})(I_s)$$

The mutual inductance is the flux through the coil divided by the current in the solenoid. Thus

$$M = \frac{\phi_{mc}}{I_s} = \frac{(1.15 \times 10^{-5} \text{ Wb/A})(I_s)}{I_s}$$
$$= 1.15 \times 10^{-5} \text{ H}$$

8. (a) 33.5 G   (b) 9.47×10^{-4} Wb   (c) 0.947 mH

9. (a) The magnetic field inside a long solenoid is

$$B = \mu_0 \left(\frac{N}{\ell}\right) I = (4\pi \times 10^{-7} \text{ N/A}^2) \left(\frac{1600}{0.20 \text{ m}}\right) (2.80 \text{ A})$$
$$= 2.81 \times 10^{-2} \text{ T}$$

(b) The inductance of a solenoid is

$$L = \mu_0 \left(\frac{N^2}{\ell}\right) A = (4\pi \times 10^{-7} \text{ N/A}^2) \left[\frac{(1600)^2}{0.20 \text{ m}}\right] (6 \times 10^{-4} \text{ m}^2)$$
$$= 9.65 \times 10^{-3} \text{ H}$$

(c) The energy stored in the solenoid is

$$U = \tfrac{1}{2}LI^2 = \tfrac{1}{2}(9.65 \times 10^{-3} \text{ H})(2.80 \text{ A})^2 = 0.0378 \text{ J}$$

(d) The interval volume of the solenoid is

$$\text{Vol} = \ell A = (20 \text{ cm})(6 \text{ cm}^2) = 120 \text{ cm}^3 = 1.2 \times 10^{-4} \text{ m}^3$$

so the stored energy density is

$$\eta = \frac{U}{\text{Vol}} = \frac{0.0378 \text{ J}}{1.2 \times 10^{-4} \text{ m}^3} = 3.15 \times 10^2 \text{ J/m}^3$$

(e) Calculating the magnetic energy density directly from the magnetic field B, we get

$$\eta = \frac{B^2}{2\mu_0} = \frac{(2.81 \times 10^{-2} \text{ T})^2}{(2)(4\pi \times 10^{-7} \text{ N/A}^2)} = 3.14 \times 10^2 \text{ J/m}^3$$

Within round-off error, this is what we got in part (d).

10. (a) 4.93 mH   (b) 5.00 Ω   (c) 0.986 ms

11. The maximum emf to be generated is 110 V:

$$\mathcal{E}_{max} = NBA\omega \quad \text{or} \quad \omega = \frac{\mathcal{E}_{max}}{NBA}$$

Here

$$NBA = (500)(0.58 \text{ T})(0.1 \text{ m})(0.25 \text{ m}) = 7.25 \text{ Wb}$$

so

$$\omega = \frac{110 \text{ V}}{7.25 \text{ Wb}} = 15.2 \text{ s}^{-1}$$

and

$$f = \frac{\omega}{2\pi} = \frac{15.2 \text{ s}^{-1}}{2\pi} = \underline{2.42 \text{ Hz}}$$

12. <u>377 V</u>

CHAPTER 23

Alternating Current Circuits

I. Key Ideas

Alternating Current Almost all the electrical energy we use is in the form of alternating current (ac) produced by electromagnetic induction in an ac generator. Such a generator produces an emf that varies sinusoidally with time. By Ohm's law, the current that an alternating emf produces in a resistor varies sinusoidally with the same frequency as the emf. The average power dissipation in a resistance is less than $\mathcal{E}_{max}I_{max}$ because both voltage and current are varying; we usually use *root-mean-square* (*rms*) values. Standard household ac power in the United States is 110 V rms.

Behavior of an Inductance The behavior of inductors and capacitors in an ac circuit is very different from that in a dc circuit. The emf produced in an inductor depends on the rate at which the current in it is changing and thus on the frequency as well as the magnitude of the current; the higher the frequency, the greater the back emf induced. The sinusoidal current in an inductance *lags behind* the voltage by one-fourth of a cycle or 90°. No power is dissipated in a pure inductance. The *inductive reactance* is a quantity, analogous in some ways to resistance, that determines the current in an inductor due to an applied emf. It is directly proportional to frequency.

Behavior of a Capacitance To change the potential difference across a capacitor, charge must be moved on and off its plates. This takes time; as a result, voltage across a capacitor lags behind the current through it by one-fourth of a cycle or 90°. The *capacitive reactance* of a capacitor is inversely proportional to frequency. Note that a continuing alternating current can exist in a circuit containing a capacitor even though charge can't flow through it. Again, in a pure capacitance, no power is dissipated.

LCR Circuits The behavior of a series circuit containing a resistance, an inductance, and a capacitance is important. In the steady state, an alternating current flows in the circuit with the frequency of the driving emf. Because of their phase differences, the voltage drops across R, L, and C in a series circuit do not simply add. Instead, they can be added vectorially on a two-dimensional diagram on which the direction angle of a vector representing each voltage drop is used to represent its phase angle. The quantity that combines resistance and reactance to determine the current that flows in the circuit is the *impedance* of the circuit.

Resonance In a series *LCR* circuit, the voltage drop across the capacitor lags and the voltage across the inductance leads the current. The drop across the resistance is in phase with the current. The reactances of inductance and capacitance thus tend to cancel. At a certain frequency, they cancel exactly, and the current in the circuit has its maximum value; this is the *resonance frequency* of the circuit. The behavior of this circuit closely resembles that of a driven harmonic oscillator; for example, the average power being delivered by the emf to the circuit has its maximum value at resonance. The sharpness of the resonance curve, which is described by the *Q factor*, depends on the resistance in the circuit. A familiar application of the series resonance circuit is in the tuning of a radio-frequency receiver.

Transformers A transformer is a device for changing ac voltages and currents by mutual inductance with little loss of power. Two coils are closely associated geometrically so that essentially all the flux of one passes through the other. In order for the flux to be the same, the coil with fewer turns must have a larger current in it and, if the power is to be the same, a lower voltage. The ability to step voltage levels up or down makes possible long-distance power transmission.

Rectification A *diode* is a device that conducts current in only one direction. Diodes can be used to *rectify* alternating current, that is, to convert it to a direct current.

Amplification A third element called a *grid* inserted into a vacuum diode produces a *vacuum triode*. Small changes in the voltage at the grid produce corresponding large changes in the plate current, making possible the *amplification* of electric signals.

II. Key Equations

Key Equations

Alternating current in a resistor

$$I = \frac{\mathcal{E}}{R} = I_{max} \sin \omega t$$

$$I_{max} = \frac{\mathcal{E}_{max}}{R}$$

Power dissipated in a resistor

$$P_{av} = (I^2 R)_{av} = \tfrac{1}{2} I^2_{max} R = I^2_{rms} R$$

$$P_{av} = \tfrac{1}{2} \mathcal{E}_{max} I_{max} = \mathcal{E}_{rms} I_{rms}$$

Root-mean-square quantities

$$I_{rms} = \frac{1}{\sqrt{2}} I_{max} \qquad \mathcal{E}_{rms} = \frac{1}{\sqrt{2}} \mathcal{E}_{max}$$

Current in an inductor

$$I = -I_{max} \cos \omega t = I_{max} \sin(\omega t - 90°)$$

$$\text{when } \mathcal{E} = \mathcal{E}_{max} \sin \omega t$$

$$\text{with } I_{max} = \frac{\mathcal{E}_{max}}{\omega L} = \frac{\mathcal{E}_{max}}{X_L}$$

Inductive reactance

$$X_L = \omega L = 2\pi f L$$

Current in a capacitor

$$I = I_{max} \cos \omega t = I_{max} \sin(\omega t + 90°)$$

$$\text{with } I_{max} = \omega C \mathcal{E}_{max} = \frac{\mathcal{E}_{max}}{X_C}$$

Capacitive reactance

$$X_C = \frac{1}{\omega C} = \frac{1}{2\pi f C}$$

Current in an *LCR* circuit

$$I = I_{max} \sin(\omega t - \phi) \qquad I_{max} = \frac{\mathcal{E}_{max}}{Z}$$

Impedance

$$Z = \sqrt{R^2 + (X_L - X_C)^2}$$

Phase angle

$$\tan \phi = \frac{X_L - X_C}{R}$$

Resonance frequency

$$f_0 = \frac{1}{2\pi \sqrt{LC}}$$

Q factor

$$Q = \frac{2\pi f_0 L}{R}$$

Power factor

$$P_{av} = \mathcal{E}_{rms} I_{rms} \cos \phi$$

Transformer voltage ratio

$$V_2 = \frac{N_2}{N_1} V_1$$

III.   Possible Pitfalls

Remember that the average value of the square of any sinusoidal function over one or more complete cycles is $\frac{1}{2}$.

Remember the difference between maximum and root-mean-square (rms) quantities. You can't multiply voltage and current to get power unless you use rms values (to get the average power).

In general, the current in an ac circuit is not proportional to the emf applied to it or to the voltage drop over a circuit element other than a resistance. They differ in *phase*—the current in a capacitor or inductor oscillates at the same frequency as the voltage, but ahead of it or behind it. Impedance or reactance is the ratio of maximum or rms voltage and current values, not the instantaneous values.

A reactance is not a resistance, although both are measured in ohms. In a pure reactance, there is no power dissipation, the ratio of voltage to current is frequency dependent, and the voltage and current are not in phase.

Capacitive and inductive reactances have opposite frequency dependences. A capacitor has a very large reactance at low frequencies and a very small reactance at high frequencies; the reverse is true for an inductor.

You can't, in general, simply add voltage drops or currents in an ac circuit because of phase differences. You must add them *vectorially*, including the phase of each quantity, as discussed in the chapter.

The behavior of a series *LCR* circuit near resonance depends on all three quantities, but the *resonance frequency* is determined by L and C independent of R.

IV.   Questions and Answers

Questions

1. Inductive and capacitative reactances in series subtract rather than add. Why?

2. Electric power for domestic use is transmitted at very high voltages and is then stepped down to 110 V by a transformer near the point of consumption. Why?

3. The reactance of a capacitor decreases with increased frequency, whereas the reactance of an inductor increases. Why?

4. At a certain instant, the current in a series *LCR* circuit is zero. In general, the power being supplied by the generator is not zero at this instant even though there is no energy being dissipated in the resistance. Where is the energy going?

5. What is meant by saying that the current through an inductor "lags behind the voltage by 90°"?

6. In a coil that has both resistance and inductance, does the phase angle between the current through the coil and the voltage drop across it depend on the frequency?

7. Is the current through a resistor in an ac circuit always in phase with the emf applied to the circuit? Why or why not?

Answers

1. In a series connection, the current through two circuit elements must be the same. But the voltage across a capacitor is 90° behind the current, whereas that across an inductor is 90° ahead of the current. The voltage across the capacitor is therefore 180° out of phase with that across the inductor, so their resultant is the difference in the magnitudes of the voltages. This difference is the (common) current times a total reactance, so the reactances have subtracted rather than added.

2. It saves on wire. To reduce the power loss in transmission lines, which is equal to I^2R, both the resistance of and current in these lines should be as low as possible. The only way to lower the resistance is to use large diameter wire, so instead power companies choose to step up voltage thus reducing current.

3. As the frequency increases, less charge has to be moved on and off the capacitor plates (the current flows in each direction for a shorter time), so the voltage required to cause a given current decreases. This means that the reactance decreases. But to produce a given amount of current in an inductor, as the frequency increases, a given magnetic field in the inductor has to be set up and taken down more rapidly. Thus, the voltage required increases and so does the reactance.

4. The power dissipated in the resistance has to equal that being supplied by the generator on the average—but not instantaneously. At the instant described, the energy being supplied is going into charging up the capacitor or setting up the magnetic field in the inductor.

5. Both the voltage and the current are sinusoidal functions of time. If the current lags behind the voltage, this means that the current reaches its maximum value one-quarter cycle (90° since a full cycle is 360°) later than the voltage across it does.

6. Yes. The phase angle between the current and the voltage across the coil (a combination of L and R) depends on the ratio of the inductive reactance to the resistance. Since the inductive react-ance depends on the frequency, so does the phase angle.

7. No. This would be true only if the net reactance of the circuit were zero, which would mean (*a*) that there is negligible induct-ance and capacitance in the circuit or (*b*) that the circuit is at reso-nance.

V.   Problems and Solutions

Problems

1. An electric generator operating at 60 Hz delivers 100 W to a purely resistive load of resistance 150 Ω. If the generator consists of a coil of 250 turns, each with an area 0.1 m^2, being rotated in a uniform magnetic field, what must be the value of the magnetic field?

How to Solve It
- Calculate the rms current to the resistor from the given power and resistance values.

- Calculate the maximum emf of the generator.

- You can use this to calculate the maximum magnetic flux through the coil and thus the magnetic field.

2. A 100-Ω resistor is connected across a 110 V ac power line. Find (*a*) I_{rms}, (*b*) I_{max}, and (*c*) the average power dissipated in the resistor.

How to Solve It
- The 110 V specified for ordinary household power is an rms value. Use Ohm's law to get the rms value of the current.

- Calculate the maximum current value from the rms value.

- The product of rms voltage and rms current is the average power being delivered to the circuit.

3. A real coil has both resistance and inductance; consider it as a 50-Ω resistor and a 0.10-H inductor in series. The coil is connected across a 120 V rms, 60 Hz power line. Find (*a*) the rms current, (*b*) the power drawn from the line, and (*c*) the power factor.

How to Solve It
- Calculate the impedance of the series *LR* combination from the resistance and reactance. Remember that reactance depends on frequency.

- The rms current can be found from the rms voltage and the impedance.

- The (average) power being drawn from the line is equal to the power being dissipated in the resistance. Be careful: it is not the product of rms voltage and current; that's what the power factor is about.

4. An emf of 40 V rms and a frequency of 100 Hz is applied to (*a*) a 0.2-H inductor and (*b*) a 50-μF capacitor. In each case, find the peak value of the current.

How to Solve It
- Since no resistance is mentioned, you must assume a pure inductance and a pure capacitance.

- Calculate the reactance of each at a frequency of 100 Hz. The rms voltage divided by the reactance gives you the rms current.

- Don't forget to convert the rms values to peak current values.

5. The circuit in Figure 23-1 consists of a 1.5-μF capacitor and a 100-Ω resistor in series with negligible inductance. (*a*) Draw a vector diagram showing the voltage drops across *C* and *R*, and find the phase difference between the emf and the current at a frequency of 300 Hz. (*b*) Repeat for a frequency of 5000 Hz.

Figure 23-1

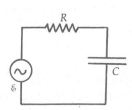

How to Solve It
- The voltage drop across the resistor is in phase with the current, and the same current flows through both *R* and *C*. Start your diagram by putting V_R on the horizontal axis.

- The voltage across the capacitor is 90° out of phase with the current (and thus with V_R). Which way? Put V_C on the vertical axis.

- Add the two vectorially, and work out the phase difference.

6. The circuit in Figure 23-2 consists of a 250-Ω resistor and a
0.04-H inductor in series with negligible capacitance. Find the voltage across the inductor (*a*) at a frequency of 50 Hz and (*b*) at a frequency of 5000 Hz. (*c*) An *LR* circuit is often used in electronic
circuits as a filter. What do you suppose it filters?

Figure 23-2

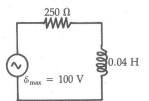

How to Solve It
- An *LR* circuit is just an *LCR* circuit with $X_C = 0$. Calculate its
  impedance from the *LCR* circuit formula.

- From the impedance, calculate the current in the circuit.

- Once you know the current and the reactance of the inductor, calculate the voltage drop across the inductor.

7. In the series *LCR* circuit in Figure 23-3, we have $L = 0.01$ H,
$R = 220\ \Omega$, and $C = 0.1\ \mu$F. The amplitude of the driving voltage
is $\mathcal{E}_{max} = 150$ V. At a frequency of 1000 Hz, what power is being
supplied by the generator?

Figure 23-3

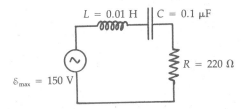

How to Solve It
- Calculate the impedance of the circuit. The driving emf divided by
  the impedance is the current in the circuit.

- Use the current in the circuit to find the power being dissipated in
  the resistor.

- This is the same as the (average) power being supplied by the generator since the reactive elements (*L* and *C*) are nondissipative.

8. A series *LCR* circuit is driven by 115 V rms from the ac power
line. The power being drawn is 65 W and the current in the circuit
is 1.0 A rms. Find (*a*) the resistance and (*b*) the net reactance of the
circuit.

How to Solve It
- The 65 W is being dissipated in the resistor; this lets you calculate
  the resistance.

- The ratio of the rms voltage to the current in the circuit is its impedance.

- Use the impedance and the resistance to find the reactance.

9. In the series *LCR* circuit in Figure 23-4, find the maximum voltages V_R, V_C, and V_L across the resistor, capacitor, and inductor, respectively, (*a*) at a frequency of 100 Hz and (*b*) at resonance.

Figure 23-4

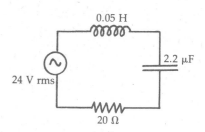

0.05 H

2.2 μF

24 V rms

20 Ω

How to Solve It
- At each frequency, first calculate the impedance of the circuit. At resonance, the impedance is equal to the resistance.

- Use the impedance to calculate the current in the circuit.

- The voltage drop across each element is its resistance or reactance times the current.

- Remember to express your answers as maximum voltages.

10. A series *LCR* circuit consists of an 8-Ω resistor, a 100-mH inductor, and a 5-μF capacitor. A signal generator applies 5 V rms to this circuit. (*a*) At what frequency will maximum power be delivered to this circuit? (*b*) What is this maximum power?

How to Solve It
- The maximum power is delivered when the generator frequency is the resonance frequency of the circuit.

- Calculate the resonance frequency of the circuit from the given *L* and *C*.

- At this frequency the impedance of the circuit is just equal to the resistance in it.

11. When a certain *LCR* series circuit is connected to the power line (110 V rms, 60 Hz), it draws 700 W of power, and the current leads the line emf by 50°. Find (*a*) the rms current in the circuit and (*b*) its resistance. (*c*) Is the resonance frequency greater than or less than 60 Hz?

How to Solve It
- You are given the phase angle between the voltage and the current; its cosine is the power factor.

- From the power, the rms voltage, and the power factor, calculate the rms current.

- The power drawn is dissipated in the resistance.

- Because the current leads the voltage, the net reactance is acting more like a capacitance than an inductance.

12. A power line with a resistance of 10^{-5} Ω/m is to be used to transmit 100 kW of power from a generating station to a city 30 km away. Find the power loss in the transmission line if the line voltage is (*a*) 1000 V and (*b*) 40,000 V.

How to Solve It
- Calculate the resistance of the power line.

- In each case, calculate the current that is to reach the city.

- Calculate the power loss from the current and resistance.

13. The primary coil of a step-down transformer has 400 turns and is connected to a 110-V rms ac line. The secondary coil is to supply 15 A at 6.3 V rms. Assuming there is no power loss in the transformer, find (*a*) the number of turns in the secondary coil and (*b*) the current in the primary coil.

How to Solve It
- The ratio of the primary and secondary voltages is the same as the turns ratio of the primary and secondary coils.

- If there are no losses in the transformer, the power drawn from the input is the same as the power drawn from the secondary.

Solutions

1. The average power dissipated in the resistor is

$$P_{av} = I_{rms}^2 R$$

so

$$I_{rms} = \left(\frac{P_{av}}{R}\right)^{1/2} = \left(\frac{100\text{ W}}{150\ \Omega}\right)^{1/2} = 0.816 \text{ A (rms)}$$

The emf of the generator is thus

$$\mathcal{E}_{max} = I_{max}R = \sqrt{2}\, I_{rms}R$$

$$= (1.414)(0.816\text{ A})(150\ \Omega) = 173\text{ V}$$

This emf is also given by

$$\mathcal{E}_{max} = NBA\omega$$

where

$$\omega = 2\pi f = (2\pi)(60 \text{ Hz}) = 377 \text{ s}^{-1}$$

so

$$B = \frac{\mathcal{E}_{max}}{NA\omega} = \frac{173 \text{ V}}{(250)(0.1 \text{ m}^2)(377 \text{ s}^{-1})} = \underline{0.0183 \text{ T}}$$

2. (*a*) <u>1.10 A (rms)</u>   (*b*) <u>1.56 A (max)</u>   (*c*) <u>121 W</u>

3. (*a*) The inductive reactance of the coil is

$$X_L = \omega L = (2\pi)(60 \text{ Hz})(0.10 \text{ H}) = 37.7 \text{ }\Omega$$

so

$$Z = \sqrt{R^2 + X_L^2} = \sqrt{(50 \text{ }\Omega)^2 + (37.7 \text{ }\Omega)^2} = 62.6 \text{ }\Omega$$

and

$$I_{rms} = \frac{\mathcal{E}_{rms}}{Z} = \frac{120 \text{ V}}{62.6 \text{ }\Omega} = \underline{1.92 \text{ A}}$$

(*b*) The power being dissipated in the resistance of the coil is

$$P_{av} = I_{rms}^2 R = (1.92 \text{ A})^2(50 \text{ }\Omega) = \underline{184 \text{ W}}$$

Pure inductance is nondissipative, so this is also the power being drawn from the source.

(*c*) The source power is given by

$$P_{av} = \mathcal{E}_{rms}I_{rms} \cos \phi$$

so the power factor is

$$\cos \phi = \frac{P_{av}}{\mathcal{E}_{rms}I_{rms}} = \frac{184 \text{ W}}{(120 \text{ V})(1.92 \text{ A})} = \underline{0.799}$$

4. (*a*) <u>0.450 A (max)</u>   (*b*) <u>1.78 A (max)</u>

5. (*a*) At 300 Hz, $\omega = 2\pi f = (2\pi)(300 \text{ Hz}) = 1885 \text{ s}^{-1}$, so

$$X_C = \frac{1}{\omega C} = \frac{1}{(1885 \text{ s}^{-1})(1.5 \times 10^{-6} \text{ F})} = 354 \text{ }\Omega$$

so

$$V_C = IX_C = (I)(354 \text{ }\Omega)$$

and

$$V_R = IR = (I)(100 \ \Omega)$$

The ratio

$$\frac{V_C}{V_R} = \frac{354 \ \Omega}{100 \ \Omega}$$

so, plainly, $V_C = 3.54V_R$. The voltage drop across a capacitor lags the current, and thus V_R, by 90°. Thus, the vector diagram looks like what is sketched in Figure 23-5. Recalling that $X_L = 0$, the phase angle ϕ is

Figure 23-5

$$\tan \phi = \frac{X_L - X_C}{R} = \frac{0 - 354 \ \Omega}{100 \ \Omega} = -3.54$$

$$\phi = -74.2°$$

(b) For $f = 5000$ Hz, exactly the same procedure gives $X_C = 21.2 \ \Omega$, so $V_C = 0.212V_R$. The vector diagram is shown in Figure 23-6, and $\underline{\phi = -12.0°}$.

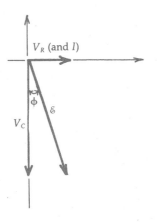

Figure 23-6

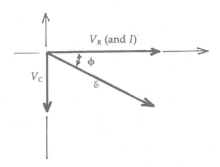

6.   (a) 5.02 V   (b) 98.1 V   (c) At high frequencies, essentially all of the input voltage appears across the inductor; at low frequencies, very little of it does. The circuit can thus be used to remove low-frequency components from a signal.

7.   At a frequency of 1000 Hz, the reactances are

$$X_L = \omega L = (2\pi)(1000 \ \text{Hz})(0.01 \ \text{H}) = 62.8 \ \Omega$$

and

$$X_C = \frac{1}{\omega C} = \frac{1}{(2\pi)(10^3 \ \text{Hz})(10^{-7} \ \text{F})} = 1592 \ \Omega$$

so the impedance is

$$Z = \sqrt{R^2 + (X_L - X_C)^2} = \sqrt{(220 \ \Omega)^2 + (62.8 \ \Omega - 1592 \ \Omega)^2}$$

$$= 1545 \ \Omega$$

Thus, the current in the circuit is

$$I_{max} = \frac{\mathcal{E}_{max}}{Z} = \frac{150 \ V}{1545 \ \Omega} = 0.0971 \ A$$

The power dissipated in the resistor, which is also the power being drawn from the generator, is therefore

$$P_{av} = \tfrac{1}{2}I_{max}^2 R$$

$$= \tfrac{1}{2}(0.971 \ A)^2(220 \ \Omega) = \underline{1.04 \ W}$$

8.   (a) $\underline{65 \ \Omega}$ (b) $\underline{94.9 \ \Omega}$

9.   (a) At f = 100 Hz, $\omega = 2\pi f = 628 \ s^{-1}$. Thus, the inductive and capacitive reactances are

$$X_L = \omega L = (628 \ s^{-1})(0.05 \ H) = 31.4 \ \Omega$$

and

$$X_C = \frac{1}{\omega C} = \frac{1}{(628 \ s^{-1})(2.2 \times 10^{-6} \ F)} = 72.4 \ \Omega$$

and the impedance of the circuit is

$$Z = \sqrt{(20 \ \Omega)^2 + (31.4 \ \Omega - 72.4 \ \Omega)^2} = 45.6 \ \Omega$$

so

$$I_{rms} = \frac{\mathcal{E}_{rms}}{Z} = \frac{24 \ V}{45.6 \ \Omega} = 0.526 \ A$$

Thus,

$$V_{R,rms} = IR = (0.526 \ A)(20 \ \Omega) = 10.5 \ V$$

$$V_{R,max} = \sqrt{2} \ V_{R,rms} = \sqrt{2}(10.5 \ V) = \underline{14.9 \ V}$$

and

$$V_{L,max} = \sqrt{2} \ IX_L = (1.414)(0.526 \ A)(31.4 \ \Omega) = \underline{23.4 \ V}$$

and $V_{C,max} = \underline{53.9 \ V}$. Notice that the three voltage drops don't add up to the generator voltage because of phase differences.

(b) The resonance frequency of this circuit is

$$f_0 = \frac{1}{2\pi\sqrt{LC}} = \frac{1}{2\pi\sqrt{(.05\ H)(2.2\ \mu F)}} = 480\ Hz$$

At resonance, $Z = R = 20\ \Omega$, so $I_{rms} = 24\ V/20\ \Omega = 1.20\ A$.
Therefore, $V_{R,rms} = 24\ V$ and $V_{R,max} = \underline{33.9\ V}$. At resonance, V_L and
V_C cancel each other, so they must be equal in magnitude. Thus,

$$V_{C,rms} = V_{L,rms} = IX_L = I\omega L$$

$$= 2\pi IfL = (2\pi)(1.2\ A)(480\ Hz)(0.05\ H) = 181\ V\ (rms)$$

so

$$V_{C,max} = V_{L,max} = \sqrt{2}V_{L,rms} = (1.414)(181\ V) = \underline{256\ V}$$

10. (a) $\underline{225\ Hz}$ (b) $\underline{3.12\ W}$

11. (a) The power factor is

$$\cos \phi = \cos 50° = 0.643$$

and we know the average power is 700 W:

$$P_{av} = \mathscr{E}_{rms}I_{rms} \cos \phi$$

so

$$I_{rms} = \frac{P_{av}}{\mathscr{E}_{rms} \cos \phi} = \frac{700\ W}{(110\ V)(0.643)} = \underline{9.90\ A}$$

(b) The power dissipated in the resistance is

$$P_{av} = I_{rms}^2 R$$

so

$$R = \frac{P_{av}}{I_{rms}^2} = \frac{700\ W}{(9.90\ A)^2} = \underline{7.14\ \Omega}$$

(c) Because the voltage in the circuit lags behind the current,
the reactance of the circuit is acting like a capacitance; that is,
the capacitative reactance exceeds the inductive reactance. Thus,
$1/\omega C > \omega L$, so $\omega^2 < 1/LC$. The frequency of 60 Hz is therefore
<u>below</u> the resonance frequency.

12. (a) $\underline{33.3\ kW}$ (b) $\underline{20.8\ W}$

13.   (*a*) The turns ratio is equal to the voltage ratio

$$\frac{N_p}{N_s} = \frac{V_p}{V_s}$$

so the number of turns in the secondary coil is

$$N_s = \frac{V_s N_p}{V_s} = \frac{(6.3 \text{ V})(400)}{110 \text{ V}} = 22.9 = \underline{23 \text{ turns}}$$

(*b*) Since there are no power losses,

$$V_p I_p = V_s I_s$$

The current in the primary coil is therefore

$$I_p = \frac{V_s I_s}{V_p} = \frac{(6.3 \text{ V})(15 \text{ A})}{110 \text{ V}} = \underline{0.859 \text{ A}}$$

CHAPTER 24

Light

I. Key Ideas

Light The nature of light was debated for centuries. Newton believed that light is a propagation of particles; on this basis, he could explain many observed phenomena, but doing so required other assumptions that turned out to be wrong. Light passing across a boundary between transparent media is reflected in part and transmitted in part. In general, the direction of the transmitted light changes at the boundary. This is called *refraction*. Newton's explanation of refraction, that it is due to attractive forces on light "particles" at the boundary, required that light speed up as it passes from air into water, for example. A wave theory of light, on the other hand, can explain reflection and refraction at a boundary by Huygens' wavelet construction, provided that light travels slower in water than in air. Newton rejected a wave theory of light because, in his time, it appeared that the diffraction of light did not occur. Many interference and diffraction experiments in the early nineteenth century, however, proved the wave nature of light; later, the speed of light in water was measured and was found to be less than that in air.

The Nature of Light Maxwell's theory of electromagnetism predicted the existence of electromagnetic waves that would propagate at the speed of light, and we identify light as one form of Maxwell's electromagnetic waves. Although a wave theory of light has been firmly established, experiments in this century have shown that, in the interaction of light with matter, light can be understood only as a propagation of discrete particles! In modern quantum theory, the fundamental entities of which light is made have aspects of both wave and particle.

Electromagnetic Waves Electromagnetic waves of any frequency and wavelength can exist, although in different wavelength ranges, their properties may be very different. We therefore speak of different "kinds" of electromagnetic radiation: radio waves, light waves, x-rays, and so forth. All move at the same speed in a vacuum. The human eye responds to wavelengths between 400 and 700 nanometres—this is *visible light*.

Production and Detection Electromagnetic waves are produced by oscillating electric charges; the frequency of oscillation is the same as the wave frequency. The electric and magnetic fields in electromagnetic waves are perpendicular to each other and to the direction of wave propagation. Half the energy density in the electromagnetic wave is that stored in the electric field, and half is that stored in the magnetic field. Electromagnetic waves at radio and television frequencies are detected by currents induced in electric circuits;

in the visible range, they can be detected by the eye or by photographic film. These respond mainly to the electric field. In many cases, electromagnetic waves have the properties of the radiation emitted by an oscillating electric dipole: the intensity is zero along the dipole axis and is maximum in the plane perpendicular to the axis.

Speed of Light   The speed of light can be measured simply by measuring the time required for light to traverse a given distance, but the speed is very high and elaborate methods are required. The first indications of the true magnitude of the speed of light came from observations of the moons of Jupiter. The first laboratory measurements were made in the nineteenth century. They involved the light reflected from a distant mirror being interrupted periodically by a toothed wheel, a rotating mirror, or the like. Foucault measured the speed of light in water and found it to be less than in air. The speed of electromagnetic waves can be derived from the fundamental electric and magnetic constants and agrees with measured values of the speed of light.

Reflection   When electromagnetic waves encounter a barrier or a boundary between transparent media, they are in general partially reflected. The angle that the reflected wave makes with a plane reflecting boundary is the same as that of the incident wave. The fraction of the intensity that is reflected at a boundary depends on the angle of incidence and the wave speeds in the two media. The speed of electromagnetic waves in a material is characterized by the *index of refraction* of the material. Images can be formed by the *specular reflection* from a smooth surface; the *diffuse reflection* from rougher surfaces does not form images. The *law of reflection* follows from Huygens' principle.

Refraction   When electromagnetic waves encounter a boundary between transparent media, some of the energy is transmitted across the boundary. In general, the propagation direction changes at the boundary; this is *refraction*. Refraction at a boundary depends on the indexes of refraction of both media according to Snell's law, which can also be derived from Huygen's principle.

Total Internal Reflection   When light traveling in a medium of higher refractive index is incident upon the boundary of a medium with a lower index, no refraction can occur for angles of incidence greater than a *critical angle* characteristic of the two media. In this case, the light is totally reflected in the first medium. This is the principle on which optical fibers are based.

Polarization   Light is a transverse wave, so it can be polarized. An electromagnetic wave is *polarized* if the plane in which the electric field oscillates is fixed or rotates in simple way and *unpolarized* if the field direction varies randomly. The electromagnetic waves produced by a single atom, an antenna, and the like are generally polarized. The polarization state of light is affected by its interactions with matter: by absorption, reflection, and scattering or by passing through a birefringent (doubly refracting) material.

II.   Numbers and Key Equations

Numbers

Speed of light in a vacuum

$$c = 2.998 \times 10^8 \text{ m/s}$$

Key Equations

Wave frequency and speed

$$f = \frac{c}{\lambda}$$

Electric and magnetic fields in an electromagnetic wave

$$E = cB$$

Wave intensity in terms of energy density or electric field

$$I = \eta c = c\epsilon_0 E_{rms}^2$$

Speed of light in terms of the electromagnetic constants

$$c = \frac{1}{\sqrt{\mu_0 \epsilon_0}}$$

Index of refraction

$$n = \frac{c}{v}$$

Reflected intensity at normal incidence

$$I = \left(\frac{n_2 - n_1}{n_2 + n_1}\right)^2 I_0$$

Snell's law

$$n_1 \sin \theta_1 = n_2 \sin \theta_2$$

Critical angle for total internal reflection

$$\sin \theta_c = \frac{n_2}{n_1}$$

Malus' law

$$I = I_0 \cos^2 \theta$$

Polarizing angle (Brewster's law)

$$\tan \theta_p = \frac{n_2}{n_1}$$

III. Possible Pitfalls

Unlike mechanical waves, such as sound, light requires no medium to propagate it; the electromagnetic field is its own medium. Thus, some wave phenomena that depend specifically on the medium of propagation, such as the doppler effect, may behave differently for light than for mechanical waves.

In a sense, the wave-particle debate about the nature of light was never settled—light is something more complex than either and has aspects of both.

The laws of reflection and refraction can be explained in terms of *either* a wave or a particle theory of light. However, to explain them in terms of a particle theory requires assumptions about the speed of light in different media that turn out to be false.

Light passes from one medium into another with its frequency unchanged; after all, the frequency is the rate at which the electric field is oscillating at each point. When the speed of propagation changes, the *wavelength* must change in proportion.

The index of refraction is the speed of light in a vacuum divided by that in a material; you are likely to get this upside down if you aren't careful. Remember that the speed of light in any material is always less than that in a vacuum, so the index of refraction is always greater than 1.

Total internal reflection occurs only for light that is propagating from a medium with a lower speed of light toward one with a higher speed of light, that is, from a medium of higher refractive index toward one of lower refractive index.

Light is polarized (or more precisely, its state of polarization is affected) by reflection at any angle. The polarizing angle given by Brewster's law is the angle at which the polarization of the reflected light is total.

The *optic axis* in a birefringent material is not a specific line but a *direction*.

IV. Questions and Answers

Questions

1. How should a dipole broadcast antenna be oriented?

2. There are situations in which light travels through a medium with a continuously varying index of refraction. The index of refraction of a gas, for example, is proportional to its density, so the index of refraction of the atmosphere decreases with increasing altitude. What would the path of a ray of light in such a situation look like?

3. Estimate the time it took light to travel the round-trip in Galileo's attempt to measure the speed of light if Galileo and his assistant stood on hills 1 km apart.

4. A timely measurement of the speed of light in a transparent medium such as glass or water compared to that of air would have settled the debate between wave and particle theories of light. How?

5. What kind of electromagnetic radiation has a wavelength of 1 m? 100 μm? 10 nm? What kind has a frequency of 10^6 Hz? 10^{12} Hz? 10^{18} Hz?

6. Which way is the transmission axis oriented in antiglare Polaroid sunglasses?

7. When you are driving, why is a wet highway on a rainy night so much harder to see than a dry one?

8. Why do goggles help you to see under water?

9. Blue light is bent more by a glass prism than red light. Which color of light travels faster in glass?

10. Why does the oar you are rowing with appear to be bent at the water's surface?

11. Why is it that we can hear but not see around corners?

Answers

1. The antenna should be oriented vertically since the intensity, or power, radiated by an oscillating dipole is greatest in the medium plane of the dipole. (This is assuming, of course, that you want to broadcast horizontally.)

2. Figure 24-1 shows a ray of light in a medium in which the refractive index decreases continuously as you go upward. Thinking of this as the earth's atmosphere, you can see that the changing index of refraction affects the way you see objects in the sky, especially those that are near the horizon.

Figure 24-1

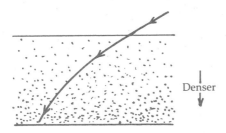

Denser

3. If the hills were 1 km apart, the round-trip would have taken about 7×10^{-6} s.

4. Both theories can be made to predict Snell's law of refraction. However, using the particle model to do so requires that the speed of light in glass or water be greater than that in air (which is, of course, not true) in order to explain the observed direction of refraction. The wave model, on the other hand, requires that the speed of light in glass be less than that in air. Thus, a measurement of the speed of light in air and glass would confirm one or the other. As it happened, this measurement could not be made until after the observation of interference had already confirmed the wave theory.

5. A 1-m wavelength is radio (UHF); 100 μm is around the boundary between microwaves and infrared light; and 10-nm waves are very long x-rays. A 10^6-Hz frequency is in the AM radio band; a 10^{12}-Hz is in the microwave region; and 10^{18}-Hz electromagnetic waves are x-rays.

6. Most of the glare you want to reduce is the reflection of the sun off horizontal surfaces—a highway or the surface of a swimming pool, for instance. The reflected light will thus be partially polarized in a horizontal plane, so you want the polarizing lenses of your sunglasses to pass light that is polarized vertically. Thus, the transmission axis should be vertical.

7. Reflection from a dry highway is diffuse, so some of the light from your headlamps is reflected back at you; this is what lets you see the road surface. When the highway is wet, the water surface reflects the light almost specularly, so little of it comes back to your eye.

8. The stuff your eye is made of is almost water, so there is very little refraction at the eye-water boundary, which means that your eye can't form a clear image of the objects outside it. When you wear goggles, you are looking through an eye-air boundary, and your eye functions more nearly normally.

9. The bending—refraction—of light by a prism occurs because of the difference between the speed of light in the glass and that in air, so the lower the speed of light in the glass, the more the light is bent. Thus, the red light, which is refracted less, travels faster in the glass.

10. Because of the refraction at the air-water boundary, you see the underwater portion of the oar at a different position than where it actually is, and the shaft of the oar appears to bend at the water's surface.

11. In part this is because, since the wavelengths of sound are so much larger, the reflection of sound is more nearly specular. The more important reason, however, is that sound, with wavelengths in metres, is diffracted around corners, doorways, and the like, whereas light, with its very much shorter wavelengths, is not.

V.   Problems and Solutions

Problems

1.   In a certain electromagnetic wave, the rms value of the electric field is 14 V/m. Find (*a*) the rms value of the magnetic field and (*b*) the intensity of the wave.

How to Solve It
- The ratio of magnitudes of the electric to the magnetic field in the wave is the speed of light *c*.

- There is a formula for the intensity of the electromagnetic field in terms of the rms value of the electric field; use this to calculate *I*.

2.   The average intensity of sunlight at the top of the atmosphere is 1350 W/m^2. Assuming the light to be a single electromagnetic wave, what are the corresponding rms values of the electric and magnetic fields?

How to Solve It
- There is a formula for the intensity of the electromagnetic field in terms of the rms value of the electric field; given *I*, use this formula to calculate E_{rms}.

• The ratio of magnitudes of the electric to the magnetic field in the wave is the speed of light c.

3. If 60 percent of the energy consumed by a 60-W light bulb goes into light energy, find (*a*) the intensity of the light at a distance of 10 m from the bulb and (*b*) the rms value of the electric field in the light wave.

How to Solve It
• Intensity is energy propagated across unit area per unit time. If the light energy radiated by the light bulb is propagated uniformly in all directions, the intensity at a distance r is the power emitted by the source divided by the area of a sphere of radius r.

• Don't forget that only 60 percent of the power consumed by the light bulb is radiated as light.

• Once you have the intensity, you can find the rms electric field just as you did in Problem 2.

4. Light is incident normally on a slab of flint glass of refractive index 1.63. What fraction of the incident light intensity passes through the slab?

How to Solve It
• Calculate the fraction of light reflected at the front surface of the glass from its index of refraction using the formula in the text.

• This fraction is reflected at each surface of the slab; whatever's left is the fraction transmitted through it.

5. A ray of light is incident at an angle of 45° on a slab of glass 1 cm thick, as shown in Figure 24-2. The index of refraction of the glass is 1.51. (*a*) Show that the emergent ray is parallel to the incident ray. (*b*) Find the lateral displacement d between the incident and emergent rays.

Figure 24-2

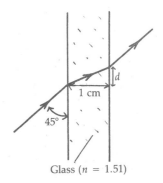

Glass (n = 1.51)

How to Solve It
• The faces of the glass are parallel; thus, if the ray that has passed through the slab makes the same angle with the normal as does the incident light, the transmitted ray is parallel to the incident ray, as advertised.

• Use Snell's law to find the angle of refraction at each surface of the slab.

• From the angles of refraction and the thickness of the glass slab, you can calculate the lateral displacement of the transmitted ray.

6. A glass surface ($n = 1.54$) is covered by a layer of water ($n = 1.33$) of uniform thickness. There is air ($n = 1.00$) on the other side of the water layer. At what minimum angle of incidence must light in the glass strike the glass-water surface in order for it to be totally reflected internally at the air-water surface?

How to Solve It
- Snell's law describes the refraction of the light at each boundary.

- Start with the fact that the ray is totally reflected at the air-water boundary, and calculate its angle of incidence.

- This angle is the angle of refraction at the glass-water boundary; use Snell's law then to calculate the angle of incidence for light coming from within the glass.

7. The object in Figure 24-3 is 85 cm under water. How far back under the dock (distance D in the drawing) must the object be if it cannot be seen from any point above the water's surface? The index of refraction of water is 1.33.

Figure 24-3

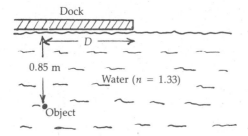

How to Solve It
- If the object cannot be seen at all from above the surface, it must be because any light that could get past the edge of the dock is coming to the water's surface at so large an angle that it is totally reflected.

- Write an expression for the angle of incidence of light that strikes the water just at the edge of the dock.

- Calculate the value of D for which this angle is the critical angle for a water-air boundary.

8. The slab of transparent plastic in Figure 24-4 has an index of refraction $n = 1.31$. Light is incident on the upper surface. For

Figure 24-4

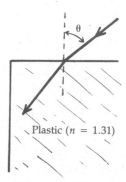

what angles θ will total internal reflection occur at the vertical face of the slab?

How to Solve It

- First calculate the critical angle for the plastic-air boundary.

- Assume that light strikes the vertical boundary at this angle. With a little geometry, you can find the angle at which light is refracted into the plastic at the upper horizontal surface.

- Using this angle, calculate in turn the angle of incidence upon the upper surface.

9. Light is incident on a piece of plastic at an angle of 48° with the normal to the surface. The light refracts into the material in a direction 32° from the normal. What is the speed of light in the plastic?

How to Solve It

- Use Snell's law and the given angles to calculate the index of refraction of the plastic.

- This index of refraction is the ratio of the speed of light in a vacuum to that in the plastic material.

10. The index of refraction of a certain piece of glass is 1.47. Red light of wavelength 635 nm in a vacuum is incident on the surface of the glass. What is the wavelength of the light in the glass?

How to Solve It

- The frequency of the light is unchanged by the boundary, so the wavelength is proportional to the speed of light.

- Thus, the wavelength is changed by a factor equal to the refractive index. Is it longer or shorter in the glass?

11. Assume that you are repeating Fizeau's measurement of the speed of light. Light is divided by a wheel with 700 teeth, travels to a mirror 8.41 km away, and reflects back. Maximum transmission of the reflected light first occurs when the wheel is turning at 1513 rev/min. What value for the speed of light do you infer from these data?

How to Solve It

- The calculations here are very simple—you find the speed of light from the time it takes to travel a measured distance.

- The distance is that to the mirror and back again.

• The time is the time it takes for the wheel to advance from one notch between teeth to the next.

12. The radius of the orbit of Mars around the sun is 2.28×10^{11} m; that of the earth is 1.50×10^{11} m. Space travelers on Mars use radio waves to communicate with the earth. What is the maximum time delay for their signal to reach the earth? What is the minimum delay?

How to Solve It

• Calculate the distance between the earth and Mars at their closest approach.

• Calculate the greatest distance between the earth and Mars, which is when they are on opposite sides of the sun.

• Calculate the travel time for radio waves in each case.

13. Unpolarized light of intensity 1 W/m² is incident on two pieces of ideal Polaroid oriented with their transmission axes at 90° to one another so that none of the light passes through them. A third Polaroid is then placed between the first two, with its transmission axis oriented at 30° to that of the first. What is the intensity of light passing through the stack of Polaroids? Could the third one (the "meat in the sandwich") be oriented differently so as to make the stack pass more light?

How to Solve It

• The function of the first Polaroid is to make plane-polarized light out of the initially unpolarized light.

• What effect does the first Polaroid have on the intensity?

• The light incident on the second and third Polaroids is plane-polarized—use Malus' law to calculate the fraction of the intensity incident on each that gets through it.

Solutions

1.   (*a*) The relation between electric and magnetic field amplitudes in an electromagnetic wave is

$$E = cB$$

Thus,

$$B_{rms} = \frac{E_{rms}}{c} = \frac{14 \text{ V/m}}{3.00 \times 10^8 \text{ m/s}} = \underline{4.67 \times 10^{-8} \text{ T}}$$

(b) The intensity is

$$I = c\epsilon_0 E_{rms}^2$$

$$= (3.00 \times 10^8 \text{ m/s})(8.85 \times 10^{-12} \text{ C}^2/\text{N·m}^2)(14 \text{ V/m})^2$$

$$= \underline{0.520 \text{ W/m}^2}$$

2. $E_{rms} = \underline{713 \text{ V/m}}$; $B_{rms} = \underline{2.38 \times 10^{-6} \text{ T}}$

3. (a) The power being emitted as light (presumably the rest is heat) is 60 percent of the total, so it is $(0.60)(60 \text{ W}) = 36 \text{ W}$. At a distance of 10 m, this power is distributed over an area of

$$A = 4\pi r^2 = 4\pi(10 \text{ m})^2 = 1.26 \times 10^3 \text{ m}^2$$

so the intensity of the light is

$$I = \frac{P}{A} = \frac{36 \text{ W}}{1.26 \times 10^3 \text{ m}^2} = \underline{0.0286 \text{ W/m}^2}$$

(b) The intensity is

$$I = c\epsilon_0 E_{rms}^2$$

so

$$E_{rms}^2 = \frac{I}{c\epsilon_0} = \frac{0.0286 \text{ W/m}^2}{(3.00 \times 10^8 \text{ m/s})(8.85 \times 10^{-12} \text{ C}^2/\text{N·m}^2)}$$

$$= 10.8 \text{ V}^2/\text{m}^2$$

$$E_{rms} = \underline{3.29 \text{ V/m}}$$

4. $\underline{0.889}$ of the incident intensity is transmitted

5. (a) See Figure 24-5 on page 344. Using Snell's law at the front surface of the slab gives

$$n_1 \sin \theta_1 = n_2 \sin \theta_2$$

$$\sin \theta_2 = \frac{n_1}{n_2} \sin \theta_1 = \frac{1.00}{1.51} \sin 45° = 0.468$$

so that $\theta_2 = 27.9°$. Applying it in the same way at the second surface gives

$$\sin \theta_3 = \frac{1.51}{1.00} \sin 27.9° = 0.707$$

Figure 24-5

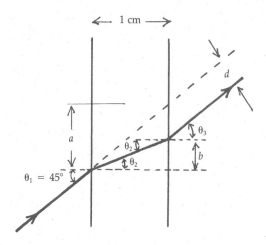

so that $\theta_3 = 45°$. Since $\theta_1 = \theta_3$, the transmitted ray is parallel to the incident ray.

(b) From the figure,

$$a = (1 \text{ cm}) \tan 45° = 1 \text{ cm}$$

and

$$b = (1 \text{ cm}) \tan 27.9° = 0.530 \text{ cm}$$

The lateral displacement is therefore

$$d = (a - b) \cos 45° = \underline{0.332 \text{ cm}}$$

6. $\underline{40.5°}$ Note this is the same as if the water layer weren't there.

7. To pass the edge of the dock, light from an underwater source must strike the water surface at an angle θ of at least that given by

$$\tan \theta = \frac{D}{0.85 \text{ m}}$$

The minimum value of θ at which light will be totally reflected at a water-air boundary is the critical angle θ_c, where

$$\theta_c = \sin^{-1}(n_1/n_2) = \sin^{-1}(1/1.33) = 48.8°$$

Thus, if the object is to be unseeable from any point above the water,

$$\tan^{-1} \frac{D}{0.85 \text{ m}} \geq 48.8°$$

so

$$D \geq (0.85 \text{ m}) \tan \theta_c = (0.85 \text{ m}) \tan 48.8° = \underline{0.971 \text{ m}}$$

8. $\theta < \underline{57.8°}$

9. According to Snell's law

$$n_1 \sin \theta_1 = n_2 \sin \theta_2$$

so the index of refraction of the plastic is

$$n_2 = \frac{n_1 \sin \theta_1}{\sin \theta_2} = \frac{(1.00) \sin 48°}{\sin 32°} = \frac{0.7431}{0.5299} = 1.402$$

The index of refraction is also

$$n_2 = \frac{c}{v}$$

so

$$v = \frac{c}{n_2} = \frac{3.00 \times 10^8 \text{ m/s}}{1.402} = \underline{2.14 \times 10^8 \text{ m/s}}$$

10. $\underline{432 \text{ nm}}$

11. The light travels 8.41 km to the main mirror and 8.41 km back in the time it takes the wheel to advance by one notch. There are 700 notches on the wheel, so the time it takes for it to advance by one notch is

$$t = \left(\frac{1}{700} \text{ rev}\right) \left(\frac{1 \text{ min}}{1513 \text{ rev}}\right) (60 \text{ s/min}) = 5.67 \times 10^{-5} \text{ s}$$

Since in this time the light has traveled $(2)(8.41 \text{ km}) = 16.82 \text{ km}$, its speed is

$$c = \frac{d}{t} = \frac{1.682 \times 10^4 \text{ m}}{5.67 \times 10^{-5} \text{ s}} = \underline{2.97 \times 10^8 \text{ m/s}}$$

12. <u>21.0 min</u>; <u>4.33 min</u>

13. An "ideal" Polaroid passes all the light that is polarized along its transmission axis. The incident light is unpolarized, which means that its polarization direction is random; as much of its energy is associated with electric field components along the transmission axis of the Polaroid as with field components transverse to it. Thus, half the incident unpolarized light intensity gets through the first Polaroid:

$$I_1 = \tfrac{1}{2}I_0$$

By Malus' law, the intensity that gets through the second Polaroid is

$$I_2 = I_1 \cos^2 30° = \tfrac{1}{2}I_0(0.8660)^2 = 0.375I_0$$

What gets through is plane-polarized along the transmission axis of the second Polaroid, so it is at an angle of 60° to that of the third. Thus, what gets through the third is

$$I_3 = I_2 \cos^2 60° = (0.375I_0)(0.500)^2 = \underline{0.0938I_0}$$

If the middle Polaroid is oriented at 45° rather than 30°, one-eighth of the incident intensity gets through the stack; this is the maximum possible.

CHAPTER 25

Geometric Optics

I. Key Ideas

Plane Mirrors If light from a point source is reflected from a smooth surface, such as a plane mirror, it comes to the eye just as if it came from a point behind the mirror. This apparent source is a *virtual image* of the actual source—"virtual" because light does not actually emanate from that point. The image formed by a plane mirror is the same distance behind the plane of the mirror as the source is in front of it, along a line perpendicular to the plane of the mirror. The image formed by one reflecting surface may in turn be imaged by another.

Spherical Mirrors Light from a point can form a *real image* in a concave spherical mirror—"real" in that light actually converges to and diverges from the image point. The real image is inverted. All rays from an object point that are sufficiently close to the symmetry axis of the mirror (*paraxial rays*) converge to a point image. Rays far from the axis are not imaged at the same point, causing a blurred image; this is called *spherical aberration*. Point-to-point imaging is reciprocal.

Focal Point The point at which parallel incident light from a very distant source is brought to a focus by a concave spherical mirror is the *focal point*. The distance of the focal point from the mirror's surface (the *focal length* of the mirror) is half the mirror's radius of curvature. An object closer to the mirror than the focal point forms a virtual image behind the mirror.

Images Formed by Spherical Mirrors The location of the image point can be calculated from the position of the object point and the focal length. If all that is needed is the approximate location of an image, it can be located geometrically by drawing a diagram of the system including certain principal rays, such as those through the focal point and the center of curvature. The same calculation, and the corresponding rays, also apply to a convex mirror if the *focal length is taken to be negative*. That is, the focal point of a convex mirror is behind the mirror.

Magnification The size of the image of an extended object is not in general the same as the size of the object. The ratio of image size to object size is called the *lateral magnification* of the image. The magnification is considered to be positive if the image is upright, negative if it is inverted.

Refraction Refraction of light at the boundary between two transparent media that have different indexes of refraction can also form images. At a spherical boundary, *paraxial rays* (those that strike the mirror very near its axis) from a point source are focused to a point image. The same sign conventions as for a mirror apply *if* a positive (real) image distance or radius of curvature is taken to mean that the image point or the center of curvature is on the side of the boundary that is actually traversed by refracted light. The location of the image can be derived from the law of refraction (Snell's law).

Lenses The most important applications of imaging by refraction involve lenses. A *lens* is a piece of transparent material with curved (spherical) surfaces. The focal length of a thin lens depends on the two radii of curvature and the refractive index of the lens material. A lens has two focal points, one on each side. The first focal point is the source point that would be imaged at infinity; the second is the point at which a parallel incident beam is focused. The focal points of a thin lens are equidistant from the lens.

Images Formed by Lenses A lens that is thicker at middle than at the edges converges parallel incident light—it has a positive focal length—and vice versa. A converging lens forms real images of sources outside its focal point. As in the case of a mirror, images can be located approximately by drawing ray diagrams. The principal rays are those through the focal points and the undeflected ray through the center or *vertex* of the lens. The same rays can be drawn for a negative lens, except that each focal point is on the *opposite side* of the lens from where the corresponding focal point of a positive lens would be.

Combinations Practical optical instruments often involve lenses and mirrors in combinations. These are analyzed by taking the image formed by each lens, refracting surface, or mirror as the object that is imaged by the next.

Power of a Lens The focusing *power* of a lens measures its ability to focus parallel light; it is the reciprocal of its focal length. If the focal length is in metres, power is in *diopters*.

Aberrations Spherical reflecting and refracting surfaces do not focus all parallel incident rays to a single focal point, but only those close to the axis. The blurring of an image due to off-axis rays is *spherical aberration*. Other effects, such as *coma*,

astigmatism, and *distortion*, result from imaging off-axis points or extended objects. These aberrations are due not to defects in the mirror or lens but to properties of the geometry of spherical surfaces. Lenses display *chromatic aberration* as well because the refractive index of the lens material depends on wavelength.

II.   Key Equations

Key Equations

Focal length of a spherical mirror

$$f = \frac{r}{2}$$

Focusing equation for a spherical refracting surface

$$\frac{n_1}{s} + \frac{n_2}{s'} = \frac{n_2 - n_1}{r}$$

Lens-maker's formula

$$\frac{1}{f} = (n - 1)\left(\frac{1}{r_1} - \frac{1}{r_2}\right)$$

Focusing equation for a mirror or a thin lens

$$\frac{1}{s} + \frac{1}{s'} = \frac{1}{f}$$

Magnification

$$m = -\frac{s'}{s}$$

Thin lenses in contact

$$\frac{1}{f} = \frac{1}{f_1} + \frac{1}{f_2}$$

Power

$$P = \frac{1}{f}$$

III.   Possible Pitfalls

The formulas for calculating the image positions for mirrors and lenses must be used with proper *signs* for all quantities, or they'll give you nonsense. Always keep the following *sign conventions* in mind:

> The object distance is positive for a real object, that is, one on the side of the lens or mirror where the actual light is traveling before it is reflected or refracted; otherwise, it is negative. Images are real and image distances are positive for images that are formed on the side of the lens or mirror where the actual light is traveling *after* it is reflected or refracted; otherwise, the image is virtual and the image distance is negative.

> The radius of curvature of a reflecting or refracting surface is positive if the center of curvature is on the side of the surface where a *real* image would be formed; otherwise, it is negative.

Conventionally, we draw ray diagrams with light proceeding to the right from a source on the left. This is not what determines signs, however. The image distance from a lens is negative when the image is on the side *not traversed by refracted light*— that is, when it is on the same side as the source— and thus is virtual, not simply when the image is to the left of the lens.

A positive magnification corresponds to an upright image; a negative magnification, to an inverted image.

When drawing ray diagrams for negative (diverging) lenses, remember that the focal points are on the "wrong" side of the lens. The first focal point, the point of "origin" for light that is refracted parallel to the axis, is *beyond* the lens. The second focal point is a point *before* the lens from which parallel incoming rays *appear* to diverge after they are refracted in the lens.

For a single lens or mirror, an upright image is always virtual and an inverted one is always real. This is *not* necessarily true when there is more than one element, however.

The focusing equation for a mirror or thin lens is usually written in terms of *inverse* distances. When you use it, don't forget the final inversion that gives you a distance for an answer.

The refracting power of a lens is the *inverse* of its focal length; a stronger lens has a smaller, not a larger, focal length.

IV.   Questions and Answers

Questions

1.   Objects under water, when seen from above the water's surface, appear to be at less than their actual depth. Why?

2.   A plane mirror seems to invert your image left and right but not up and down. Why is this?

3.   What is chromatic aberration? What causes it?

4.   Cross-sectional sketches of several thin lenses appear in Figure 25-1 on page 350. Which of these are converging lenses and which are diverging?

Figure 25-1

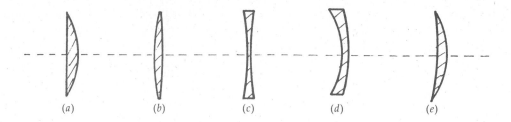

(a)   (b)   (c)   (d)   (e)

5. What is the focal length of a plane mirror?

6. Is it possible for a convex mirror to produce a real image?

7. Both radii of curvature of a certain lens are positive. Is it a converging lens or a diverging lens? Or can't you tell?

8. What is the minimum length of a plane mirror that is necessary in order for you to be able to see your whole body reflected in it?

9. A shaving mirror should produce an image of your face that is larger than your face and upright. Should the mirror be concave or convex?

10. Under what circumstances do an object and its image formed by a spherical mirror coincide? Is the image real or virtual? Is it upright or inverted?

11. Answer Question 10 for a thin lens.

Answers

1. Underwater objects appear to be at less than their actual depth because the index of refraction of water is greater than that of air, so light leaving the water is bent away from the normal. See the ray diagram in Figure 25-2.

Figure 25-2

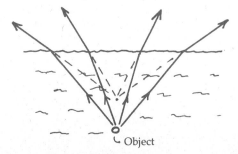

Object

2. Well, really, it doesn't do either one. Look at Figure 25-3. If your head is up, so is your image's head; if your left hand—the one with the ring—is pointing west, so is your image's left hand. What actually happens is that the image is inverted back to front: you are facing north; your image is facing south. Such an image is sometimes called *perverted*.

Figure 25-3

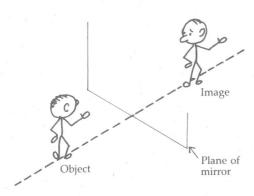

Image

Plane of mirror

Object

3. Chromatic aberration is the appearance of colors not present in the object in images formed by white light. It is due to the dispersion of the light by the lens material. The speed of light in any transparent material is a function of the wavelength, and wavelength is color. Thus, the focal length of a given hunk of glass for red light is a little different than it is for blue, for example.

4. A converging lens is thicker in the middle than it is at the edges; a diverging lens is thinner in the middle. The converging lenses here are (*a*), (*b*), and (*e*). The others are diverging lenses.

5. A plane surface has an infinite radius of curvature, so the focal length of a plane mirror is infinite.

6. A convex mirror has a negative focal length, and for a real image, the image distance s' is positive. Thus, when the object distance in the focusing equation for a convex mirror

$$\frac{1}{s} = -\frac{1}{s'} + \frac{1}{f}$$

is negative, a real image will be formed. This means that the object is virtual. This case is illustrated in Figure 25-4 on page 352.

Figure 25-4

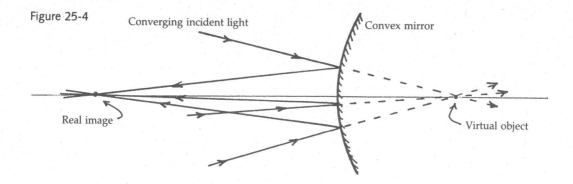

Converging incident light

Convex mirror

Real image

Virtual object

7. You can't tell. If the radius of the front surface has a larger magnitude, it's a diverging lens; otherwise, it's a converging lens.

8. The mirror has to be half your height; see Figure 25-5. Perhaps surprisingly, the answer doesn't depend on how far from the mirror you stand.

Figure 25-5

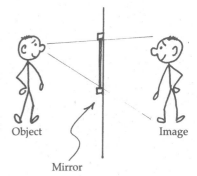

Object

Image

Mirror

9. To get an upright (virtual) image that is larger than the object, you must be using a concave mirror with the object (your face) inside the focal point.

10. For a mirror, object and image coincide if $s' = s = 2f = r$; that is, the object is at the center of curvature of the mirror. The image is real, since s' is positive, and inverted, since $m = -s'/s = -1$ is negative.

11. For a lens, real image space is on the opposite side of the lens from object space, so for object and image to coincide means that $s' = -s$. From the focusing equation for a mirror or thin lens, this means that $1/f = 0$. In this case, the lens is a flat piece of glass (or perhaps a meniscus with parallel surfaces), and the image is the object.

V. Problems and Solutions

Problems

1. An object is 25 cm in front of a convex spherical mirror. The image of the object in the mirror is half the size of the object. What is the radius of curvature of the mirror?

How to Solve It

- You're given the lateral magnification of the image; remember that it is also the ratio of image to object distance.

- Be careful—is the image in front of the mirror or behind it?

- Use the focusing equation for a mirror and the image and object distances to calculate the focal length of the mirror, which is half its radius of curvature.

2. The opposite walls of a barbershop are covered by plane mirrors such that you see many reflected images of yourself, receding to infinity, due to multiple reflections. The width of the shop is 6.5 m, and you are standing 2 m from the north wall. How far apart are your first two images behind the mirror on the north wall?

How to Solve It

- Each image formed in either mirror acts as an object that is reflected in the other mirror. If you have trouble seeing this, draw a ray diagram.

- The image formed by a plane mirror is the same distance behind the mirror plane as the object is in front of it.

- The first two images behind the north mirror are (1) the first image of you and (2) the image of the first image formed in the south mirror.

3. An object is 40 cm in front of a concave spherical mirror whose radius of curvature is 32 cm. Locate the image formed by the mirror (*a*) by calculating the image distance and (*b*) by drawing a ray diagram.

How to Solve It

- First, calculate the image distance directly from the focusing equation for a mirror. Is the image real or virtual? Upright or inverted? Larger or smaller than the object?

- To construct the ray diagram, first draw a ray from an off-axis object point parallel to the axis of the mirror. This ray is reflected through the focal point *F*.

- Then draw a ray from the same object point that passes through *F* before hitting the mirror. This ray will be reflected parallel to the axis.

- The third principal ray for mirror diagrams is one from the object point through the center of curvature *C*; it reflects back upon itself. This one's a little hard to fit into the diagram in this particular case.

4. A concave spherical mirror has a radius of curvature of 60 cm. Where should an object be placed if its image is to be virtual and three times the size of the object?

How to Solve It

- The fact that the image is to be three times the size of the object means that it must be three times as far from the mirror.

- The fact that the image is to be virtual means that s' must be negative. Thus, the object is inside the focal length f.

- The focal length of a spherical mirror is half its radius of curvature. Knowing s' and f, you can use the focusing equation for a mirror to find s.

5. A fish is 12 cm from the front surface of a spherical fish bowl of radius 20 cm. A person 1 m outside the surface of the bowl is looking at the fish. (*a*) Where does the fish appear to be according to the person viewing it from outside? (*b*) What does the fish see of the person? The index of refraction of water is 1.33.

How to Solve It

- Use the equation for focusing by a single spherical refracting surface.

- Remember the sign convention: If the surface is concave toward the object, its radius of curvature is negative; if the surface is convex, its radius is positive.

- In (*a*), the fish is the object, and we want to know where its image is.

- In (*b*), the person outside is the object and we want to know what the fish sees. Don't forget that the refractive indexes interchange!

6. Consider the front surface of the eye to be a spherical refracting surface of a material with an index of refraction of 1.35 that forms an image of very distant objects on the retina 2.7 cm behind it. What must be the radius of curvature of the surface?

How to Solve It

- Use the equation for focusing by a single spherical refracting surface.

- Remember the sign convention: If the surface is concave toward the object, its radius of curvature is negative; if the surface is convex, its radius is positive.

- Here we know s, s', n_1, and n_2, and we want to find r.

7. The focal length of a glass lens (index of refraction 1.55) is +16 cm. If the radius of curvature of its front surface is +12 cm, what is the radius of curvature of its rear surface?

How to Solve It

- Use the lens-maker's formula, which gives the focal length of a lens from its refractive index and radii of curvature.

- In this case, you know all the quantities in the lens-maker's formula except for r_2, so you can solve for that.

8. A thin lens made of glass of refractive index 1.60 has radii of curvature of magnitude 12 and 18 cm. What are the possible values for its focal length?

How to Solve It

- Use the lens-maker's formula.

- The only thing that makes for more than one possibility is the signs of the radii of curvature. You get one value for the magnitude of the focal length if the radii are of like sign and a different value if they are of opposite sign.

9. A light source is 1 m from a screen. A lens placed at either of two points between the source and the screen focuses an image of the source onto the screen. The image formed with the lens in one of these two positions is three times the size of the image formed with the lens in the other position. What is the focal length of the lens?

How to Solve It

- You can set this up and solve for the two lens positions by brute force, but this isn't really necessary. Remember that imaging is reversible.

- The two different lens positions must just have the object and image distances interchanged.

- Be careful—what you're told here isn't the magnification in either case, but rather the ratio of the magnifications in the two cases.

10. An object is 32 cm from a converging lens whose focal length is 40 cm. Locate the image formed by the lens (*a*) by calculating the image distance and (*b*) by drawing a ray diagram.

How to Solve It

- First, calculate the image distance directly from the focusing equation for a thin lens. Is the image real or virtual? Upright or inverted? Larger or smaller than the object?

- To construct the ray diagram, first draw a ray from an off-axis object point parallel to the axis of the lens. This ray is refracted through the second focal point.

- Then draw a ray from the same object point that passes through the first focal point before hitting the lens. This ray will be reflected parallel to the axis.

- The third principal ray for lens diagrams is one from the object point through the vertex of the lens (the point where the symmetry axis passes through the lens). This ray is undeflected.

11. When a bright light source is placed 36 cm in front of a lens, there is an upright image 14 cm from the lens. There is also a faint inverted image 13.8 cm in front of the lens due to reflection from the front surface of the lens. When the lens is turned around, this weaker inverted image is 25.7 cm in front of the lens. What is the index of refraction of the lens material?

How to Solve It
- The object of the game here is to calculate n from the lens-maker's formula. To do that, you need the focal length and the two radii of curvature of the lens.

- You can get the focal length from the information about the image formed by the lens. Be careful—is it real or virtual?

- You can get the radius of curvature of each surface from the image formed by reflection from it, just as if it were a spherical mirror.

12. Two lenses, one of focal length $+15$ cm and one of focal length -15 cm, are 10 cm apart. Where is the final image of an object that is 30 cm in front of the positive lens?

How to Solve It
- First, ignore the second lens and, using the focusing equation for a thin lens, find out where the first (positive) lens acting alone would form its image.

- The next thing that happens to light from the source is that it passes through the second (negative) lens. Thus, the image that is formed by the first lens is the object that is imaged by the second.

- Actually, the image that would be formed by the first lens is behind the position of the second, so it is a virtual object for the second. This makes no difference in applying the focusing equation, but the value of s is negative.

Solutions

1.   Because the mirror is convex, it forms an upright virtual image behind the mirror. An image "half the size of the object" means that the magnification of the lens is 0.5:

$$m = -\frac{s'}{s}$$

$$s' = -ms = -(0.5)(25 \text{ cm}) = -12.5 \text{ cm}$$

We can now use the focusing equation for a mirror to find the focal length:

$$\frac{1}{s} + \frac{1}{s'} = \frac{1}{f}$$

$$\frac{1}{25 \text{ cm}} + \frac{1}{-12.5 \text{ cm}} = \frac{1}{f}$$

so

$$f = -25 \text{ cm}$$

The focal length is half the radius of curvature, so

$$r = 2f = \underline{-50 \text{ cm}}$$

The mirror is convex with a radius of curvature of 50 cm.

2.   <u>9.0 m apart</u>

3.   (*a*) Image formation is described by

$$\frac{1}{s} + \frac{1}{s'} = \frac{1}{f}$$

The radius of curvature of the mirror is 32 cm, so its focal length $f = \frac{1}{2}r = 16$ cm, and

$$\frac{1}{40 \text{ cm}} + \frac{1}{s'} = \frac{1}{16 \text{ cm}}$$

$$\frac{1}{s'} = 0.0625 \text{ cm}^{-1} - 0.0250 \text{ cm}^{-1} = 0.0375 \text{ cm}^{-1}$$

so

$$s' = \underline{26.7 \text{ cm}}$$

The image is real, inverted, and 26.7 cm in front of the mirror.

(b) The three principal rays that should appear on your ray diagram are as follows: (1) A ray that leaves the object parallel to the axis and is reflected through the focal point F. (2) A ray from the object that comes to the mirror through the focal point F and is reflected parallel to the axis. (3) A ray from the object that comes to the mirror through its center of curvature C and is reflected back upon itself. These three rays are shown on the ray diagram in Figure 25-6.

Figure 25-6

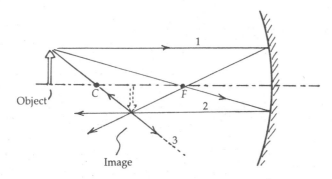

4. 20.0 cm in front of the mirror

5. The focusing equation for a single spherical refracting surface is

$$\frac{n_1}{s} + \frac{n_2}{s'} = \frac{n_2 - n_1}{r}$$

(a) The person outside is looking at the fish. The refracting surface is concave toward the fish (the object). Thus, $s = 12$ cm and $r = -20$ cm, and we want the image distance s'

$$\frac{1.33}{12 \text{ cm}} + \frac{1}{s'} = \frac{(1 - 1.33)}{-20 \text{ cm}}$$

$$\frac{1}{s'} = \frac{0.33}{20 \text{ cm}} - \frac{1.33}{12 \text{ cm}} = -0.0943 \text{ cm}^{-1}$$

$$s' = -10.6 \text{ cm}$$

The man sees the fish 10.6 cm behind the surface of the bowl.

(b) Now the person outside is the object; consider how the

fish sees him. The surface—concave away from the object—has a radius of curvature $+20$ cm. Thus,

$$\frac{1}{100 \text{ cm}} + \frac{1.33}{s'} = \frac{1.33 - 1}{+20 \text{ cm}}$$

giving $s' = \underline{205 \text{ cm}}$. The image of the man formed by the surface of the fishbowl is far behind the position of the fish, so the fish doesn't see it. The fish sees the man directly, badly out of focus.

6. $\underline{7.0 \text{ mm}}$

7. The focal length of the lens is given by the lens-maker's equation:

$$\frac{1}{f} = (n - 1)\left(\frac{1}{r_1} - \frac{1}{r_2}\right)$$

so here

$$\frac{1}{r_1} - \frac{1}{r_2} = \frac{1}{(n - 1)(f)} = \frac{1}{(0.55)(16 \text{ cm})} = 0.1136 \text{ cm}^{-1}$$

We know that $r_1 = +12$ cm, so

$$\frac{1}{12 \text{ cm}} - \frac{1}{r_2} = 0.1136 \text{ cm}^{-1}$$

which gives $1/r_2 = -0.0303 \text{ cm}^{-1}$ or $r_2 = \underline{-33.0 \text{ cm}}$. The negative sign means that the second surface is convex also. Thus, the lens is biconvex, with radii of 12 cm and 33 cm.

8. The focal length is $\underline{\pm 12 \text{ cm}}$ if both radii of curvature have the same sign, $\underline{\pm 60 \text{ cm}}$ if they are of opposite signs.

9. The magnification of a single lens is $m = -s'/s$, and here we are given that $m_2 = 3m_1$. Thus,

$$\frac{s_2'}{s_2} = 3\frac{s_1'}{s_1}$$

But imaging is a reversible process: if point A is imaged at B, then B is imaged by the same lens at A. As there are only two focusing positions, the image distance in one case must be the object distance in the other and vice versa. Thus, $s_2' = s_1$ and $s_2 = s_1'$, so

$$\frac{s_1}{s_1'} = 3\frac{s_1'}{s_1}$$

Cross-multiplying,

$$3s_1'^2 = s_1^2$$

$$s_1' = \frac{s_1}{\sqrt{3}} = 0.577s_1$$

We know that the distance from source to screen is 1 m, so

$$s_1 + s_1' = 1 \text{ m}$$

Substituting in the value for s_1',

$$s_1 + 0.577s_1 = 1\text{m}$$

$$s_1 = \frac{1 \text{ m}}{1.577} = 0.634 \text{ m}$$

Then

$$s_1' = 0.577s_1 = (0.577)(0.634 \text{ m}) = 0.366 \text{ m}$$

The focal length of the lens is therefore

$$\frac{1}{f} = \frac{1}{s} + \frac{1}{s'} = \frac{1}{0.634 \text{ m}} + \frac{1}{0.366 \text{ m}}$$

$$f = \underline{23.2 \text{ cm}}$$

10. The image is virtual, and <u>160 cm from the lens.</u>

11. Let the radius of curvature of the original front surface of the lens be r_1. This will be positive if the front surface is convex; thus, the radius of curvature of the front surface considered as a spherical mirror is $-r_1$. Using the focusing equation for a mirror, then,

$$\frac{1}{s} + \frac{1}{s'} = \frac{1}{f} = \frac{2}{r}$$

$$\frac{1}{36 \text{ cm}} + \frac{1}{13.8 \text{ cm}} = \frac{2}{-r_1}$$

giving $r_1 = -20.0$ cm. The negative sign means that the lens surface is concave. The same analysis with the lens turned around shows that its other surface is also concave, with a radius of curvature of 30.0 cm. The lens is thus biconcave and is therefore a diverging lens; the lens image must therefore be virtual, so its image distance is negative. Thus, for the lens, $s' = -14$ cm. We get the lens focal length from the focusing equation for a thin lens:

$$\frac{1}{f_{lens}} = \frac{1}{s} + \frac{1}{s'} = \frac{1}{36 \text{ cm}} + \frac{1}{-14 \text{ cm}} = -0.04365 \text{ cm}^{-1}$$

$$f_{lens} = -22.9 \text{ cm}$$

At last, we have the focal length and both radii of curvature of the lens; thus, we can get the refractive index from the lens-maker's formula:

$$\frac{1}{f_{lens}} = (n - 1)\left(\frac{1}{r_1} - \frac{1}{r_2}\right)$$

$$\frac{1}{-22.9 \text{ cm}} = (n - 1)\left(\frac{1}{-20.0 \text{ cm}} - \frac{1}{+30.0 \text{ cm}}\right)$$

$$-0.04365 \text{ cm}^{-1} = (n - 1)(-0.08333 \text{ cm}^{-1})$$

giving

$$n - 1 = 0.524$$

$$n = \underline{1.524}$$

12. 50 cm in front of the first lens

CHAPTER 26

Optical Instruments

I. Key Ideas

The Eye The eye is our most familiar and most important optical instrument. Light coming to the eye is focused primarily by the front surface (the *cornea*) onto the photosensitive rear surface (the *retina*) about 2.5 cm away. To image objects at different distances, slight changes in focal length are made by changing the shape of the lens (accommodation). Vision is most distinct at the shortest distance for clear vision (the *near point*). Focusing errors must often be corrected with external lenses.

Apparent Size The apparent size of an object seen by the eye is determined by the size of the image on the retina, which is determined by the *angle* the object subtends at the eye.

Simple Magnifier A short-focal-length positive lens allows one to look at a very near object by forming an image of it farther away. The object appears larger because, being nearer, it subtends a larger angle at the eye. In compound optical instruments, the *eyepiece* is a simple magnifier used to look at the image formed by other lenses.

The Camera A camera functions very much like the eye: a positive lens forms an image of objects outside the camera on photosensitive film at its focal plane. The focal length of the lens is fixed;

accommodation is accomplished by moving the lens toward or away from the film. As in the eye, a variable aperture adjusts the amount of light admitted. The maximum aperture size is determined by the size of the lens. In good cameras, the optical systems use combinations of several lenses.

The Microscope A compound microscope is a two-lens system used to look at very small objects. The *objective* lens is a short-focus positive lens that forms a large real image of a small object that is just outside its focal point. A second positive lens, the eyepiece, is used as a simple magnifier to look closely at this image.

The Telescope A telescope is a two-lens system used to view distant objects. The objective is a positive lens with a long focal length that forms a real image of the object at its focal point. The eyepiece is a simple magnifier used to view this image. The angular magnification is the ratio of the objective and eyepiece focal lengths. The main purpose of an astronomical telescope is not magnification, as many astronomical objects are so far away as to still appear as points, but to gather more light. Thus, a large objective is desirable. The largest telescopes use mirrors rather than lenses as objectives. Terrestrial telescopes are designed slightly differently in order to produce an upright final image.

II. Key Equations

Key Equations

Angular magnification of a simple magnifier

$$M = \frac{x_{np}}{f}$$

Camera aperture

$$f/\text{number} = \frac{f}{D}$$

Magnification of a compound microscope

$$m = -\frac{lx_{np}}{f_o f_e}$$

Angular magnification of a telescope

$$M = -\frac{f_o}{f_e}$$

III. Possible Pitfalls

The cornea—the front surface of the eye—does most of the eye's focusing. The lens, despite its name, is only a fine-tuning device. It is nearly correct to think of the eye as a single refracting surface.

The magnification of a simple magnifier is an *apparent* magnification. The angle that the image subtends at the eye is approximately the same as it would be without the lens—except that, without the lens, the object would have to be held much farther away in order for the eye to focus on it. This is why the near point distance of the eye figures in the formulas for magnification.

The most important attribute of an astronomical telescope is not its magnification—individual stars remain points—but its light-gathering power.

The magnification of a telescope is greater for a longer objective focal length; that of a microscope is greater for a shorter objective focal length.

IV. Questions and Answers

Questions

1. When you buy a decent 35-mm camera, it will usually come with an $f/1.8$ or $f/2$ lens; a "faster" $f/1.4$ lens will cost you a lot more—probably as much as the camera body. How come?

2. The objectives of really large astronomical telescopes are always mirrors rather than lenses. Why is this?

3. As we have described it, the final image of a telescope is virtual. How then is it possible to take telescopic (that is, astronomical) photographs by allowing the image to fall on photographic film?

4. I am very nearsighted; my personal near point is 14 cm from my eyes. Do I need a converging or a diverging lens to correct this defect?

5. What's the point of bifocals?

6. Eyeglasses of power +1.25 diopters are prescribed for a certain person. Is she nearsighted or farsighted?

7. When might you want to use a small camera aperture rather than a large one?

Answers

1. The f/number refers to the diameter of the lens relative to its focal length; for a given focal length, there is twice the area of glass in an f/1.4 as in an f/2 lens. Not only do the precision optical elements have to be twice as big, but the lens design is probably more complex as well: it is much harder to correct aberrations in the larger lens.

2. A mirror can be ground with a paraboloidal rather than a spherical surface and so be free of spherical aberration, and mirrors do not exhibit chromatic aberration in any case. The real advantage, however, is mechanical rather than optical: only one surface of a mirror matters, so it can be supported from the back. A lens can be supported only at its edges, so mounting large ones is very difficult. Very large lenses sag under their own weight enough to affect their optical properties.

3. The eyepiece is dispensed with, and the film is placed at the position of the real image formed by the objective.

4. Since I am nearsighted, my uncorrected eye brings light from distant objects to a focus in front of my retina and light from very near objects on the retina. To correct this, I use diverging lenses in front of my eyes to cancel out a little of the positive focusing power of my uncorrected eye.

5. The aging eye loses some of its ability to accommodate because the crystalline lens becomes less flexible. Bifocals help the eye to accommodate by giving it the choice of two corrective lenses of slightly different focal length. (Some of us decrepit types, in fact, need trifocals!)

6. She is farsighted—her eye needs the help of a converging (positive) lens to bring light from nearby objects to a focus on the retina.

7.   You would choose a small aperture when there is plenty of light and depth of field is important.

V.   Problems and Solutions

Problems

1.   The most distant object on which my (uncorrected) eyes will focus clearly is about 40 cm away and the nearest about 14 cm away. (*a*) If glasses are to correct my vision so that I can focus at infinity, what focal length is required? (*b*) With corrective lenses of this focal length, what will be my near point?

How to Solve It
- Assume the cornea-to-retina distance is 2.5 cm—this is taken as standard several times in the text.

- The lenses must image very distant objects at a distance of 40 cm from my eyes.

- The corrected near point will be the position of an object that the lenses image at a distance of 14 cm from my eyes.

2.   A farsighted secretary needs to read from a word-processor screen 50 cm from her eyes; her uncorrected near point distance is 110 cm. What must the focal length of the lenses in her reading glasses be to form an image of the screen 110 cm away?

How to Solve It
- You can find the power of the secretary's uncorrected eye at its near point. (This is the shortest focal length or the maximum power that her eye will provide.)

- The extra power required to focus on a closer object must be provided by her glasses lenses.

- For lenses in contact, which is nearly true here, the focusing powers just add. The power of a lens is $1/f$.

3.   A simple magnifier is a positive lens with a focal length of 4.5 cm. What apparent magnification does it provide (*a*) for a person with "normal" vision and (*b*) for the farsighted secretary in Problem 2, whose near point distance was 110 cm?

How to Solve It
- The apparent magnification of a simple magnifier is just the ratio of the user's near point distance to the focal length of the lens.

- The "normal" value of near point distance is 25 cm.

4. The instructions that came with your film suggest an aperture stop of $f/5.6$ at 1/200 s for today's conditions. (*a*) If you want to photograph a high-jumper at 1/1500 s, what aperture stop should you use? (*b*) If you want to photograph a landscape using an aperture of $f/16$ for extreme depth of field, what shutter speed should you use?

How to Solve It
- Assume that the film will respond the same to the same amount of light, no matter how long an exposure is required to get it. Actually, as photographers know, this isn't quite true, but it's near enough.

- The amount of light that gets through the lens is proportional to the area of the aperture—that is, to the square of the aperture diameter.

- To take the picture faster, a larger aperture (a smaller $f/$number) is needed.

5. An astronomical telescope consists of two positive lenses 1 m apart and has a magnification of 20. What are the focal lengths of the two lenses?

How to Solve It
- In an astronomical telescope, the two interior focal points coincide, so the distance between the two lenses is the sum of their focal lengths.

- The magnification (never mind the sign) of an astronomical telescope is the ratio of the focal lengths.

- Now you know the ratio of f_o and f_e and you know their sum; solve for the individual values.

6. The objective lens of a large astronomical telescope has a focal length of 14 m. It is used to view Jupiter when that planet is at its closest approach to the earth. (Under these conditions Jupiter, whose diameter is 143,000 km, is 630,000,000 km away.) What is the diameter of the image of Jupiter formed by the objective?

How to Solve It
- The image of a single lens subtends the same angle at the lens as does the object.

- Thus, the sizes of Jupiter and its image formed by the objective lens are in the same ratio as their distances from the lens.

7.   A 35-mm camera with interchangeable lenses is used to take a picture of a hawk 30 m away that has a wing span of 1.2 m. What focal length lens should be used to make an image of the hawk 2.5 cm wide on the film?

How to Solve It
- You know the sizes of both the image and the object; calculate the lateral magnification of the lens.

- The magnification is also the ratio of image and object distances from the lens.

- Use the image and object distances to calculate the focal length.

8.   A compound microscope has an overall length of 26 cm and an overall magnification of 40. If the focal length of the eyepiece is 4 cm, what is the focal length of the objective?

How to Solve It
- Assume that the viewer has the "normal" near point distance of 25 cm.

- The tube length l of the microscope is the distance between the interior focal points; thus, you can find it by subtracting the two focal lengths from the overall length of the instrument.

- Unfortunately you don't know f_o, so you have to leave l in terms of the unknown f_o. Solve the equation that results for f_o.

9. . A compound microscope has a tube length of 20 cm and an objective lens of focal length 8 mm. (*a*) If it is to have an overall magnification of 200, what should be the focal length of the eyepiece? (*b*) If the final image is viewed at infinity, how far from the objective should the object be placed?

How to Solve It
- Assume that the viewer has the "normal" near point distance of 25 cm.

- You can calculate the focal length of the eyepiece from the formula for the magnification of a compound microscope.

- The image distance from the objective is the focal length of the objective plus the tube length. Using this and the focal length, calculate the object distance.

Solutions

1. (*a*) The most distant image on which my eye will focus is at 40 cm. The correcting lens should therefore image very distant objects at this distance. The point at which very distant objects are imaged is (by definition) the focal point. Since this focal point is in front of the lens, f is negative. Thus, $f = -40$ cm.

(*b*) The actual near point of my eye is at 14 cm distance, so the nearest object on which I can focus with the corrective lens is one that is imaged 14 cm from my eye. This is 14 cm behind the lens, so $s' = -14$ cm. With $f = -40$ cm, my corrected near point s is

$$\frac{1}{s} + \frac{1}{s'} = \frac{1}{f}$$

$$\frac{1}{s} = \frac{1}{f} - \frac{1}{s'} = \frac{1}{-40 \text{ cm}} - \frac{1}{-14 \text{ cm}} = 0.0464 \text{ cm}^{-1}$$

or

$$s = 21.5 \text{ cm}$$

2. $+92$ cm

3. The apparent magnification of a simple magnifier is

$$M = \frac{x_{np}}{f}$$

if the image is viewed at infinity.

(*a*) For the person with normal vision,

$$M = \frac{25 \text{ cm}}{4.5 \text{ cm}} = 5.6$$

(*b*) For the farsighted secretary,

$$M = \frac{110 \text{ cm}}{4.5 \text{ cm}} = 24$$

4. (*a*) $f/2.0$ (*b*) $1/25$ s

5. When the astronomical telescope is focused at infinity, its interior focal points coincide. Therefore,

$$f_o + f_e = 1 \text{ m}$$

The magnification (disregarding the sign) of the telescope is

$$M = f_o/f_e = 20$$

so

$$f_e = 0.05f_o$$

Thus,

$$f_o + 0.05f_o = 1 \text{ m}$$

$$f_o = \frac{1 \text{ m}}{1.05} = \underline{0.952 \text{ m}}$$

and

$$f_e = 0.05f_o = \underline{0.0476 \text{ m}}$$

6. <u>3.18 mm</u>

7. The camera has a single lens that casts a real image, which is inverted, on the film plane. From the known image and object sizes,

$$m = \frac{y'}{y} = \frac{(-0.025 \text{ m})}{1.20 \text{ m}} = -0.0208$$

This is also the ratio of image and object distances:

$$m = -\frac{s'}{s}$$

$$s' = -ms = -(-0.0208)(30 \text{ m}) = 0.625 \text{ m}$$

From the equation for a thin lens,

$$\frac{1}{f} = \frac{1}{s} + \frac{1}{s'} = \frac{1}{30 \text{ m}} + \frac{1}{0.625 \text{ m}} = 1.633 \text{ m}^{-1}$$

$$f = \underline{0.612 \text{ m}}$$

This is just over two feet. A lens with a focal length this long is not at all impossible, although a practical version would probably by "folded" optically so as to be more convenient to handle.

8. <u>2.97 cm</u>

9. (a) The tube length l of the microscope is the interior distance

between the focal points of the two lenses. The overall magnification is

$$m = \frac{x_{np}l}{f_o f_e}$$

so, if the user has "normal" vision ($x_{np} = 25$ cm),

$$f_e = \frac{x_{np}l}{mf_o} = \frac{(25 \text{ cm})(20 \text{ cm})}{(200)(0.80 \text{ cm})} = \underline{3.12 \text{ cm}}$$

(b) For the final image to be at infinity, the image formed by the objective lens must be at the focal point of the eyepiece, so the image distance is

$$s' = 20 \text{ cm} + 0.8 \text{ cm} = 20.8 \text{ cm}$$

Now

$$\frac{1}{s} + \frac{1}{s'} = \frac{1}{f_o}$$

$$\frac{1}{s} = \frac{1}{f_o} - \frac{1}{s'} = \frac{1}{0.8 \text{ cm}} - \frac{1}{20.8 \text{ cm}} = 1.20 \text{ cm}^{-1}$$

$$s = \underline{0.832 \text{ cm}}$$

The object should be 8.32 mm from the objective, that is, 0.32 mm outside the focal point of the objective.

CHAPTER 27

Physical Optics: Interference and Diffraction

I. Key Ideas

Phase Difference When harmonic waves of the same frequency and wavelength combine, the resultant is another wave of the same frequency and wavelength but with an amplitude that depends on the phase difference between the component waves. If the component waves are in phase (or any multiple of 360° out of phase, which is the same thing) they interfere *constructively*; that is, their amplitudes simply add. If they are 180° out of phase, they cancel each other; this is *destructive interference*.

Sources of Phase Difference If light reaches a point by two or more different paths, the difference in path length produces a phase difference. A path length difference of one wavelength corresponds to a phase difference of one cycle or 360°, which is the same as no phase difference. Reflection of light at a boundary may also introduce a phase difference of 180° if the speed of light is greater in the first (incident) medium than in the second.

Coherence Waves that differ in phase by a constant amount are said to be *coherent*. Light waves from separate sources are incoherent; coherence in optics is achieved by dividing the light from a single source.

Thin-Film Interference We observe interference in the light reflected from the two surfaces of a thin film of transparent material. There is a path length difference because the light reflected by the farther surface has traveled twice the thickness of the film farther than that reflected from the near surface. Note that this extra path length is in the transparent material. The reflection at either surface or both may introduce another 180° phase difference. When the total phase difference is a whole number of cycles, the interference is constructive and the intensity of the reflected light is a maximum; when it is an odd multiple of 180°, the interference is destructive and the reflected intensity is a minimum. If the film thickness varies, the interference is alternately constructive and destructive, and we see "fringes" in the reflected light. If white light is incident, the interference maxima occur at different thicknesses for different wavelengths, and we see various colors in the reflected light.

Interferometry An *interferometer* is an instrument that uses interference phenomena to make precise measurements. In a *Michelson interferometer*, light from a source is split by a half-silvered mirror, reflected back along two perpendicular paths, and recombined. When the mirror in one arm moves by a quarter wavelength, the interference phase changes by one-half cycle, and the interference pattern changes by half a fringe; that is, a dark fringe replaces a bright one. Thus, distance measurements to a fraction of one wavelength of light can be made.

Two-Slit Interference Young's experiment, which demonstrated conclusively the wave nature of light, was a two-slit interference experiment. In this experiment, the "two sources" are two narrow parallel slits in a barrier that are illuminated by a single light source. The light from the slits on a screen far away displays an interference pattern determined by the difference in path from the slits to each point on the screen. The pattern is determined by the ratio of the slit spacing to the wavelength of the light.

Interference of Three or More Sources In the interference pattern produced by three or more identical sources, the main interference maxima (there are secondary maxima in between) are in the same place no matter how many sources there are, but they are much sharper and brighter when there are many sources.

Diffraction Diffraction is the propagation of light waves in other than straight lines. It occurs whenever part of a wavefront is limited, as by an aperture or obstacle. Because of diffraction, light passing through a single slit in an opaque barrier spreads outside the geometrical shadow of the aperture. The spreading angle is approximately the ratio of wavelength to slit width. The intensity pattern far from the slit (*Fraunhofer diffraction*) can be derived from Huygens' principle. Diffraction patterns near the aperture (*Fresnel diffraction*) are much more difficult to calculate.

Interference and Diffraction The actual pattern produced in the two-slit experiment is a combination of the two-slit interference pattern *and* the diffraction pattern of the light passing through each slit.

Aperture and Resolution In any optical instrument in which light enters through a finite aperture, diffraction at the aperture limits the instrument's *resolution*. The central maximum in the pattern from a circular aperture is within an angle given approximately by $1.22\ \lambda/d$. The images of objects that are separated by angles smaller than this will not be distinguishable. Astronomical instruments need to have large apertures for the sake of resolution as well as light-gathering power.

The Diffraction Grating A diffraction grating consists of very many equally spaced lines or slits. The interference maxima in the light passing through the grating are extremely sharp. A diffraction grating is used to analyze the *spectrum* (the distribution of energy among different wavelengths) of light from various sources. The *resolving power* of the grating is determined by the total number of slits illuminated.

II. Key Equations

Key Equations

Phase difference due to path length difference

$$\delta = \frac{\Delta x}{\lambda}\, 360°$$

Thin-film interference maxima with one 180° phase change

$$\frac{2t}{\lambda'} = m + \frac{1}{2} \quad \text{for } m = 0, 1, 2, 3, \ldots$$

$$\text{with } \lambda' = \frac{\lambda}{n}$$

Thin-film interference maxima with no or two 180° phase changes

$$\frac{2t}{\lambda'} = m \quad \text{for } m = 0, 1, 2, 3, \ldots$$

$$\text{with } \lambda' = \frac{\lambda}{n}$$

Two-slit interference maxima

$$m\lambda = d \sin \theta \quad \text{for } m = 0, 1, 2, 3, \ldots$$

$$y_m = m \left(\frac{\lambda L}{d} \right)$$

Single-slit diffraction minima

$$a \sin \theta = m\lambda \quad \text{for } m = 1, 2, 3, \ldots$$

Circular aperture diffraction—first minimum

$$\sin \theta = 1.22 \frac{\lambda}{D}$$

Resolving power of a diffraction grating

$$R = \frac{\lambda}{|\Delta \lambda|} = mN$$

III.   Possible Pitfalls

Remember that a difference in path length is only one factor in determining the phase difference between light waves that have come by two different paths from a source.

In thin-film interference, which formula you must use to determine where maxima and minima occur depends not on the material of the film but on whether the reflections at the two boundaries are of the same kind or of opposite kinds. If they are of opposite kinds, a net phase difference of 180° is introduced.

The refractive index involved in the thin-film interference formulas is always that of the film material, not the material(s) surrounding it.

The equations for the maxima in the two-slit interference pattern and for the minima of the single-slit diffraction pattern are very similar. Be sure to remember that one is for maxima and one is for minima.

In the interference of three or more equally spaced sources or slits, the location of the *main* interference maxima depends only on the spacing between sources, not on how many of them there are. The *sharpness* of the maxima and the number of secondary maxima in between depend on the total number of sources or slits.

Diffraction as well as interference is always present in the two-slit intensity pattern because there is diffraction of the light that comes from each slit.

IV.   Questions and Answers

Questions

1.   What is the origin of the colors we see in the light reflected from a thin film?

2.   The maximum intensity of light in the interference pattern produced by two sources is four times, not twice, the intensity that we would see from one source alone. Where does the extra energy come from?

3.   The main reason that fairly smart guys like Newton thought that light is a propagation of particles was that diffraction of light had not been observed to happen. Why not?

4. What's the advantage of the electron microscope over a light microscope?

5. It often seems to me that a "diffraction grating" would more properly be called an "interference grating." What do you think?

6. The surface of a piece of glass is wet with water. As it dries, the glossy reflections from its surface become progressively less noticeable and then reappear. What causes this?

7. At night, looking out at a streetlamp through a sheer curtain or a window screen, you can see a cross-shaped pattern of blobs of light with the lamp at its center. What causes this?

8. Is there a maximum wavelength for which the maxima and minima of the diffraction pattern from a single aperture can be observed? Is there a minimum wavelength?

9. The gadget sketched in Figure 27-1 is Fresnel's double mirror. Like Lloyd's mirror, it is a means of realizing two coherent light sources without using slits. Explain how it works.

Figure 27-1

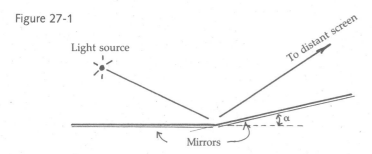

10. Describe the two-slit inteference pattern you'd get using white light.

Answers

1. The colored bands are due to the interference of light reflected at the top and bottom surfaces of a thin film of, say, oil. The different colors are due to the varying thickness of the film. At one point on the film, for instance, there will be an interference maximum for blue reflected light and a minimum for red light, so the film looks blue. For the same reason, at another point where it is slightly thicker, the film may reflect red light preferentially. Thus, when we look at sunlight reflecting from an oil slick, we see a shifting rainbow of colors in the reflected light.

2.   The energy isn't "extra"; it comes from the places in the pattern where the intensity is less than what one source alone would produce. The energy is rearranged in the interference pattern.

3.   Because the wavelengths of visible light are so small, the diffraction of light is usually too small to see.

4.   Electrons have wavelengths that are much smaller than those of light, so much smaller objects can be resolved distinctly by an electron microscope.

5.   Both interference and diffraction are basic to the operation of a grating. The pattern produced by the grating is just a very-many-slit interference pattern; on the other hand, the wide-angle pattern wouldn't exist if it weren't for the diffraction of the light at each slit.

6.   As the film of water evaporates, its thickness reaches and then passes a value at which the thin-film interference is predominantly destructive; it forms a temporary nonreflective coating.

7.   The threads of the curtain or the wires of the screen act as a two-dimensional diffraction grating; you are looking at an interference pattern.

8.   The first minimum in the single-slit diffraction pattern occurs at

$$\sin \theta = \lambda/a$$

where a is the width of the slit. As $\sin \theta$ can't be greater than 1, plainly this minimum can only be observed if $\lambda < a$. The minimum wavelength limitation is a practical one; as the wavelength becomes very much smaller than a, diffraction effects become unobservably small, and all you see is the geometric shadow of the slit.

9.   The two images of the actual light source in the two plane mirrors act as two point sources. The two "sources" are not in phase because they are at different distances from the mirror, but they are coherent since all the light actually comes from one source. The two image sources are separated by a distance that you can easily show to be $2R \sin \alpha$.

10. White light is a combination of all the colors of the rainbow. The different colors correspond to different wavelengths, which interfere constructively at different angles θ. Straight ahead, the central ($m = 0$) interference maximum is white, but as you go out, the pattern dissolves into colored bands rather than intensity maxima and minima.

V.   Problems and Solutions

Problems

1.   Light of wavelength 500 nm falls at near-normal incidence on two flat glass plates 8 cm long that are spaced apart at one end by a sheet of paper of thickness 0.05 mm (see Fig. 27-2). Find the spacing of the interference fringes along the length of the plates. (This is sometimes a practical way of measuring very small thicknesses.)

Figure 27-2

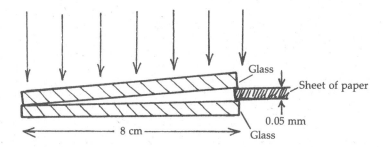

How to Solve It

- There are interference fringes because the thickness of the air wedge varies as you look along the plates.

- Find a formula for the thickness at a distance x along the plates.

- Use this expression for the thickness and the formula for interference maxima (be careful—which formula do you use?) to find an expression for the distances x at which interference maxima occur and, from this, the spacing of the fringes.

2.   A wedge-shaped air film is made by placing a small slip of paper between the edges of two glass flats 12.5 cm long. Light of wavelength 400 nm is incident normally on the glass plates. If interference fringes with a spacing of 0.20 mm are observed along the plate, how thick is the paper?

How to Solve It

- The physical setup here is the same as in Problem 1.

- When the thickness of the air film changes by half a wavelength, the interference pattern goes through one full fringe.

- Use this to calculate the angle between the glass plates and, from the angle, how far apart their edges are.

3.   A glass lens ($n = 1.42$) is coated with a thin layer of transparent material ($n = 1.59$). Light of wavelengths 495 nm and 660 nm is absent from the light reflected from the lens surface. If the coating is the thinnest possible that meets the conditions given, what wavelengths are brightest in the reflected light?

How to Solve It

- Write out the condition for destructive interference in the light reflected from a thin film. Be careful—what phase shifts occur at the two surfaces?

- If the film is the thinnest possible, there are no other destructive wavelengths between the two given. Use these wavelengths to find the thickness of the film.

- Now that you know the thickness of the film, use it to calculate the wavelengths for constructive interference.

4.   A thin film of soap solution ($n = 1.33$) surrounded by air is illuminated normally with white light. There are interference minima in the reflected light at wavelengths (in air) of 400, 480, and 600 nm. What is the thickness of the film?

How to Solve It

- The wavelengths given are the only ones in the visible region in which destructive interference is observed. Notice that the corresponding wavelengths in the soap film are different by a factor of n from those in air.

- Consider the kinds of reflection at the two surfaces to make sure you use the correct condition for destructive interference.

- To what values of the integer m do these wavelengths correspond?

- From a value of m and the corresponding wavelength, you can calculate the thickness of the film.

5. For a classroom demonstration of Young's experiment, helium-neon laser light of wavelength 632.8 nm passes through two parallel slits and falls on a screen 10.5 m away. Interference maxima on the screen are spaced 4.0 cm apart. (*a*) What is the separation of the slits? (*b*) The instructor slips a piece of clear plastic film 10^{-4} in $(2.54 \times 10^{-4}$ cm) thick with $n = 1.493$ over one slit. How does the interference pattern change?

How to Solve It

- You know how to calculate the angles, that is, the directions from the barrier, at which the light will interfere constructively.

- How are these related to the positions of the maxima on a faraway screen?

- When the instructor slips the film in front of one slit, the light from that slit is slowed down because light travels more slowly in the plastic film than it does in air.

- Calculate the time lag introduced and, from it, the phase change in the interference pattern.

6. In a two-slit interference pattern, when light of wavelength 589 nm is used and the screen is 3 m from the slits, there are 28 bright fringes per centimetre on the screen. What is the separation between the slits?

How to Solve It

- The first interference maximum ($m = 1$) is 1/28 cm from the central maximum on the screen. Use this to calculate the angle θ that corresponds to $m = 1$.

- Calculate the slit spacing d from the wavelength and the angle θ.

7. A two-slit interference pattern using 460-nm blue light is thrown on a screen 5.0 m from the slits; bright interference fringes spaced 5.0 cm apart are observed. The fourth maximum in each direction from the central maximum is missing from the pattern. What are the dimensions of the slits?

How to Solve It

- You can get the separation of the slits from the spacing of the maxima on the screen, just as in the last couple of problems.

- The reason the fourth maximum is missing is that it coincides with the first minimum in the diffraction pattern of light from each slit.

- Thus, you have the location of the first diffraction minimum, from which you can calculate the width of each slit.

8. When a thin film of a material whose index of refraction is 1.40 for 650-nm red light is inserted into one arm of a Michelson interferometer, the interference pattern shifts by 10 fringes. How thick is the film?

How to Solve It

- The effect of the transparent material is to slow down the light in that arm of the interferometer.

- Since the effect is to shift the interference pattern by 10 complete cycles, the delay introduced by the plastic film corresponds to the time it takes the light to travel 10 wavelengths.

- Remember that light traverses each arm of a Michelson interferometer twice.

9. A cell 15 cm long with transparent end windows is placed in one arm of a Michelson interferometer illuminated by 632.8-nm laser light. As all the air is pumped back out of the cell, the interference pattern shifts through 139 fringes. What value for the index of refraction of air can you infer from these data?

How to Solve It

- The light traveled more slowly when there was air in the cell because the speed of light in air is less than it is in a vacuum.

- Thus, when you pumped the air out of the cell, you removed a time lag from the light traveling in that arm of the interferometer. How much was that time lag?

- This time lag corresponds to a phase difference of 139 full cycles. From this, calculate the index of refraction of air.

10. In a lecture demonstration of diffraction, a laser beam of wavelength 632.8 nm passes through a vertical slit 0.25 mm wide and strikes a screen 12.0 m away. How wide is the central maximum of the diffraction pattern on the screen?

How to Solve It

- Calculate the angle θ corresponding to the first diffraction minimum.

- On the screen, how far is this minimum from the center of the pattern?

- Twice this value is the width of the central maximum.

11. Two light sources of wavelength 500 nm are photographed from a distance of 100 m. If the camera aperture is 1.05 cm in diameter, how far apart must the two sources be in order for them to be just resolved in the photograph according to the Rayleigh criterion?

How to Solve It

• The resolution here is limited by the diameter of the camera lens aperture.

• Use the Rayleigh criterion to find the minimum angular separation of two source points that can just be resolved.

• What is the corresponding linear separation at a distance of 100 m.

12. The world's largest refracting telescope is at Yerkes Observatory; its objective is 1.02 m in diameter. Suppose you could mount the telescope on a spy satellite 200 km above the ground. What minimum separation of two objects on the ground could it resolve? Take 500 nm as a representative wavelength for visible light.

How to Solve It

• The resolution here is limited by the diameter of the telescope aperture.

• Use the Rayleigh criterion to find the minimum angular separation of two objects that can just be resolved.

• What is the corresponding linear separation at a distance of 200 km?

13. A diffraction grating with 10,000 lines per centimetre is used to analyze the spectrum of mercury. (*a*) Find the angular separation in first order of the two spectral lines of wavelength 577 and 579 nm. (*b*) How wide must the beam of light be on the grating in order to resolve these two spectral lines?

How to Solve It

• Use the grating formula to calculate the angle at which each of these two wavelengths appears in first order. (You'll need three or four significant figures to get a meaningful value for the difference.)

• Resolution is determined by the total number of lines involved in producing the interference.

• You know the spacing; find the width of grating that contains the required number of lines.

14. The spectrum of sodium is dominated by yellow light of wavelength 589.00 and 589.59 nm. By what angle will these two lines be separated in first order by a diffraction grating of 5000 lines/inch?

How to Solve It

- What you're calculating here is not the resolution but the dispersion of the grating—the angular separation of different wavelengths.

- Just use the grating equation to calculate the angle for each wavelength value and the difference between the two.

Solutions

1. Let x be the distance measured along the plates from the point of contact. The thickness y of the air layer between the plates at position x is

$$y = \frac{(0.05 \text{ mm})}{(8 \text{ cm})}\, x = \frac{x}{1600}$$

There is glass on both sides of the air layer; thus, the reflections are of opposite kinds, and interference maxima occur for

$$\frac{2y}{\lambda'} = m + \frac{1}{2} \qquad \text{for } m = 0, 1, 2, 3, \ldots$$

where λ' is the wavelength of light in the "film." Here, the film is air, so $n = 1.00$. Thus, $\lambda' = \lambda$ and

$$y = \left(m + \frac{1}{2}\right)\frac{\lambda}{2}$$

The interference maxima thus occur at positions along the plates of

$$x = 1600y = (1600)\left(m + \frac{1}{2}\right)\left(\frac{5.0 \times 10^{-7} \text{ m}}{2}\right)$$

$$= (0.40 \text{ mm})\left(m + \frac{1}{2}\right) \qquad \text{for } m = 0, 1, 2, 3, \ldots$$

That is, the spacing of the interference maxima (the "fringes") is 0.40 mm.

2. 0.125 mm

3. The coating has a higher index of refraction ($n = 1.59$) than either the glass ($n = 1.42$) or the air on either side of it, so there is one reflection phase change of 180°. Therefore, for destructive interference between the two reflected waves,

$$\frac{2t}{\lambda/n} = m$$

or

$$\lambda = \frac{2nt}{m} \qquad \text{for } m = 1, 2, 3, \ldots$$

Putting in the two given wavelengths, we get

$$6.60 \times 10^{-7} \text{ m} = \frac{(2)(1.59)(t)}{m} = 3.18 \, \frac{t}{m}$$

$$4.95 \times 10^{-7} \text{ m} = \frac{(2)(1.59)(t)}{m'} = 3.18 \, \frac{t}{m'}$$

Dividing these two equations gives

$$m' = \frac{6.60}{4.95} \, m = \frac{4}{3} \, m$$

Thus, m and m' are two integers that are in the ratio of 4 to 3. They might be 4 and 3 or 12 and 9 or whatever. We are asked for the thinnest layer possible, however; thus, they must correspond to the smallest integers possible. So $m = 3$ and $m' = 4$ and

$$6.60 \times 10^{-7} \text{ m} = 3.18 \, \frac{t}{3}$$

or

$$t = 6.23 \times 10^{-7} \text{ m}$$

The condition for constructive interference is

$$\lambda = \frac{2nt}{m + 1/2} = \frac{(2)(1.59)(6.23 \times 10^{-7} \text{ m})}{m + 1/2} = \frac{1.98 \times 10^{-6} \text{ m}}{m + 1/2}$$

For $m = 0, 1, 2, 3, 4, 5, \ldots$, this formula gives $\lambda = 3960, 1980, 792, 566, 440, 360, \ldots$ nm. The members of this series that lie in the visible region are 566 nm ($m = 3$) and 440 nm ($m = 4$), so we expect these to be most strongly reflected. These wavelengths are, respectively, in the yellow-green and in the blue regions of the spectrum, so the reflected light would probably appear blue-green.

4. 902 nm

5.   (*a*) From the diagram in Figure 27-3,

$$\tan \theta = \frac{x}{s}$$

Figure 27-3

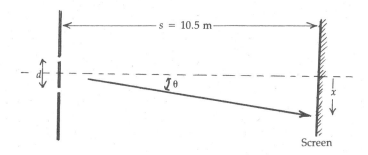

$s = 10.5$ m

Screen

For constructive interference

$$d \sin \theta = m\lambda \qquad \text{for } m = 1, 2, 3, \ldots$$

For small θ, $\sin \theta \approx \tan \theta$, so

$$x = s \tan \theta \approx s \sin \theta$$

$$= m \frac{s\lambda}{d} \qquad \text{for } m = 1, 2, 3, \ldots$$

The spacing of successive interference maxima is therefore

$$\Delta x = \frac{s\lambda}{d}$$

and thus

$$d = \frac{s\lambda}{\Delta x} = \frac{(10.5 \text{ m})(6.328 \times 10^{-7} \text{ m})}{0.04 \text{ m}} = \underline{1.66 \times 10^{-4} \text{ m}}$$

(*b*) A piece of plastic film in front of one of the slits causes a time lag in the light coming from that slit because light travels more slowly in the film than it would in air. If a is the thickness of the film, this time lag is given by

$$\Delta t = t' - t = \frac{a}{v} - \frac{a}{c} = (n - 1)\frac{a}{c}$$

$$= \frac{(1.493 - 1)(2.54 \times 10^{-4} \text{ cm})}{3.00 \times 10^{10} \text{ cm/s}} = 4.18 \times 10^{-15} \text{ s}$$

This time lag corresponds to a path length difference in a vacuum given by

$$c \, \Delta t = (3.00 \times 10^8 \text{ m/s})(4.18 \times 10^{-15} \text{ s}) = 1.25 \times 10^{-6} \text{ m}$$

This is 1.98 wavelengths, almost exactly a whole number. Thus, there is <u>very little visible change</u> in the interference pattern.

6. <u>4.95 mm</u>

7. The spacing of maxima on a screen a distance s away is

$$\Delta x = \frac{s\lambda}{d}$$

so the separation between the slits is

$$d = \frac{s\lambda}{\Delta x} = \frac{(5.0 \text{ m})(4.60 \times 10^{-7} \text{ m})}{0.050 \text{ m}} = 4.60 \times 10^{-5} \text{ m}$$

The reason that the fourth interference maximum is missing is that it coincides with the first diffraction minimum in the diffraction pattern in the light from each slit. Thus, if a is the width of each slit,

$$\sin \theta = \frac{\lambda}{a} = m \frac{\lambda}{d}$$

Thus, with $m = 4$,

$$a = \frac{d}{m} = \frac{4.60 \times 10^{-5} \text{ m}}{4} = \underline{1.15 \times 10^{-5} \text{ m}}$$

8. <u>8.1 μm</u>

9. The light travels through the cell and back again—a total distance of $(2)(15 \text{ cm}) = 0.30 \text{ m}$. With air in the cell, the speed of light in it is

$$v = \frac{c}{n_a}$$

Thus, the time it takes light to travel twice the length of the cell is

$$t_a = \frac{2L}{v} = \frac{2L}{c} n_a$$

When the air is pumped out, the light travel time in the cell is

$$t_0 = \frac{2L}{c}.$$

Pumping the air out therefore changes the light travel time in that arm of the interferometer by

$$\Delta t = t_a - t_0 = (n_a - 1)(2L/c)$$

This time lag corresponds to a path length difference in a vacuum of

$$c\,\Delta t = (n_a - 1)(2L) = 139\lambda$$

since the phase difference is 139 complete cycles. Thus,

$$(n_a - 1) = \frac{139\lambda}{2L} = \frac{(139)(6.328 \times 10^{-7}\text{ m})}{(0.30\text{ m})} = 0.000293$$

$$n_a = 1.000293$$

10. 6.1 cm

11. The Rayleigh criterion is

$$\theta \approx 1.22\,\frac{\lambda}{d} = 1.22\,\frac{5.00 \times 10^{-7}\text{ m}}{1.05 \times 10^{-2}\text{ m}} = 5.8 \times 10^{-5}$$

This is the smallest angular separation, in radians, that can be resolved by the camera. If the objects are a distance s away, their linear separation is

$$d \approx s\theta = (100\text{ m})(5.8 \times 10^{-5}) = 5.8\text{ mm}$$

or about a quarter of an inch. Of course, this is a theoretical limitation that may or may not be approachable in practice.

12. 12 cm

13.   (a) The first-order interference maximum for light of wavelength λ occurs at an angle θ given by

$$\sin\theta = \frac{\lambda}{d} = \frac{5.77 \times 10^{-7}\text{ m}}{10^{-6}\text{ m}} = 0.5770$$

$$\theta = 35.240°$$

for the 577-nm light. In just the same way, $\theta = 35.380°$ for the 579-nm light. Their angular separation is therefore

$$\Delta\theta = 35.380° - 35.240° = \underline{0.140°}$$

(b) The resolving power of a grating of N lines is $R = \lambda/\Delta\lambda \approx mN$. Here we have $m = 1$, and we use the intermediate value of 578 nm for λ. Thus,

$$N \approx \frac{578 \text{ nm}}{2 \text{ nm}} = 289$$

This corresponds to a width of $(289)(10^{-4} \text{ cm}) = \underline{0.29 \text{ mm}}$.

14. $\underline{0.0067°}$

CHAPTER 28

Relativity

I. Key Ideas

Frames of Reference Coordinate systems and clocks that are at rest with respect to a particular observer make up that observer's *frame of reference*. A frame of reference in which Newton's laws hold without modification is called an *inertial frame*. Frames of reference that are moving with uniform velocity relative to an inertial frame are also inertial frames. All inertial frames of reference are *equivalent*: there is no experiment we can do that can tell us which of two inertial frames is "moving" and which "at rest." There is no such thing as absolute motion. This is the *principle of newtonian relativity*.

The Ether Because all mechanical waves require a physical medium to transmit them, it was assumed that space must be filled with a medium that propagates light waves; this hypothetical medium was called the *ether*.

The Michelson-Morley Experiment In 1887, Michelson and Morley tried to determine the velocity of the earth with respect to the ether. Their method was to attempt to measure the difference in the travel time of light along the two arms of a Michelson interferometer, one along and one across the direction of the earth's motion through the ether. No difference was observed. Because the motion of the earth with respect to the ether

has never been detected, the ether idea has been abandoned as unnecessary.

Special Relativity Einstein postulated in a 1905 paper that (1) absolute uniform motion is not detectable in any circumstances and (2) the speed of light is independent of the motion of the source and observer. The first is just the newtonian principle of relativity, but the two taken together contradict older ideas about relative velocity. The Michelson-Morley null result is to be expected from the second postulate. The consequences of these assumptions make up the special theory of relativity. Einstein recognized that the postulates implied that measurement of space and time intervals must depend on the motion of the frame of reference in which they are observed.

Time Dilation The time interval between two events is not an absolute but depends on the motion of the observer. We can see this by simple thought experiments using a clock, that are based on light travel time. The time interval between two events in the frame of reference in which they occur in the *same place* is called the *proper time* interval. It is shorter than that measured in any other reference frame. The fact that the corresponding intervals in other reference frames are longer is called *time dilation*. One consequence of time dilation is that a clock observed in motion runs slower than would the same clock observed at rest by the factor γ:

$$\gamma = \frac{1}{\sqrt{1 - v^2/c^2}}$$

Length Contraction In the same way, we find that the length of an object measured in a frame of reference in which it is moving is shorter along the direction of motion than when it is measured at rest by the same factor γ.

Velocity Transformation The speeds of an object measured in two different frames of reference do not differ just by the relative speed of the two reference frames, as would be the case according to "classical" relativity, but are related by a more complex transformation. Again this is to be expected, as the postulates of relativity clearly alter the concepts of relative motion. One consequence is that if an object is moving at a speed less than c in one inertial frame of reference, its speed is less than c in all inertial frames.

Simultaneity Time dilation and length contraction seem self-contradictory. How, for instance, can each of two metresticks measure the other as shorter than itself? However, the apparent contradictions involve assumptions about the *simultaneity* of events that are not compatible with Einstein's postulates. Events that occur simultaneously but at different places as observed in one frame of reference are *not* simultaneous as observed in another frame that is moving relative to the first. This result, although quite counterintuitive, removes all the apparent contradictions.

The Twin Paradox The twin paradox is a famous thought experiment in special relativity. One of twins stays home while the other makes a long journey at very high speed and then returns. The traveler comes back younger than the twin who stayed home because of time dilation. This appears paradoxical because the argument seems to apply to either twin, yet clearly each cannot be younger than the other. The paradox is resolved when we realize that the roles of the twins are not symmetric after all. In fact, events corresponding to what happens to the twins are routinely observed for subatomic particles, which have longer lifetimes in high-speed accelerator orbits than they do when at rest.

Relativistic Momentum Not surprisingly, the fundamental dynamic quantities also require reformulation to be consistent with the Einstein postulates. The primary usefulness of momentum is that it is conserved in isolated systems. Momentum is redefined in such a way that it retains this property in relativistic mechanics and reduces to mv at low speeds. One consequence of the redefinition of momentum is that it can be interpreted to mean that the mass of an object depends on its speed.

Mass and Energy Likewise, we must redefine work and energy if the work-energy theorem is to apply in relativistic mechanics. One result is that we must include the *rest energy mc^2* of a particle as a part of its total energy that is interchangeable with its kinetic energy or with other forms. This equivalence of mass and energy applies to any kind of interaction between particles, but the mass change is ordinarily not noticeable except in nuclear reactions.

Stability and Binding Energy Composite particles are stable—that is, they cannot spontaneously transform into isolated particles—if the rest energy of the composite is less than the total rest energy of its component particles. The difference between these two energies is the *binding energy* of the composite. Typical values of binding energy are a few electronvolts (eV) for electrons in atoms and molecules and several megaelectronvolts (MeV) per particle in nuclei.

General Relativity Einstein generalized the theory of relativity to noninertial frames of reference beginning in 1916. His general theory is based on the *principle of equivalence,* which states that a homogeneous gravitational field is indistinguishable from a uniformly accelerated frame of reference. No experiment, Einstein assumed, can tell the difference. Consequences of the theory include the deflection of light in a gravitational field, which has been observed, and changes in time intervals and frequencies of light in a gravitational field.

Black Holes The gravitational field created by a sufficiently dense object can be so strong that nothing, not even electromagnetic radiation, can escape from it. Such an object is called a *black hole*. The critical size—the *Schwarzschild radius*—for an object with the mass of the sun to be a black hole is about 3 km.

II.   Numbers and Key Equations

Numbers

$$1 \text{ electronvolt (eV)} = 1.60 \times 10^{-19} \text{ J}$$

$$1 \text{ megaelectronvolt (MeV)} = 10^6 \text{ eV}$$

Speed of light in a vacuum

$$c = 3.00 \times 10^8 \text{ m/s}$$

Key Equations

Michelson-Morley time difference

$$\Delta t = t_1 - t_2 \approx \frac{Lv^2}{c^3}$$

Time dilation

$$\Delta t = \gamma \, \Delta t' = \frac{\Delta t'}{\sqrt{1 - v^2/c^2}}$$

$$\gamma = \frac{1}{\sqrt{1 - v^2/c^2}}$$

Lorentz-FitzGerald contraction

$$L' = \frac{L_0}{\gamma}$$

Velocity transformation

$$u_x = \frac{u_x' + v}{1 + u_x' v/c^2} \qquad u_y = \frac{u_y'}{\gamma(1 + u_x' v/c^2)}$$

Momentum of a particle

$$\mathbf{p} = \frac{m\mathbf{u}}{\sqrt{1 - u^2/c^2}} = \gamma m \mathbf{u}$$

Kinetic energy

$$E_k = (\gamma - 1)mc^2$$

Total energy of a free particle

$$E = E_k + mc^2 = \gamma mc^2$$

Energy and momentum

$$E^2 = (pc)^2 + (mc^2)^2$$

Gravitational effect on clocks

$$\frac{\Delta t_2 - \Delta t_1}{\Delta t} = \frac{\Delta \phi}{c^2}$$

Schwarzschild radius

$$R_G = \frac{2GM}{c^2}$$

III.   Possible Pitfalls

All the relativistic formulas reduce to the classical kinematic or dynamic formulas when v is very much less than c. When you should use the relativistic formulas depends on how precise you want to be, but in ordinary circumstances, you would probably want to use them if anything is moving faster than about $0.05c$. At this speed, γ is about 1.001.

Each of two metresticks in relative motion *is* shorter than the other—as measured by an observer moving with the other. This seems contradictory but isn't.

Don't fall into the trap of thinking that time dilation, length contraction, and so on are illusions of some kind, that they *appear* to occur because of time delays for light to travel from here to there. They are real: the moving metrestick *is* shorter than it would be if observed at rest.

In time dilation problems, it's very easy to get confused about which time interval is which and which clock is running slow. Remember that the time interval measured between two events in a frame of reference *in which they occur at the same place* is the *proper time* interval between them. It is shorter than the interval measured between the same events in any other frame.

We speak of the twin "paradox" because it appears that each is in the same situation relative to the other, and we assume that they must age "symmetrically." But the situation isn't symmetric because one of the twins has remained in an inertial frame of reference while the other has accelerated—at the very least in the process of turning around to come home.

The equivalence of mass and energy is *general;* all energy is interchangeable with mass. Rest mass changes with the creation or destruction of other kinds of energy in nuclear reactions, but this also happens in chemical reactions, when a rock falls from a height, or in any other circumstances. It is routinely taken into account in nuclear reactions because that's about the only case in which the mass change is large enough to observe.

IV.   Questions and Answers

Questions

1. Under what circumstances do two observers in relative motion agree that two events are simultaneous?

2. The murderer fires (event *A*) before his victim falls dead (event *B*)—in my frame of reference, at least. Can there be another reference frame in which *B* happens before *A*?

3. Human lifetimes are ordinarily under 100 years, and nothing can travel faster than light. Even so, it is possible (in theory) for you to make a journey of 200 light-years. How?

4. How is it possible that each of two observers moving relative to one another finds that the other's metrestick is shorter than her own?

5. You are standing on a street corner as your friend drives past and honks his horn twice, a few seconds apart. Both you and he measure the time interval between honks. Which of you has measured the proper time interval?

6. Describe the shape of a cube as it would be measured by observers with respect to whom it is moving at a speed near that of light.

7. Under what circumstances do two observers in relative motion agree as to the velocity of an object that is moving relative to both of them?

8. In Figure 28-1, a farmer carrying a ladder is running into his barn at a speed near that of light. The proper length of the ladder is the same as that of the barn. Clearly, because the ladder is contracted, it fits inside the barn (in the frame of reference of the picture). You could prove this by closing the barn doors (very fast) or taking pictures simultaneously at each end of the barn. But according to the farmer, it's the barn that is moving (and is thus contracted), and his ladder, which has its proper length, can't possibly fit inside it. Who's right? Does the ladder fit inside the barn or not?

Figure 28-1

9. Does the conversion of mass into energy occur only in nuclear reactions or processes?

10. Does the Michelson–Morley experiment prove that there is no ether?

11. Two globs of putty, each of rest mass 1 g, are thrown directly toward each other (see Figure 28-2). Their speed is such that the kinetic energy and the rest energy of each is the same. After they collide and stick together, the combined glob of putty must be at rest. Why? Is the rest mass of this resulting glob 2 g or 4 g? Explain.

Figure 28-2

12. The Empire State Building is 381 m high. Estimate the difference in clock rates between the top of the building and the street below.

Answers

1. If two simultaneous events occur at the same place as well as the same time, they are simultaneous for all observers. (We call such events *coincident* events.)

2. It isn't easy to prove, but the special theory of relativity is free of this kind of self-contradiction. If, in my frame of reference, *B* happens long enough after *A* so that the bullet has time to get from the murderer to the victim, no frame of reference exists in which *B* happens before *A*. Clearly this must be true, if we believe in cause and effect; if the theory allows this kind of thing to happen, there's something wrong with the theory!

3. All you have to do (!) is travel fast enough so that the 200-light-year distance is contracted enough so that it can move past you at a speed less than that of light within your lifetime. From the point of view of an observer on the earth, the distance is still 200 light-years; but your clocks are running slow due to time dilation, so you age more slowly.

4. There's nothing wrong with this, if we can just get past the tacit assumption of absolute simultaneity. Each observer would conclude that the other did something wrong in measuring her stick—timed its passage with a clock that was running slow, say. And each would be right, in her own frame of reference.

5. He has. The proper time interval is that measured by an observer with respect to whom the two events happen at the same place—in this case, the fellow that's riding along with the horn.

Figure 28-3

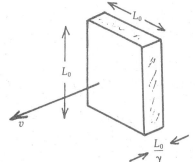

6. It would be contracted, of course, but only along the direction of motion; dimensions perpendicular to the direction of motion are unaffected. The cube in motion is a flattened, square plaque. Its shape is something like that in Figure 28-3. (Note, however, that this is not what a viewer would see because of light travel time.)

7. When—and this is the only instance—whatever it is moves at the speed of light.

8. Well, yes and no—it depends on your frame of reference. (Sorry.) Again, the apparent paradox results from our unquestioning assumption that the simultaneity of events is absolute. Suppose that in the barn frame of reference you took pictures simultaneously at the front and back doors and found that at that moment the whole ladder was in the barn (see Figure 28-4). Then, in the

Figure 28-4

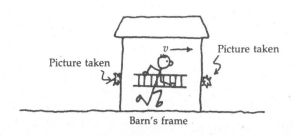

Barn's frame

Figure 28-5

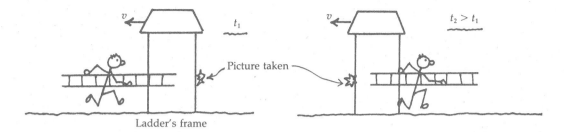

Ladder's frame

frame of reference of the moving ladder, the (contracted) barn came at you. In this frame, the front camera took its picture after the camera at the rear did, and the whole ladder was never in the barn at any one moment (see Figure 28-5).

9. No, it's quite general—mass and energy are equivalent. It's just that in almost all circumstances except for nuclear reactions, the change in the rest mass is so small as to be undetectable.

10. Maybe not, technically; if the ether were entrained by the moving earth, say, and moved with it, then the Michelson-Morley experiment would still give a null result. But it certainly makes the ether hypothesis unnecessary.

11. The rest mass of the final glob is 4 g, not 2 g. It's total energy, rather than either rest or kinetic energy separately, that is conserved. The final glob is known (by the conservation of momentum) to be at rest, so its total energy is all rest energy. (Microscopically, of course, some of this energy is the rest mass of the atoms, and some of it is internal energy.)

12. The difference is gravitational potential (potential energy per unit mass) is

$$\Delta\phi = \frac{mgh}{m} = gh = (9.81 \text{ m/s}^2)(381 \text{ m}) = 3740 \text{ m}^2/\text{s}^2$$

$$= 3740 \text{ J/kg}$$

The fractional difference in clock rates is given by

$$\frac{\Delta\phi}{c^2} = \frac{3740 \text{ m}^2/\text{s}^2}{(3.00 \times 10^8 \text{ m/s})} = 4.2 \times 10^{-14}$$

Perhaps surprisingly, there are circumstances in which a difference in clock rates as small as this one can be observed.

V.   Problems and Solutions

Problems

1. The Michelson-Morley experiment is done using an interferometer for which the length L of each arm is 4.0 m. Assume that the smallest nonzero fringe shift ΔN that can be observed with certainty is 0.01 fringe. Assuming there to be an ether, what would be the least velocity of the earth through it that can be detected in this experiment? How does this compare to the earth's orbital velocity?

How to Solve It
- The wavelength of light isn't specified, so you could choose anything. Let's use 550 nm (yellow-green, near the middle of the visible range).

- Calculate the velocity that will produce a shift of just 0.01 fringe—this is the least velocity that could be observed.

2. The earth's orbital velocity around the sun is 3.0×10^4 m/s; in Michelson and Morley's 1887 experiment, the interferometer arm length L was 11 m. If an identical interferometer is oriented with one arm along the direction of the earth's motion and one arm perpendicular to it, find the round-trip time for light along each arm and the time difference Δt to be expected from the ether theory.

How to Solve It
- There are formulas for the two travel times in the text.

3. The average lifetime of a free neutron is 15 min. If neutrons are made in nuclear reactions taking place in the sun, 1.50×10^{11} m away, how fast must an average neutron travel in order to reach the earth before disintegrating?

How to Solve It
- The neutrons must be moving fast enough so that they reach the earth in 15 min—but on their own clocks.

- Here is another way to do the problems: in a frame of reference that moves with the neutrons, the sun-earth distance is contracted.

4. Suppose you have a perfect timepiece. How much time would it gain or lose in a year at the equator due to the earth's rotation?

How to Solve It
- Use the length of the day and the circumference of the earth to find the earth's rotational velocity at the equator.

- As seen by observers outside the earth, your clocks, which are moving at this speed, are moving slow. They are slowed by the factor γ.

5. A spaceship travels from earth to a star 70 light-years away at a speed of $0.80c$. (*a*) From the point of view of observers on earth, how long does the ship take to get there? (*b*) In this interval, how much time has passed according to observers on board the ship? (*c*) From the point of view of observers on board the ship, then, what was the distance from earth to the star?

How to Solve It
- From the point of view of observers on earth, the time is just the distance divided by the speed.

- Less time than this has passed on shipboard clocks, which measure proper time between the beginning and end of the trip.

- Correspondingly, the distance between earth and star is less. (It's contracted.)

6. How fast is a spaceship moving if it is contracted (as seen by an observer "at rest") to half its length at rest?

How to Solve It
- The factor by which a moving object is contracted is γ.

- Thus, what we want is just the value of v for which $\gamma = 2$.

7. Your twin sister Barbarella travels to a star 12 light-years away and back at constant speed. When she returns, she is 1.2 years younger (according to clocks that have traveled with her) than you are. At which speed did she travel?

How to Solve It
- Neglect the time it takes Barbarella to turn around at the end of the trip so that a constant time dilation factor applies to her whole trip.

- Write expressions for the time the trip took according to you and the time it took according to her.

- The problem says that these times differ by 1.2 years.

- This is enough information to solve for her speed, although a little algebra is involved.

8. A certain spaceship whose length at rest is 250 m is traveling away from the earth at a speed of $0.6c$. A signal is sent from the

earth to the ship at the speed of light. How long does it take the
light signal to pass the length of the spaceship according to (*a*) observers on board the ship and (*b*) observers on earth?

How to Solve It

- The first question is trivial if we remember the Einstein postulates.

- The second takes a little more thought. How long is the ship according to the observers on earth?

9. Two spaceships leave the earth as indicated in Figure 28-6. According to observers on earth, they depart in directions that make a
30° angle with each other. What is the velocity (magnitude and direction) of ship *B* as measured by ship *A*?

Figure 28-6

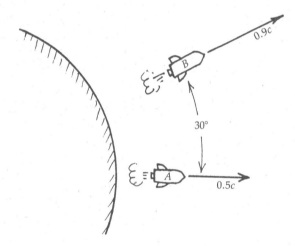

How to Solve It

- Apply the velocity transformation equations.

- The earth is one frame of reference; ship *A* is the other. Use the
components of ship *B*'s velocity in the earth's frame to calculate its
velocity components in *A*'s.

- Be careful with signs when you use the velocity transformation
equations.

10. Two spacecraft are traveling each at a speed of 2.82×10^8 m/s
relative to the earth in opposite directions. What speed does each

measure for the other if they are (a) approaching and (b) receding from each other.

How to Solve It
● The transformation equations apply not to an observer at one fixed position but to an entire frame of reference.

● Think about the problem for a moment—does it matter whether the ships are approaching or receding?

● Be careful with signs when you use the velocity transformation equations.

11. The K mesons and π mesons are subnuclear particles with rest masses of 498 MeV/c^2 and 135 MeV/c^2, respectively. A K meson with total energy of 680 MeV/c^2 decays in flight, producing two π mesons moving along the same line as the original K meson. Find the momentum and the kinetic energy of each π meson.

How to Solve It
● The original total energy and momentum (of the K meson before the decay) remain the total energy and momentum of the two π mesons after.

● The conservation of momentum, the conservation of energy, and the momentum-energy relationship for each π meson provide four equations that can be solved for the two momenta and the two energies.

● The rest is algebra (but there's quite a bit of it).

12. An object of rest mass M_0 explodes, breaking apart into two fragments, each of rest mass $0.32M_0$. What is the speed of each fragment?

How to Solve It
● By symmetry (or by the conservation of momentum, if you like) the two fragments have the same velocity.

● Because of this, you know the rest energy and the kinetic energy of each fragment. Solving for v is easy.

Solutions

1.   The fringe shift in the Michelson-Morley experiment is

$$\Delta N = \frac{2c\,\Delta t}{\lambda} = \frac{2L}{\lambda}\frac{v^2}{c^2}$$

Solving for the speed v that corresponds to 0.01 fringes gives

$$\frac{v}{c} = \left(\frac{\lambda\,\Delta N}{2L}\right)^{1/2} = \left(\frac{(5.50 \times 10^{-7}\text{ m})(0.01)}{(2)(4.0\text{ m})}\right)^{1/2} = 2.62 \times 10^{-5}$$

Thus,

$$v = (2.62 \times 10^{-5})(3.00 \times 10^8\text{ m/s}) = \underline{7.9 \times 10^3\text{ m/s}}$$

For comparison's sake, this is about one-fourth the speed of the earth in its orbit around the sun.

2. $t_1 = 1.000000010(2L/c) = \underline{7.33 \times 10^{-8}\text{ s}}$;
$t_2 = 1.000000005(2L/c) \approx t_1$; $\Delta t = \underline{3.7 \times 10^{-16}\text{ s}}$

3. If τ is the neutron's lifetime at rest, its lifetime in the earth's frame of reference is $\gamma\tau$. Let d be the earth-sun distance; then for the neutrons to reach the earth,

$$v > \frac{d}{\gamma\tau}$$

We must solve this for v; a little algebra is called for. We have

$$\gamma = \frac{1}{\sqrt{1 - v^2/c^2}}$$

so

$$v > \frac{(d)(1 - v^2/c^2)^{1/2}}{\tau}$$

$$(v\tau/d)^2 > 1 - v^2/c^2$$

$$v^2[(\tau/d)^2 + (1/c)^2] > 1$$

or

$$v/c > [1 + (c\tau/d)^2]^{-1/2}$$

$$= \left[1 + \left(\frac{(3.00 \times 10^8\text{ m/s})(900\text{ s})}{(1.50 \times 10^{11}\text{ m})}\right)^2\right]^{-1/2} = 0.486$$

so

$$v > 0.486c = \underline{1.46 \times 10^8\text{ m/s}}$$

4. About $\underline{38\ \mu s}$

5. (a) From the point of view of observers on earth

$$t = \frac{d}{v} = \frac{(70 \text{ light-years})}{0.80c} = \frac{(70)(9.46 \times 10^{15} \text{ m})}{(0.80)(3.00 \times 10^8 \text{ m/s})}$$

$$= \underline{2.76 \times 10^9 \text{ s}} = \underline{87.5 \text{ y}}$$

(b) On board the ship, the clocks are "moving slow" (according to earthbound observers) by the factor

$$\gamma = \frac{1}{\sqrt{1 - (v/c)^2}} = \frac{1}{\sqrt{1 - (0.80)^2}} = 1.667$$

Thus, on the ship clocks the trip took

$$t' = \frac{t}{\gamma} = \frac{2.76 \times 10^9 \text{ s}}{1.667} = \underline{1.66 \times 10^9 \text{ s}} = \underline{52.5 \text{ y}}$$

(c) The distance traveled by the ship was thus

$$d' = vt' = (0.80c)(52.5 \text{ y}) = \underline{42.0 \text{ light-years}}$$

6. $\underline{0.866c}$ or $\underline{2.60 \times 10^8 \text{ m/s}}$

7. Let Barbarella's speed be v. Then the time it takes her to go and return as measured on earth is

$$t' = \frac{(2)(12 \text{ light-years})}{v} = \frac{24 \text{ y}}{v/c}$$

(We neglect the time it takes her to turn around at the other end of the trip.) On board ship, less time has passed (ship time is proper here):

$$t = t'/\gamma = (1 - v^2/c^2)^{1/2} \left(\frac{24 \text{ y}}{v/c} \right)$$

$$= (24 \text{ y})(c^2/v^2 - 1)^{1/2}$$

According to the problem,

$$t = t' - 1.2 \text{ y}$$

$$(24 \text{ y})(c^2/v^2 - 1)^{1/2} = (24 \text{ y})(c/v) - (1.2 \text{ y})$$

so

$$(c^2/v^2 - 1)^{1/2} = (c/v) - 0.05$$

Squaring both sides:

$$c^2/v^2 - 1 = (c/v)^2 - (2)(0.05)(c/v) + (0.05)^2$$

so

$$c/v = 10.025$$

and

$$v = 0.09975c = 2.99 \times 10^7 \text{ m/s}$$

8.   (a) $\underline{0.83 \text{ } \mu\text{s}}$ (b) $\underline{1.67 \text{ } \mu\text{s}}$

9.   Ship A's frame of reference moves with respect to the earth's in the (common) x direction at $v = 0.5c$. In the earth's frame, the components of the velocity u of B are

$$u_x = u \cos 30° = (0.9c) \cos 30° = 0.779c$$

$$u_y = (0.9c) \sin 30° = 0.450c$$

We simply have to transform these into A's frame of reference:

$$u_x' = \frac{u_x - v}{1 - u_x v/c^2} = \frac{(0.779c) - (0.5c)}{1 - (0.779c)(0.5c)/c^2} = 0.458c$$

and

$$u_y' = \frac{u_y}{\gamma(1 - u_x v/c^2)}$$

where

$$\gamma = \frac{1}{\sqrt{1 - (0.50)^2}} = 1.155$$

so

$$u_y' = \frac{0.450c}{(1.155)[1 - (0.779)(0.5)]} = 0.638c$$

Figure 28-7

Thus,

$$u' = \sqrt{(0.458c)^2 + (0.638c)^2} = \underline{0.785c} = \underline{2.35 \times 10^8 \text{ m/s}}$$

is B's speed as observed by A. The direction angle (see Figure 28-7) is $\tan^{-1}(0.638/0.458) = \underline{54.3° \text{ from the } x \text{ axis.}}$

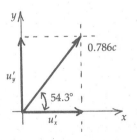

10. $\underline{0.998c \text{ in either case}}$

11. The total energy available to the two mesons is 680 MeV. The total momentum is given by

$$E^2 = (pc)^2 + (mc^2)^2$$

$$(680 \text{ MeV})^2 = (pc)^2 + (498 \text{ MeV})^2$$

or

$$p = 463 \text{ MeV}/c$$

Thus, for the π mesons

$$E_1 + E_2 = 680 \text{ MeV}$$

$$p_1 + p_2 = 463 \text{ MeV}/c$$

from the conservation of momentum and energy. Likewise, for each π meson we have

$$E^2 = (pc)^2 + (135 \text{ MeV})^2$$

Thus, altogether there are four equations to solve for E_1, E_2, p_1, and p_2. Writing the energy-momentum formula for the π mesons gives

$$E_1^2 = (p_1c)^2 + (135 \text{ MeV})^2$$

and

$$E_2^2 = (p_2c)^2 + (135 \text{ MeV})^2$$

We have

$$E_2 = 680 \text{ MeV} - E_1$$

and

$$p_2 = 463 \text{ MeV}/c - p_1$$

so

$$(680 \text{ MeV} - E_1)^2 = (463 \text{ MeV} - p_1c)^2 + (135 \text{ MeV})^2$$

Squaring out the second of these equations and subtracting them gives

$$462{,}400 - 1360E_1 = 214{,}400 - 926p_1c$$

or

$$E_1 = 0.681p_1c + 182.4 \text{ MeV}$$

Substituting this back into the equation for E_1^2 above to eliminate E_1 gives

$$(p_1c)^2 - (463 \text{ MeV})(p_1c) - 28{,}030 \text{ MeV}^2 = 0$$

The solutions of this quadratic are $p_1 = \underline{-54 \text{ MeV}/c}$ or $\underline{517 \text{ MeV}/c}$.

Whichever one of these solutions we choose, the other is the value of p_2. Thus, these are the momenta of the two π mesons; the corresponding kinetic energies are $E_k = \underline{11 \text{ MeV}}$ or $\underline{399 \text{ MeV}}$.

12. $\underline{0.769c}$ or $\underline{2.31 \times 10^8 \text{ m/s}}$

CHAPTER 29

The Origins of Quantum Theory

I. Key Ideas

Blackbody Radiation All objects emit electromagnetic radiation; the intensity of the radiation increases with increasing temperature. Above about 1000 K, some of the radiation begins to be in the visible region, and the object begins to glow. The radiated power is maximum at a wavelength that is inversely proportional to T; this is the *Wien displacement law*. A *blackbody* is an idealized physical system that absorbs *all* the radiation that is incident on it. The calculation of its radiation spectrum is a fundamental problem; it should follow directly from classical thermodynamics, but such treatments were not consistent with the properties of the observed spectrum.

Planck's Theory Max Planck, in 1900, was able to account for the observed blackbody spectrum by assuming that electromagnetic radiation can be emitted or absorbed *only* in fixed amounts or *quanta* of energy that are proportional to the frequency of the radiation. Planck found no way to fit this idea into the classical scheme of things.

Photoelectric Effect When light is incident on a clean metal surface, electrons may be emitted from the surface; this is the *photoelectric effect*.

Photons Einstein applied Planck's idea of energy quantization to explain the photoelectric effect. He assumed that a single Planck quantum of radiation, which he called a *photon*, is absorbed by an electron and gives it enough energy to escape from the metal. The maximum kinetic energy of the emitted electrons is thus the energy of a photon less the energy needed for the electron to escape. The latter is called the *work function*; it is a characteristic of the metal. The success of this treatment, which correctly predicted yet-unknown properties, established that Planck's quanta of light were localized, indivisible particles.

X-Rays X-rays were discovered by Roentgen in 1895. They are electromagnetic radiation with frequencies very much greater (and thus wavelengths very much shorter) than those of visible light. They can be produced by accelerating electrons to high speed and then stopping them on a target of some material in a vacuum tube. This produces a continuous spectrum of x-rays whose minimum wavelength corresponds to the maximum kinetic energy of the incident electrons. The wavelengths are of the order of the spacing between atoms in crystals, and x-ray diffraction has been a supremely important tool in investigating crystal structure.

Characteristic X-Ray Lines Superimposed on the continuous spectrum produced in an x-ray tube is a series of sharp "lines" (radiation at discrete wavelengths) that is characteristic of the target material.

Compton Scattering The scattering of photons from atomic electrons can be likened to collisions between billiard balls. The energy transferred to the electron reduces the energy of the photon, so the scattered radiation has longer wavelengths. This process is called *Compton scattering*; it is strong evidence for the concept of the photon as a localized, particle-like entity.

Quantization of Atomic Energies Bohr applied the concept of energy quantization to the hydrogen atom to produce the first successful model of atomic structure. He successfully predicted the observed line spectrum of hydrogen by assuming that the atom radiates only in the process of making a transition from one to another of a discrete set of stationary states and that, when such a transition takes place, a photon is emitted or absorbed to conserve total energy. Only certain values of energy and angular momentum are possible in Bohr's model of hydrogen. In particular, the allowed states are those in which the angular momentum is an integer multiple of Planck's constant divided by 2π.

II. Numbers and Key Equations

Numbers

Planck's constant
$$h = 6.626 \times 10^{-34} \text{ J·s}$$
$$= 4.136 \times 10^{-15} \text{ eV·s}$$
$$hc = 1.240 \times 10^{-6} \text{ eV·m}$$

Rydberg constant
$$R_\infty = 1.0974 \times 10^7 \text{ m}^{-1}$$

Bohr radius
$$a_0 = 0.0529 \text{ nm}$$

Bohr energy value
$$E_0 \approx 13.60 \text{ eV}$$

Key Equations

Rayleigh-Jeans law
$$P(\lambda, T) = 2\pi c k T \lambda^{-4}$$

Wien displacement law
$$\lambda_m = \frac{2.898 \text{ mm·K}}{T}$$

Electromagnetic energy quantum
$$E = hf$$

Einstein's photoelectric equation
$$(\tfrac{1}{2}mv^2)_{max} = eV_0 = hf - \phi$$

Work function
$$\phi = hf_t = \frac{hc}{\lambda_t}$$

Bragg's condition
$$2d \sin \theta = m\lambda \qquad \text{for m} = 1, 2, 3, \ldots$$

Bremsstrahlung cutoff wavelength
$$\lambda_m = \frac{hc}{eV_0}$$

Photon energy and momentum
$$E = pc$$

Compton scattering wavelengths
$$\lambda_2 - \lambda_1 = \frac{h}{mc}(1 - \cos \theta)$$

Balmer wavelengths
$$\lambda = \frac{364.6n^2}{n^2 - 4} \text{ nm} \qquad \text{for } n = 3, 4, 5, \ldots$$

Bohr's second postulate
$$f = \frac{E_i - E_f}{h}$$

Bohr's allowed energy levels
$$E_n = -\left(\frac{2\pi^2 m k^2 e^4}{h^2}\right)\frac{Z^2}{n^2} = -Z^2\frac{E_0}{n^2}$$

III. Possible Pitfalls

Notice that the long argument about the nature of light—waves versus particles—really ends in a draw; light is both, and more than either. Don't think that the discovery of photons wipes out two centuries' worth of observations of light *waves*.

Planck's theory of blackbody radiation introduced the idea that electromagnetic radiation is absorbed and emitted in discrete amounts, but our picture of the *photon* as a discrete, localized *particle* of light originated with Einstein's theory of the photoelectric effect.

In the photoelectric effect, the energy of the emitted electrons is determined by the frequency and not the intensity of the light. An electron can acquire either a "whole photon's worth" of energy from the light, or none.

The wavelength change in Compton scattering is independent of the wavelength of the scattered light, but the photon energy change is not.

The total energies of bound atomic states, as in the Bohr model of the hydrogen atom, are negative because we call the potential energy zero when the electron is infinitely far away from the nucleus.

IV. Questions and Answers

Questions

1. How does the intensity of light from a blackbody change when its temperature is increased? What changes occur in the body's radiation spectrum?

2. Everything emits thermal radiation—you, too. At approximately what wavelength is your blackbody radiation spectrum concentrated?

3. According to classical electromagnetic theory, an accelerated charge emits electromagnetic radiation. What would this mean for the electron in the Bohr atom? What would happen to its orbit?

4. Are there quantities in classical physics that are quantized?

5. What kind of electromagnetic radiation is that with the shortest wavelength that can be emitted by a hydrogen atom?

6. Consider the photoelectric emission of electrons induced by monochromatic incident light. The incoming photons all have the same energy, but the emitted electrons have a range of kinetic energies. Why is this?

7. The Compton effect is practically unobservable for visible light. Why?

8. When an atom making a transition from a higher to a lower quantum state emits a photon, the photon energy is actually a (very) little bit less than the energy difference between the initial and final atomic states. Why?

9. A markedly nonclassical feature of the photoelectric effect is the fact that the energy of the emitted electrons doesn't increase as you increase the intensity of the light striking the metal surface. What change does occur as the intensity is increased?

Answers

1. Intensity is proportional to the fourth power of the temperature (this is the Stefan-Boltzmann law; see Chapter 13), so a small increase in T produces a large increase in radiated power. The spectrum is shifted toward shorter wavelengths—higher frequencies—approximately in proportion to the change in absolute temperature.

2. Using $T \approx 300$ K in the Wien displacement law gives $\lambda \approx 10$ μm.

3. The emission of radiation would have to be at the expense of the orbiting electron s mechanical energy—there's nowhere else for it to come from. Thus, the electron's orbital radius would decrease with time, and the electron would eventually spiral into the nucleus. The atom would collapse.

4. Yes; electric charge is the most obvious example, but all the mechanical properties of standing waves on a string would qualify, too.

5. The photon energy would be equal to the binding energy of the ground state, which is 13.6 eV. At 91 nm, this is in the far ultraviolet.

6. Some of the electrons have to knock around in the metal, thereby losing a variable amount of kinetic energy in collisions, before they make it to the surface and escape. The emitted kinetic energies range anywhere from zero to a maximum given by the incoming photon energy less the work function of the metal.

7. From the Compton scattering formula, the maximum possible wavelength change is $2h/mc$, which is about 0.005 nm. This is so small relative to the wavelength of visible light as to be practically unobservable.

8. Because if momentum is to be conserved, the emitting atom must recoil in the direction opposite that of the photon. This recoil takes a very small part of the total energy released.

9. Increasing the intensity means delivering photons to the surface of the metal at a greater rate; correspondingly, the photocurrent—the rate at which the electrons are emitted—increases.

V. Problems and Solutions

Problems

1. A light bulb emits 40 W of light uniformly in all directions. Take 500 nm as the average wavelength of the light. At what rate do photons from this light bulb enter your eye (pupil diameter 6 mm) if you are a mile away from it?

How to Solve It
- Calculate the energy of each photon and, from that, the rate at which the source emits photons.

- At a distance of 1 mi, calculate the rate at which photons are received per unit area. (They are emitted uniformly in all directions.)

- Calculate the area of the pupil of your eye and, from that, the rate at which photons enter it.

2. What range of photon energies corresponds to the visible range of wavelengths?

How to Solve It
- Wavelengths in the range 400–700 nm are visible.

- Calculate the corresponding photon energies.

3. The surface temperature of the sun is about 5900 K. If the sun were a perfect blackbody at this temperature, what would be the peak wavelength λ_{max} of its spectrum?

How to solve it
- The peak wavelength and the temperature are inversely proportional.

- Use the Wien displacement law.

4. What is the minimum wavelength of x-rays produced by a tube in which the electron beam has been accelerated through a potential difference of 30,000 V?

How to Solve It
- The minimum wavelength corresponds to the maximum photon energy.

- The maximum photon energy that an electron can produce is equal to the kinetic energy of the electron.

- The kinetic energy of these electrons is determined by the accelerating voltage of the x-ray tube.

5. Ultraviolet light of wavelength 260 nm incident upon a certain metal ejects electrons whose maximum kinetic energy is 1.34 eV. What is the maximum wavelength of light that will eject electrons from this same metal?

How to Solve It
- Use Einstein's photoelectric equation to calculate the work function ϕ of the metal.

- The work function determines the minimum energy that must be delivered by the photon in order to get an electron free of the surface.

- This minimum photon energy corresponds to the maximum wavelength that can cause photoemission.

6. The longest wavelength of light that will cause photoelectric emission of electrons from a certain metal is 522 nm. If light of wavelength 348 nm is incident upon this same metal, what are the maximum and minimum kinetic energies of the emitted electrons?

How to Solve It
- The longest wavelength that can cause photoemission corresponds to a photon energy equal to the work function of the metal.

- Use Einstein's photoelectric equation to determine the maximum kinetic energy of emitted electrons.

- What is the minimum kinetic energy of the emitted electrons?

7. The difference between initial and final photon energy in the Compton effect is the kinetic energy of the recoiling electron. Photons of energy 1.16×10^5 eV (this would be in the γ-ray region) undergo Compton scattering. What are the maximum and minimum values possible for the kinetic energy of the recoiling electron?

How to Solve It

• The maximum kinetic energy of the recoiling electron corresponds to the minimum energy of the scattered photon—this occurs when the photon is scattered straight back at 180°.

• Use the Compton formula to calculate the energy of the scattered photon; the difference between this and the incident photon energy is the energy of the recoiling electron.

• Likewise, the minimum kinetic energy of the recoiling electron corresponds to the maximum kinetic energy of the scattered photon.

8. X-ray photons are Compton-scattered by electrons in carbon. Each photon that is scattered through 90° loses energy of 139 eV. What is the wavelength of the incident x-rays?

How to Solve It

• Use $\theta = 90°$ or $\cos \theta = 1$ to simplify the Compton scattering formula.

• Write the incident and scattered wavelengths in the Compton scattering formula in terms of the corresponding photon energies. The difference between the initial and scattered photon energies is known.

• From this you get an equation that you can solve for the incident photon energy and, in turn, for the incident wavelength.

9. A hydrogen atom is in the tenth ($n = 10$) Bohr state. (*a*) Find the orbital frequency of the electron. (*b*) Find the frequency of the light that is emitted when the electron drops from the $n = 10$ to the $n = 9$ state.

How to Solve It

• You know the angular momentum of the electron in the $n = 10$ orbit. From this you can calculate the velocity.

• From the velocity and the known orbital radius, calculate the orbital frequency.

• Determine the frequency of the emitted light from the energy difference between the $n = 9$ and $n = 10$ states.

10. Find the four longest wavelengths in the Balmer series of hydrogen.

How to Solve It
- The Balmer series consists of the transitions $n > 2 \to n = 2$.

- The longest wavelengths correspond to the lowest transition energies and thus to the lowest integers n.

11. Hydrogen atoms in their ground state ($n = 1$) are illuminated by ultraviolet light of wavelength 50 nm. Electrons absorb photons and are ejected from the atoms. What is the kinetic energy of the emitted electrons?

How to Solve It
- The energy absorbed by an electron is that of the incoming photon.

- The energy of the electron emitted from the atom is that absorbed from the photon less the binding energy of the hydrogen atom in its ground state.

Solutions

1. At 1 mi = 1609 m, the energy from the 40-W source is distributed uniformly over an area $4\pi r^2 = (4\pi)(1609 \text{ m})^2 = 3.25 \times 10^7 \text{ m}^2$. The energy of a photon of wavelength 500 nm is

$$E = hc/\lambda = \frac{1240 \text{ eV·nm}}{500 \text{ nm}} = 2.48 \text{ eV} = 3.97 \times 10^{-19} \text{ J}$$

so the rate at which the photons are being emitted is

$$\frac{40 \text{ W}}{3.97 \times 10^{-19} \text{ J/photon}} = 1.01 \times 10^{20} \text{ photons/s}$$

The number per unit area per unit time a mile away is

$$\frac{1.01 \times 10^{20} \text{ photons/s}}{3.25 \times 10^7 \text{ m}^2} = 3.10 \times 10^{12} \text{ photons/s·m}^2$$

If the diameter of your pupil is 6 mm, its area is $(\pi)(3.0 \times 10^{-3} \text{ m})^2 = 2.83 \times 10^{-5} \text{ m}$, so the rate at which photons enter your eye is

$$(3.10 \times 10^{12} \text{ photons/s·m}^2)(2.83 \times 10^{-5} \text{ m}^2)$$
$$= \underline{8.8 \times 10^7 \text{ photons/s}}$$

2. $\underline{1.77\text{–}3.10 \text{ eV}}$ (700–400 nm)

3. From the Wien displacement law,

$$\lambda_{max} = \frac{2.898 \text{ mm·K}}{T} = \frac{2.898 \text{ mm·k}}{5900 \text{ K}}$$

$$= 4.91 \times 10^{-4} \text{ mm} = \underline{491 \text{ nm}}$$

4. <u>0.0413 nm</u>

5. Using Einstein's photoelectric equation,

$$E_{max} = hf - \phi = \frac{hc}{\lambda} - \phi$$

$$\phi = \frac{hc}{\lambda} - E_{max} = \frac{1240 \text{ eV·nm}}{260 \text{ nm}} - 1.34 \text{ eV} = 3.43 \text{ eV}$$

The maximum wavelength of light (that is, the light with the lowest incident photon energy) that can make this happen is that corresponding to $E_{max} = 0$:

$$\frac{hc}{\lambda_t} = \phi$$

$$\lambda_t = \frac{hc}{\phi} = \frac{1240 \text{ eV·nm}}{3.43 \text{ eV}}$$

$$= \underline{362 \text{ nm}}$$

6. <u>1.19 eV</u> and <u>0</u>

7. The energy of a photon is hc/λ, so the energy transferred to the electron is

$$E_k = \frac{hc}{\lambda} - \frac{hc}{\lambda'} = \frac{hc(\lambda' - \lambda)}{\lambda\lambda'}$$

where λ' is the wavelength of the scattered photon. The Compton scattering formula is

$$\lambda' - \lambda = \lambda_e(1 - \cos \theta)$$

or

$$\lambda' = \lambda + \lambda_e(1 - \cos \theta)$$

where $\lambda_e = h/mc$. Thus,

$$E_k = \frac{hc\lambda_e(1 - \cos\theta)}{\lambda[\lambda + \lambda_e(1 - \cos\theta)]}$$

The maximum energy transfer occurs for $\theta = 180°$, in which case $\cos\theta = -1$ and

$$E_{k,\text{max}} = \frac{2hc\lambda_e}{\lambda(\lambda + 2\lambda_e)}$$

Dividing by $2\lambda_e$ gives

$$E_{k,\text{max}} = \frac{hc/\lambda}{1 + \lambda/2\lambda_e}$$

Now

$$\lambda = hc/E$$

and

$$\lambda/2\lambda_e = \frac{hc/E}{2h/mc} = \frac{mc^2}{2E}$$

so

$$E_{k,\text{max}} = \frac{E}{1 + mc^2/2E}$$

The incident photon energy $E = 116$ keV, and $mc^2 = 511$ keV, so the maximum energy transfer is

$$E_{k,\text{max}} = \frac{116 \text{ keV}}{1 + 511/232} = 36.2 \text{ keV}$$

From the Compton scattering equation, clearly zero energy is transferred when $\theta = 0$, so the energy transferred to the electron ranges from 0 to 36.2 keV.

8.   0.1459 nm

9.   (a) The easiest way to get the velocity of the orbiting electron is from its angular momentum:

$$L = mvr = \frac{nh}{2\pi} \qquad \text{so} \qquad v = \frac{nh}{2\pi mr}$$

The orbital frequency is thus

$$f_o = \frac{v}{2\pi r} = \frac{nh}{4\pi^2 mr^2}$$

For $n = 10$ the orbital radius is

$$r = n^2 a_0 = (10)^2(0.0529 \text{ nm}) = 5.29 \text{ nm}$$

Thus,

$$f_o = \frac{(10)(6.626 \times 10^{-34} \text{ J·s})}{(4\pi^2)(9.11 \times 10^{-31} \text{ kg})(5.29 \times 10^{-9} \text{ m})^2} = \underline{6.58 \times 10^{12} \text{ s}^{-1}}$$

(b) The radiation frequency is

$$f_r = \frac{E_{10} - E_9}{h} = \frac{\left(\dfrac{-13.60 \text{ eV}}{10^2}\right) - \left(\dfrac{-13.60 \text{ eV}}{9^2}\right)}{4.14 \times 10^{-15} \text{ eV·s}}$$

$$= \underline{7.71 \times 10^{12} \text{ s}^{-1}}$$

The orbital and radiation frequencies are about the same. In the limit of very large n, they are the same.

10. $\underline{656.3, 486.1, 434.0, \text{ and } 410.2 \text{ nm}}$

11. The energy of an incoming photon is

$$E = hc/\lambda = \frac{1240 \text{ eV nm}}{50 \text{ nm}} = 24.80 \text{ eV}$$

The binding energy of the electron in the ground state is 13.60 eV, so absorbing a photon leaves the electron with a final kinetic energy of

$$E_f = 24.80 \text{ eV} - 13.60 \text{ eV} = \underline{11.20 \text{ eV}}$$

CHAPTER 30

Electron Waves and Quantum Theory

I. Key Ideas

De Broglie Waves In 1924, de Broglie suggested that electrons and other "material" particles share the dual nature of electromagnetic radiation, possessing both wave and particle properties; that is, that wave propagation of some sort is associated with the motion of the particle. He proposed that the same wavelength-to-momentum relationship applies as for electromagnetic waves.

Wave Mechanics De Broglie's idea provides a satisfying rationale for quantization in the Bohr model of the hydrogen atom: the allowed states are those in which the circumference of the electron orbit is a whole number of electron wavelengths. This idea of quantization as a kind of standing-wave condition was developed by Schrödinger into a general theory of *wave mechanics*.

Electron Diffraction Electrons accelerated through rather low potential differences have wavelengths approximately the size of atoms, leading to the possibility of *diffraction* of electrons by orderly arrays of atoms in crystals. The first observation of electron diffraction by Davisson and Germer in 1927 confirmed de Broglie's hypothesis. In the years since, electron diffraction has developed into a very important tool in crystallography.

The Electron Microscope Because the wavelengths of electrons that have been accelerated through not very large voltages are orders of magnitude smaller than those of visible light, electron waves can be used to make a microscope with a resolution very much finer than that of optical microscopes.

Wave-Particle Duality It is problematical to consider the same entity as at once a wave and particle: a particle is discrete and localized; a wave is nonlocal and displays interference and diffraction. However, the classical notions of waves and particles as separate entities are inadequate; everything propagates as a wave but exchanges energy as if it were a classical particle. In macroscopic situations, the wavelength is very small, and the wave nature of particles makes no difference. In microscopic situations, both aspects can be important.

Uncertainty Principle It's assumed in classical mechanics that experimental uncertainties can, in principle, be made as small as desired. In fact, it is impossible to reduce uncertainties beyond limits imposed by wave-particle duality. The position and velocity of a particle cannot be measured simultaneously with unlimited precision. Because of quantization, the observation of a system disturbs it by at least a minimum amount. Heisenberg's *uncertainty principle* states that, in a given situation, the *product* of the uncertainties in the position and in the corresponding momentum of a particle cannot be less than a limiting value of the order $h/2\pi$.

Zero-Point Energy Because of wave-particle duality (or, which is the same thing, because of the uncertainty principle), a particle that is confined to a finite region of space cannot have zero kinetic energy. The minimum kinetic energy possible is called the *zero-point energy*.

Wave Function The state of a particle wave, such as that of an electron, is described by its *wave function*. The intensity of the wave, which is the square of the wave function, is proportional to the probability of finding the particle at some particular place. The wave function displays interference and diffraction of itself, whether many particles are present to interfere or not.

Particle in a Box The case of a particle that is free except for being confined to a finite region of space is illustrative. The wave function of the particle is a simple sine wave, but it must vanish at the ends of the interval, as the particle is not to get out. Thus, the allowed wave functions are standing sine waves in which the length of the interval is an integral number of half wavelengths of the particle. Note that this predicts a minimum kinetic energy that is of the order of magnitude predicted by the uncertainty principle.

Correspondence Principle In the classical limit, quantum calculations must reduce to classical mechanics. This is Bohr's *correspondence principle*. In the limit of very large quantum numbers, classical and quantum calculations must give the same results. In the "particle in a box" problem, for example, for very large values of n, the particle has uniform probability of being anywhere inside.

II. Key Equations

Key Equations

De Broglie equations

$$f = \frac{E}{h} \qquad \lambda = \frac{h}{p}$$

Electron wavelength

$$\lambda = \frac{1.226 \text{ nm} \cdot \text{eV}^{1/2}}{\sqrt{E_k}}$$

Uncertainty principle

$$\Delta x \, \Delta p \geq \frac{h}{4\pi}$$

Standing-wave condition

$$n\frac{\lambda}{2} = L$$

Energy levels for a particle in a box

$$E_n = n^2 E_1 \qquad E_1 = \frac{h^2}{8mL^2}$$

III. Possible Pitfalls

In the two-slit interference experiment for electrons, one electron is *not* interfering with another. This is a hard point to get hold of, but the interference effects will occur even if the electron intensity is so low that there is never more than one electron in the apparatus at any given time. The interference is an attribute of the wave function.

You calculate the de Broglie wavelength of a particle in just the same way regardless of its speed, but at high speeds, the correct (that is, the relativistic) formulation for the momentum must be used.

The fact of the "associated wave" doesn't imply any oscillatory motion of the electron itself. Don't let sinusoidal graphs of wave functions put the picture into your mind of the electron itself jiggling up and down somehow.

Watch out for the electronvolt. It is such a convenient unit for atomic-scale phenomena that you'll certainly be using it, but its name is misleading in that it's a unit of energy, not potential difference. As it is not an SI unit, you'll need to convert it to joules for lots of calculations.

Remember that the Heisenberg uncertainty principle doesn't put any limitation at all on how precisely you can determine the position of a particle *or* on how precisely you can measure its momentum. It only limits how precisely *both* quantities can be measured simultaneously.

IV.   Questions and Answers

Questions

1. Is Bohr's picture of the hydrogen atom consistent with the uncertainty principle? Explain.

2. An electron and a proton have the same kinetic energy. Which has the longer wavelength?

3. Is the wavelength of an electron the same as that of a photon of the same total energy?

4. Why does the uncertainty principle alone imply the existence of a zero-point energy for a confined particle?

5. The speed of a photon is known precisely in any circumstance—it is c. Does it follow from the uncertainty principle that nothing whatever can be known about the photon's position? Explain.

6. Why do we never observe the "wave nature of particles" for everyday objects—for buckshot or bumblebees, for example.

7. Does the de Broglie wavelength of a particle increase or decrease as its kinetic energy increases?

8. What does the wave function ψ for a free particle in a box look like for very large n? How does this "correspond" (as Bohr's correspondence principle says it ought to) to the classical situation?

Answers

1. No. The very idea of a sharply defined orbit implies simultaneous specification of the electron's position and its momentum, which is just what the uncertainty principle forbids. Limitations intrinsic to the model imply minimum uncertainties in the electron's position that are of the order of the size of the whole atom. All that's left of Bohr's picture is an atom-sized fuzzy region with the electron in it somewhere.

2. For given kinetic energy, the proton has the greater momentum and thus the shorter wavelength.

3. No; for given total energy, the electron has the greater momentum and thus the shorter wavelength.

4. Because if the particle were at rest and were known to be at rest, then we know its momentum to be precisely zero. If there is perfect knowledge of the momentum, then from the uncertainty principle there can clearly be no knowledge whatever of position. The particle must be able to be literally anywhere, not confined to a finite region.

5. No; our exact knowledge of the photon's speed tells us nothing about its momentum, which can be anything, depending on the wavelength. It is simultaneous position and momentum measurements that are related by the uncertainty principle.

6. Because the wavelength predicted by the de Broglie formula for the momenta of such large objects is far too small to observe.

7. As the kinetic energy of a particle increases, so does its momentum; thus, its wavelength, which is inversely proportional to momentum, decreases.

8. For very large n, the wavelength, which is equal to $2L/n$, is very small, and the wave function is sinusoidal, with very small "pitch" (see Figure 30-1). When n is large enough, the probability of finding the particle within some interval is independent of where that interval is; this corresponds to the classical situation, in which the particle is equally likely to be anywhere.

Figure 30-1

V. **Problems and Solutions**

Problems

1. The kinetic energy of an electron in the third Bohr orbit of hydrogen is 1.511 eV; the radius of the orbit is 0.476 nm. Find the de Broglie wavelength of the electron.

How to Solve It
- There are a couple of different ways to do this. You can calculate the wavelength directly from the given kinetic energy.

- An alternative is to use the standing-wave condition and the radius of the orbit to get the wavelength.

2.   Thermal neutrons from nuclear reactors are used in neutron diffraction experiments in crystallography. Neutrons in thermal equilibrium in a nuclear reactor have kinetic energy of 0.026 eV. What is the de Broglie wavelength of such a neutron? The mass of a neutron is 1.67×10^{-27} kg.

How to Solve It
- These neutrons are very slow.

- Use the kinetic energy to calculate the momentum and the momentum to calculate the de Broglie wavelength.

3.   A typical atom may be 0.2 nm in diameter. An electron that is known to be part of an atom thus has a position uncertainty that is no larger than this. (*a*) To what minimum uncertainty in momentum does this correspond? (*b*) What would be the kinetic energy of an electron with a momentum of this value?

How to Solve It
- Use the known maximum uncertainty in the position of the electron to estimate a minimum value for Δp.

- Take the momentum of an electron to be numerically equal to the value you calculated in part (*a*), and calculate the corresponding kinetic energy.

- This is a version of the zero-point energy calculation.

4.   (*a*) What is the momentum of an electron whose de Broglie wavelength is equal to that of laser light with a wavelength of 632.8 nm? (*b*) If the wavelength is known to within 0.1 nm, what is the minimum uncertainty in the electron's position?

How to Solve It
- Calculate the electron's momentum from its known wavelength.

- Use the given uncertainty in the wavelength to calculate the uncertainty in the electron's momentum.

- Use the uncertainty principle to calculate the minimum uncertainty in the electron's position.

5.   An electron is confined to a one-dimensional box of length 0.25 nm. (This is roughly atom-sized.) (*a*) Calculate the first four energy levels of the electron. (*b*) An electron in the $n = 4$ state makes transitions, emitting photons, that eventually leave it in the ground state ($n = 1$). What wavelengths of light could be emitted in the process?

How to Solve It
- The first part is just a plug—use the equation for the energy levels of a particle in a box.

- The energy of the emitted photon is equal to the difference in energy between initial and final electron states.

- Calculate all the possible photon energies and, from them, the wavelengths.

6. The zero-point energy result says that, just because you confine a particle to a finite region of space, it cannot be at rest. This seems to be in contradiction to our everyday experience; but is it? Calculate the minimum speed of a buckshot (mass 0.92 g) that is "confined" to a one-dimensional cigar box (length 20 cm).

How to Solve It
- From the quantization condition for a free particle in box, derive an expression for the minimum speed of the shot. Of course, minimum speed corresponds to minimum kinetic energy.

- In using the values given, be careful of units: the data given aren't in SI!

Solutions

1. The momentum of an electron with kinetic energy of 1.511 eV is

$$p = \sqrt{2mE} = \sqrt{(2)(5.11 \times 10^5 \text{ eV}/c^2)(1.511 \text{ eV})} = 1243 \text{ eV}/c$$

and the de Broglie wavelength is

$$\lambda = \frac{h}{p} = \frac{hc}{pc} = \frac{1.240 \times 10^{-6} \text{ eV·m}}{1243 \text{ eV}}$$
$$= 0.998 \times 10^{-9} \text{ m} = \underline{0.998 \text{ nm}}$$

An easier way to do this is to remember that the circumference of the nth Bohr orbit is n wavelengths. Thus,

$$\lambda = \frac{2\pi r}{n} = \frac{(2\pi)(0.476 \text{ nm})}{3} = \underline{0.997 \text{ nm}}$$

The difference is just round-off error.

2. $\underline{0.178 \text{ nm}}$

3. (*a*) From the uncertainty principle, the minimum uncertainty in the momentum of the electron is given by

$$\Delta p\, \Delta x > h/4\pi$$

$$\Delta p > \frac{h}{4\pi\, \Delta x} = \frac{6.626 \times 10^{-34}\ \text{J·s}}{(4\pi)(2 \times 10^{-10}\ \text{m})} = \underline{2.64 \times 10^{-25}\ \text{kg·m/s}}$$

(*b*) Assuming the electron is nonrelativistic, its kinetic energy is

$$E_k = \frac{p^2}{2m}$$

For a momentum equal to the momentum uncertainty found in (*a*),

$$E_k = \frac{(2.64 \times 10^{-25}\ \text{kg·m/s})^2}{(2)(9.11 \times 10^{-31}\ \text{kg})} = 3.82 \times 10^{-20}\ \text{J} = \underline{0.238\ \text{eV}}$$

4. (*a*) $\underline{1.05 \times 10^{-27}\ \text{kg·m/s}}$ (*b*) $\underline{0.32\ \text{mm}}$

5. (*a*) For a box of length L, the lowest energy is

$$E_1 = \frac{h^2}{8mL^2}$$

$$= \frac{(6.626 \times 10^{-34}\ \text{J·s})^2}{(8)(9.11 \times 10^{-31}\ \text{kg})(2.5 \times 10^{-10}\ \text{m})^2} = 9.64 \times 10^{-19}\ \text{J}$$

$$= (9.64 \times 10^{-19}\ \text{J})(6.24 \times 10^{18}\ \text{eV/J}) = 6.02\ \text{eV}$$

The energy levels are

$$E_n = n^2 E_1$$

The first four, therefore, are

$$E_1 = \underline{6.02\ \text{eV}}$$

$$E_2 = 4E_1 = \underline{24.1\ \text{eV}}$$

$$E_3 = 9E_1 = \underline{54.2\ \text{eV}}$$

$$E_4 = 16E_1 = \underline{96.3\ \text{ev}}$$

(*b*) There are six possible transitions:

$n = 4 \to 3$	$\Delta E = E_4 - E_3 = 42.1\ \text{eV}$	$\lambda = \underline{29.4\ \text{nm}}$
$n = 4 \to 2$	$\Delta E = 72.2\ \text{eV}$	$\lambda = \underline{17.2\ \text{nm}}$
$n = 4 \to 1$	$\Delta E = 90.2\ \text{eV}$	$\lambda = \underline{13.7\ \text{nm}}$
$n = 3 \to 2$	$\Delta E = 30.1\ \text{eV}$	$\lambda = \underline{41.2\ \text{nm}}$

$n = 3 \rightarrow 1 \quad \Delta E = 48.2 \text{ eV} \qquad \lambda = \underline{25.7 \text{ nm}}$

$n = 2 \rightarrow 1 \quad \Delta E = 18.1 \text{ eV} \qquad \lambda = \underline{68.6 \text{ nm}}$

6. $\underline{1.80 \times 10^{-30} \text{ m/s}}$

CHAPTER 31

Atoms, Molecules, and Solids

I. Key Ideas

Wave Mechanics The Bohr model of the hydrogen atom was a success, but it was unsatisfactory in that it was an *ad hoc* treatment, and there was no clear way to extend it to atoms of more than one electron. A more satisfactory theory is based on Schrödinger's wave equation. The standing-wave solutions of this equation correspond to the allowed states in the Bohr model and the energy values agree.

Hydrogen-Atom Wave Functions The state of an electron in an atom is described by a wave function ψ. The square of the wave function evaluated at some position measures the probability of finding the electron there. Only certain discrete values of the electron's energy and angular momentum can exist; that is, physically acceptable solutions of Schrödinger's equation exist only for certain values of the dynamic variables. The energy and so forth are said to be *quantized*. The states of an electron in a hydrogen atom are specified by three *quantum numbers: n, ℓ,* and *m.* Energies correspond to the Bohr results, and the most probable distances of the electron from the nucleus correspond to the Bohr atom radii.

Electron Spin Discrete spectral lines result from the transition of an electron between discrete energy levels. The predictions of wave mechanics

for the hydrogen atom agree with the spectral series predicted by the Bohr model, but each of the lines, on close inspection, is really two or more very closely spaced lines. This *fine structure* is the result of an intrinsic magnetic moment of the electron that corresponds to an intrinsic angular momentum called *electron spin*. The magnitude of the electron spin can have only one value; it is a fixed property of an electron, like its charge. The z component of the spin has just two possible values, $\frac{1}{2}\hbar$ and $-\frac{1}{2}\hbar$.

Atoms with More than One Electron With the spin quantum number, the states of the electron in hydrogen require four quantum numbers to specify them completely. In atoms with more than one electron, these four quantum numbers describe the state of each electron. The *Pauli exclusion principle* says that no two electrons in an atom can be in precisely the same state; that is, no two electrons can have the same values for all four of the quantum numbers. The specification of which states the electrons are in is called the *electron configuration*. The electron configuration for the lowest-energy state of an atom determines its chemical properties.

The Periodic Table Because of the Pauli exclusion principle, the electrons in multielectron atoms fill up the available states in *energy order*. The energy increases with increasing n and, for given n, with increasing ℓ. This leads to a *periodicity* in the chemical properties of the elements. For instance, when an atom (in its ground state) has just one electron in the outermost "shell" (that is, in the state with the largest n value), this electron is very loosely bound. As a result, such atoms interact readily with others; these elements (Li, Na, K, Rb, and Cs) are all chemically very active metals.

Atomic Spectra The wavelengths of light emitted by an isolated atom are determined by the allowed transition energies between the quantum states of the atom. Because atoms have discrete energy levels, they emit light of discrete wavelengths. In most cases, an excited state of an atom involves the excitation of just one of its electrons. In particular, the alkali metal elements (Li, Na, K, and so forth) have only one electron outside of a closed shell, and their optical spectra are very similar to that of hydrogen.

X-Ray Spectra At much higher energies, an inner electron can be knocked out of the atom, as by electron bombardment in an x-ray tube, for example. When an outer electron drops into the vacant state, a photon, usually in the keV region, is emitted. This is the origin of the *characteristic x-ray lines* of each element.

Molecular Bonding Single atoms are not usually what we encounter; we normally see atoms bonded together in molecules. Various mechanisms involving interactions of the outer electrons of atoms are involved in bonding. In *ionic bonds,* which are important in simple salts, atoms interchange one or more electrons to produce positive and negative ions that attract each other and bond together due to electrostatic attraction. This mechanism clearly doesn't explain the bonding of molecules like H_2 or N_2. Such atoms bond by *sharing* electrons in such a way as to reduce the repulsion between their nuclei. This is a *covalent bond*. The *hydrogen bond* that is responsible for shaping and linking large organic molecules is a similar mechanism.

Van der Waals and Metallic Bonds The *van der Waals bonds* are weaker bonds that hold solids and liquids together by attraction between the instantaneous electric dipole moments induced by molecules in one another. In the metallic bond, the valence electrons are shared by many atoms and so hold the lattice of metal ions together.

Molecular Spectra Electromagnetic radiation is emitted by molecules not only due to transitions between electron states but also due to transitions between vibrational or rotational states of the molecule as a whole. Vibrational and rotational spectral bands may be superimposed on the spectra of electron transitions, or they may occur alone at much lower energies, appearing in the infrared region. Much information about molecular structure is derived from the study of vibrational and rotational spectra.

Absorption and Emission of Light We get most of our information about the states of atoms and molecules from the electromagnetic radiation that they emit and absorb. Absorption lines in stellar spectra show particular wavelengths at which photons are absorbed by elements in the atmospheres of the stars. Incident photons of the proper energy can also *stimulate* emission of photons of the same energy. When this occurs, the emitted light is coherent with the incident light; this is the principle of the laser.

Solids A prominent characteristic of solids is the enormous range in their electrical conductivity. In a conductor, there are free electrons that can move through the material. Atomic energy levels in a solid form bands of very many, very closely spaced energy levels. These bands may be separated, or they may overlap. If the last filled band (the valence band) is followed by an energy gap before the first empty states (the conduction band), there are no free electrons and the material is an insulator; if these bands overlap, there are available states and the material is a conductor. If there is a gap between the bands but it is very small, thermal excitation excites some electrons into the empty conduction band (leaving "holes" in the valence band). This is a *semiconductor* material; its conductivity increases strongly with temperature.

Impurity Semiconductors By *doping* a semiconductor with an appropriate impurity, one can add conduction electrons or holes and thereby control the electrical properties of the material. Conduction can be either primarily by electrons (an *n-type semiconductor*) or primarily by positive holes (*p-type*).

Semiconductor Devices The *pn junction* (an interface between *p*-type and *n*-type semiconductor materials) is the basis of modern electronic technology. A simple *pn* junction rectifies current—that is, it passes current freely in one direction but not in the other. Many other applications exist. A material that contains two nearby *pn* junctions, forming a *pnp* or *npn* "sandwich," is a *transistor*. Current into the middle layer controls the flow from one end to the other. The transistor can amplify a signal or act as a switch.

Emitted photon energy

$$hf = \frac{hc}{\lambda} = E_i - E_f$$

Characteristic x-ray lines

$$\lambda = \frac{hc}{E_n - E_1} = \frac{hc}{(Z - 1)^2(E_0)(1 - 1/n^2)}$$

Rotational kinetic energy

$$E = \frac{1}{2}I\omega^2 = \frac{L^2}{2I}$$

Rotational energy levels

$$E = \ell(\ell + 1) E_{0r} = \ell(\ell + 1)\frac{\hbar^2}{2I}$$

Vibrational energy levels

$$E_n = \left(n + \frac{1}{2} \right) hf$$

II. Numbers and Key Equations

Numbers

$$\hbar = h/2\pi = 1.055 \times 10^{-34} \text{ J·s}$$

Bohr energy

$$E_0 = 13.60 \text{ eV}$$

H atom quantum numbers

$$n = 1, 2, 3, \ldots$$

$$\ell = 0, 1, 2, \ldots, n - 1$$

$$m = -\ell, -\ell + 1, \ldots, -1, 0, 1, \ldots,$$
$$\ell - 1, \ell$$

$$m_s = \pm\frac{1}{2}$$

Key Equations

Angular momentum

$$L = \sqrt{\ell(\ell + 1)} \, \hbar \qquad L_z = m\hbar$$

III. Possible Pitfalls

Don't mix up the quantum number m with the electron mass m! There are only so many letters in the alphabet.

The quantum numbers m and m_s, which measure *components* of angular momentum quantities (in units of $\hbar$), can be negative; n and ℓ are never negative.

Angular momentum quantum numbers like ℓ and m always come in pairs in which one (ℓ, for example) determines the magnitude of the angular momentum and the other (m), its projection along a given axis. Spin angular momentum, likewise, is described by two quantum numbers, s and m_s; but since the magnitude of quantum number s is *always* ½—the magnitude of the spin is an *intrinsic* property of the electron, just like its charge—it is usually not specified.

When we say (with Pauli) that no two electrons in a multielectron atom can be in the same quantum state, this means *exactly* the same—that is, all four quantum numbers cannot be the same. There's nothing wrong with saying that the two valence electrons in magnesium, for example, are in the 3s state, but they must have different values of m or m_s.

The vibrational and rotational spectra of simple molecules are in the infrared region if the transitions involve *only* change of vibrational or rotational state. If electron transitions, which are of much larger energy, are also involved, the radiation will be in the visible or ultraviolet regions, but with the lines split into closely spaced vibration-rotation *bands*.

Remember that the states in multielectron atoms don't fill simply in order of n; the energy of the electron states depend on ℓ as well. Thus, after argon, the nineteenth electron goes into the 4s state in potassium even though the 3d state is still unfilled.

Not all the radiation produced by an x-ray tube is characteristic radiation of the target atoms. The bulk of the radiation is the continuous bremsstrahlung spectrum, which is produced by a process that is largely independent of the target material.

The conductivity of most semiconductors, in practice, is determined not by electrons excited thermally across the band gap but by the density and type of impurities in the semiconductor material.

IV.   Questions and Answers

Questions

1. How does the periodic table demonstrate the need for electron spin? What properties would helium have if there were only the three quantum numbers n, ℓ, and m for electron states?

2. Would you expect the optical spectrum of magnesium to be more like that of hydrogen or that of helium? Explain.

3. The angular momentum of the electron in hydrogen cannot be directed along the z axis. Why not?

4. Why do we say that electron states in the hydrogen atom have definite values of L_z but not of L_x? What is it that's special about the z axis?

5. The optical spectra of certain elements are dominated by doublets—closely spaced pairs of spectral lines. Sodium (Na; 11 electrons) is an example. The very next element, magnesium (Mg; 12 electrons), is quite different. Its spectrum contains single lines and triplets, not doublets. Why is this?

6. When the $n = 2$ shell is just filled, there are 10 electrons in the atom; $Z = 10$ is neon, an "inert gas." To fill the $n = 3$ shell would take 18 more electrons; $Z = 28$ is nickel. Why isn't nickel the next inert gas?

7. The chemical and physical properties of sodium (Na) and those of lithium (Li) are very similar. Why?

8. Except for the inert gases, the elements near either end of a "row" in the periodic table tend to be chemically more active than those toward the middle. Why is this?

9. When the active electron in a helium atom makes a transition from the 2s to the 1s state, an ultraviolet photon with $\lambda \approx 60$ nm is emitted. When the same transition occurs in the He^+ ion, the wavelength is just about half this value. Why?

10. For a given value of the principal quantum number n, electron states in multielectron atoms increase with increasing ℓ, whereas for hydrogen they are independent of ℓ. How come?

11. Would you expect sodium fluoride (Na—F) to be an example of primarily covalent or ionic bonding? What about fluorine gas (F—F)? If your answers are not the same, why not?

12. How can you learn the bond length (the equilibrium inter-nuclear distance) in a diatomic molecule from measurements of its infrared spectra?

13. Give an example of an element that would act as a donor impurity in a semiconductor. Give an example of an acceptor impurity element. Explain the difference in behavior of the two elements you suggested.

14. Would you expect the resistivity of a semiconductor to increase, to decrease, or to be mostly unaffected by an increase in temperature? Explain your answer.

15. In actual molecular spectra, the spacing of rotational lines in a band tends not to be quite constant; instead, it decreases slightly with increasing ℓ. Why?

16. The periodic table really isn't, quite. The "periods" are of different lengths. Why?

Answers

1. The structure of the periodic table is determined by the order in which the electron states fill up. Since the Pauli exclusion principle allows only one electron for each different set of quantum numbers, all the available quantum numbers affect the periodic table. If there were no electron spin, only half of the available sets of different quantum numbers would exist. Helium's electron configuration would be $1s2s$ rather than $1s^2$—I guess it would have properties like those of lithium.

2.   Since the electron configuration of the outer, "active" electrons in magnesium is $3s^2$, its optical spectrum would be more like helium ($1s^2$) than hydrogen ($1s$).

3.   If the electron is in a state that *has* angular momentum ($\ell > 0$), then

$$L = \sqrt{\ell(\ell + 1)}\, \hbar > \ell\hbar$$

but

$$L_z = m\hbar \leq \ell\hbar$$

For **L** to be parallel to the z axis would mean $L_z = L$. Clearly, this is impossible.

4.   Nothing at all is special about the z axis; the values of the angular momentum component along any axis you like are the same: $\pm\hbar$, $\pm 2\hbar$, and so on. We specify a particular axis partly for mathematical reasons and partly to emphasize that only a single component of **L** can be known.

5.   Doublets are characteristic of the fine-structure splitting of the states of atoms with a single electron in the outermost shell, such as H or Na. In the next element, Mg, there are two active electrons, which makes its fine structure quite different.

6.   Because when the outer $3s$ and $3p$ states are filled (electrons 11 through 18), the next state that is energetically available is not the $3d$ but the $4s$ state. What makes the "inert gases" almost entirely inert is that to excite an electron out of the ground state takes a great deal of energy, so the atom is very hard to ionize. Thus $Z = 10 + 8 = 18$ (Ar) is the next inert gas.

7.   Because the outer-shell electron configurations are just the same, each having one electron in an s state.

8.   Because the elements at the ends of the rows either have a single electron outside of a closed-shell configuration, so that they readily form positive ions, or they are one electron short of a full shell, so that they readily form negative ions.

9. The photon energy is larger for the He^+ ion because the effective Z value "seen" by the electron making the transition is larger, and so its wavelength is shorter.

10. For multielectron atoms, the valence electron states are around a "core" comprising the nucleus and the interior electrons. The energy of each state depends on the effective Z value "seen" by the electron and thus on the degree to which the electron in each state "penetrates" the core. This, in turn, is determined by the value of ℓ, which determines (speaking crudely) the shape of the orbit. In hydrogen, of course, there are no interior electrons, so the energy of an electron orbit is independent of ℓ.

11. Sodium fluoride consists of a strongly positive ion (Na^+) and a strongly negative ion (F^-), so we expect the bonding to be ionic in character. Symmetric molecules such as F_2 are examples of covalent bonding.

12. By determining the molecule's rotational inertia and, from this, the bond length from its characteristic rotation energy.

13. Elements like phosphorus ($Z = 15$) that have five valence electrons act as donor impurities since they have an "extra" electron (relative to the tetravalent crystal lattice). Trivalent elements like aluminum ($Z = 13$) act as acceptor impurities and make a p-type impurity semiconductor.

14. For an intrinsic semiconductor, the resistivity decreases strongly as temperature increases because, with increased thermal energy, more electrons are excited across the band gap.

15. This results from "centrifugal stretching" of the molecule as it rotates, which increases its moment of inertia somewhat.

16. The "periods" correspond to the filling of major groups of electron states (and thus, roughly, to principal quantum number values), and there are more electron states (more ℓ values, for instance) available in the higher groups.

V. Problems and Solutions

Problems

1. The electron in a hydrogen atom is in an $\ell = 2$ state. What are the possible angles between the angular momentum vector **L** and the z axis?

How to Solve It

- The values of L and L_z are determined by the quantization condition.

- The ratio of L_z to L is the cosine of the angle between **L** and the z axis.

2. The light emitted by an ordinary He–Ne laser has $\lambda = 632.8$ nm, which is produced as an electron in neon moves from the $5s$ state to the $3p$ state. (In the ground state, the electron is in the $3s$ state.) The $3p$ state is at an energy of 18.62 eV above the ground state. What is the energy of the $5s$ state?

How to Solve It

- Calculate the photon energy of the laser light from the given wavelength.

- The photon energy is equal to the difference between the energies of the initial and final states.

- Does the $5s$ state lie above or below the $3p$ state?

3. The electron beam in a certain x-ray tube is accelerated through a potential difference of 9 kV and strikes a nickel target. (*a*) What is the shortest wavelength of the radiation that will be produced? (*b*) Will the characteristic K x-ray lines of nickel be present in the spectrum produced by the tube?

How to Solve It

- From the accelerating voltage, what is the kinetic energy of the electrons?

- The shortest wavelength corresponds to the largest photon energy. What is the maximum energy that a photon could have in this case?

- To produce K-series radiation, you must first knock out a K electron to produce a vacancy in the K shell. Is there enough energy to do that here?

4. You determine that the K_α x-ray wavelength for a certain element is 0.151 nm. What element is it?

How to Solve It

- Use the equation for the characteristic x-ray wavelengths to determine Z.

- This is the atomic number of the target element; what is the element?

5. The moment of inertia of a certain phonograph record is about 0.0012 kg·m². It is rotating at $33\frac{1}{3}$ rev/min. Find the (approximate) value of its angular momentum quantum number ℓ.

How to Solve It
- Calculate the angular momentum of the record.

- Use the formula for the magnitude of angular momentum in quantum mechanics to find ℓ.

- Note that the formula simplifies when $\ell \gg 1$.

6. The electron configuration of a certain atomic state is $1s^22s^22p^63s^13p^4$. (a) What element is this? (b) Is the atom in its ground state?

How to Solve It
- To find the element, you need its atomic number, which is equal to the number of electrons in the neutral atom.

- In the ground-state configuration, every electron is in the lowest possible energy state consistent with the Pauli exclusion principle.

7. The outermost electron in potassium is in the 4s state. If it did not penetrate the inner electron core at all, its energy would be the same as for the 4s state in hydrogen or 0.850 eV. Because it does penetrate to some degree, the effective value of Z for the 4s state in potassium is somewhat higher than 1. If the first ionization energy of potassium is 4.32 eV, what is the effective Z value for this state?

How to Solve It
- The ionization energy is just a way of telling you the 4s electron energy in potassium.

- The effective value of Z is the nuclear charge of a hypothetical one-electron atom that has its 4s state at this energy.

8. According to Equation 31-9 on page 802 of the text, the K_α line for zirconium ($Z = 40$) would have a wavelength of 0.0799 nm. What would be the wavelengths of the next two lines in the K series of Zr?

How to Solve It

- The equation in the text is approximate—it just uses the single-electron atom energies.

- The K_α line results from the transition from the $n = 2$ to the $n = 1$ state. What states are involved in the next two transitions?

9. The separation of the oxygen atoms in the O_2 molecule is 0.121 nm. Calculate the characteristic rotational energy E_{0r} for O_2.

How to Solve It

- Use the interatomic distance to calculate the moment of inertia of an O_2 molecule. You will have to look up the mass of an oxygen atom somewhere.

- Calculate the characteristic rotation energy E_{0r} from the moment of inertia.

10. The optical spectrum of sodium is dominated by a bright yellow doublet (a pair of spectral lines) with wavelengths of 588.99 and 589.95 nm. These lines result from a valence-electron transition from the $3p$ states (split by the spin-orbit effect) to the $3s$ ground state. The ground-state energy is -5.140 eV. At what energies do the $3p$ states lie?

How to Solve It

- Calculate the two photon energies.

- The photon energy is the difference between initial and final states of the atom. You know the wavelength for each transition and the energy of the final ($3s$) state; calculate the energies of the initial states.

11. A photon of wavelength 3.90 μm has just enough energy to excite an electron from the valence band to the conduction band in a certain semiconductor material. At what temperature T would kT (where k is Boltzmann's constant) be equal to the energy gap?

How to Solve It

- You can calculate the photon energy from the given wavelength.

- If the photon has just enough energy to excite an electron across the gap, its energy is equal to the gap energy.

- Thus, we want the temperature at which the kT is equal to the photon energy.

12. The energy gap between the valence and conduction bands in silicon is 1.1 eV. What maximum wavelength photon can excite an electron from the valence band to the conduction band?

How to Solve It

- The maximum wavelength corresponds to the minimum photon energy.

- The minimum photon energy that can excite an electron across the gap is, clearly, just equal to the gap energy.

13. The equilibrium separation of the K^+ and Br^- ions in KBr is 0.28 nm. (*a*) Calculate the potential energy of electrostatic attraction between the ions, assuming them to be point charges this distance apart. (*b*) The ionization energy of potassium is 4.32 eV, and the dissociation energy of KBr is 3.97 eV. Find the electron affinity of bromine.

How to Solve It

- To calculate the potential energy of the electrostatic attraction, just use the formula for the potential energy of two point charges.

- The difference between this potential and the net molecular energy (the dissociation energy) is the net energy cost of transferring an electron from K to Br.

- This net energy is the ionization energy of K less the electron affinity of Br.

14. The ionization energy of sodium is 5.14 eV and the electron affinity of fluorine is 4.07 eV. Thus, it costs 1.07 eV (net) to transfer an electron and make Na^+ and F^- ions. (*a*) At what separation distance does the electrostatic potential energy of the ions just compensate for the energy needed to exchange the electron? (*b*) The actual equilibrium interatomic distance is 0.193 nm. What is the dissociation energy of NaF?

How to Solve It

- Use the formula for the electrostatic potential energy of two point charges ($\pm e$) to find the distance at which the potential energy is -1.07 eV.

- At the separation found in part (*a*), the total energy is zero. When the ions are closer, the total energy is negative, so it takes an external energy input to dissociate the molecule.

- Thus, the dissociation energy is equal to the difference between the potential energy at the separation distance found in (*a*) and that at the equilibrium separation.

Solutions

1. The angle θ between the angular momentum vector and the z axis is given by

$$\cos \theta = \frac{L_z}{L} = \frac{m\hbar}{\sqrt{\ell(\ell + 1)}\,\hbar} = \frac{m}{\sqrt{\ell(\ell + 1)}}$$

for $\ell = 2$, $m = -2, -1, 0, 1$, and 2. For example, for $m = 2$

$$\cos \theta = \frac{2}{\sqrt{2(2 + 1)}} = \frac{2}{\sqrt{(2)(3)}} = 0.8165$$

so

$$\theta = \underline{35.3°}$$

The other possible values are found in the same way: $\theta = \underline{65.9°}$, $\underline{90.0°}$, $\underline{114.1°}$, and $\underline{144.7°}$.

2. $\underline{20.58 \text{ eV}}$

3. (*a*) The shortest-wavelength (highest-energy) radiation is produced when all the kinetic energy of an incoming electron goes to a single photon. Thus,

$$\lambda_{min} = \frac{hc}{E_k} = \frac{1240 \text{ eV·nm}}{9000 \text{ eV}} = \underline{0.138 \text{ nm}}$$

(*b*) To produce the characteristic K lines, a vacancy must be created in the K ($n = 1$) shell of the atom. For Ni ($Z = 28$), this takes energy

$$E = Z^2 \frac{E_0}{n^2} = (28)^2 \frac{(13.60 \text{ eV})}{(1)^2} = 10.66 \text{ keV}$$

The incoming 9-keV electrons do not have enough energy to knock out a K electron, so the K lines will <u>not</u> be produced

4. The equation gives $Z = 29.4$, so the element is presumably <u>copper</u>.

5. The magnitude of the angular momentum of a quantum system is given by

$$L = \sqrt{\ell(\ell + 1)}\,\hbar$$

For the rotating disk, $L = I\omega$. Here,

$$\omega = \frac{(33.33 \text{ rev/min})(2\pi \text{ rad/rev})}{60 \text{ s/min}} = 3.49 \text{ rad/s}$$

so

$$L = I\omega = (1.2 \times 10^{-3} \text{ kg·m}^2)(3.49 \text{ s}^{-1})$$

$$= 4.2 \times 10^{-3} \text{ kg·m}^2/\text{s}$$

Since this is much greater than $\hbar$, $\ell \gg 1$ and $\sqrt{(\ell + 1)} \approx \ell$, so $L \approx \ell\hbar$. Thus,

$$\ell \approx \frac{L}{\hbar} = \frac{4.2 \times 10^{-3} \text{ J·s}}{1.05 \times 10^{-34} \text{ J·s}} = \underline{4.0 \times 10^{31}}$$

6. (a) Phosphorus ($Z = 15$)   (b) No

7. Saying that the first ionization energy is 4.32 eV is the same as saying that the ground-state energy of the outer $4s$ electron is -4.32 eV. The Z_{eff} we want is the Z value of a (hypothetical) one-electron atom whose $4s$ state would be at this energy. Thus

$$E_{4s} = -(Z_{eff})^2 \frac{13.60 \text{ eV}}{n^2} = -4.32 \text{ eV}$$

so

$$(Z_{eff})^2 = (4)^2 \frac{4.32}{13.60} = 5.08$$

$$Z_{eff} = \underline{2.25}$$

8. 0.0674 nm and 0.0639 nm

9. First calculate the moment of inertia of the oxygen molecule:

$$I = 2M \left(\frac{1}{2}d\right)^2 = \frac{1}{2}Md^2$$

The mass of an oxygen atom is 2.66×10^{-26} kg. Thus,

$$I = (0.5)(2.66 \times 10^{-26} \text{ kg})(1.21 \times 10^{-10} \text{ m})^2$$

$$= 1.95 \times 10^{-46} \text{ kg·m}^2$$

and

$$E_{0r} = \frac{\hbar^2}{2I} = \frac{(1.055 \times 10^{-34} \text{ J·s})^2}{(2)(1.94 \times 10^{-46} \text{ kg·m}^2)} = 2.86 \times 10^{-23} \text{ J}$$

$$= 1.78 \times 10^{-4} \text{ eV}$$

10. -3.035 eV and -3.038 eV

11. Boltzmann's constant $k = 1.38 \times 10^{-23}$ J/K $= 8.62 \times 10^{-5}$ eV/K. We want the temperature T at which

$$kT = \frac{hc}{\lambda} = \frac{1.240 \times 10^{-6} \text{ eV·m}}{3.90 \times 10^{-6} \text{ m}} = 0.318 \text{ eV}$$

$$T = \frac{0.318 \text{ eV}}{k} = \frac{0.318 \text{ eV}}{8.62 \times 10^{-5} \text{ eV/K}} = 3690 \text{ K}$$

12. 1127 nm

13. (a) The electrostatic potential energy of two charges $+e$ and $-e$ separated by a distance r is

$$U = -\frac{ke^2}{r} = -\frac{(8.99 \times 10^9 \text{ N·m}^2/\text{C}^2)(1.602 \times 10^{-19} \text{ C})^2}{2.8 \times 10^{-10} \text{ m}}$$

$$= -8.24 \times 10^{-19} \text{ J} = -5.14 \text{ eV}$$

(b) To ionize the potassium atom requires 4.32 eV. Let the energy regained by adding an electron to bromine be A (this is the electron affinity of Br). The total energy of the molecule is thus

$$4.32 \text{ eV} - A - 5.14 \text{ eV} = E = -3.97 \text{ eV}$$

from which $A = 3.15$ eV

14. (a) 1.35 nm   (b) 6.39 eV

CHAPTER 32

Nuclear Physics

I. Key Ideas

The Nucleus The positive charge and virtually all the mass of the atom are in a tiny central region called the *nucleus*, which is about 10^5 times smaller than the atom. It is composed of protons (hydrogen nuclei) and neutrons (similar but uncharged particles), held together by the *strong nuclear force*. The total number of nucleons (neutrons and protons) is the mass number A, the number of protons is the atomic number Z of the element, and the number of neutrons is N.

Size and Shape of Nuclei We get information as to the size and shape of nuclei from scattering experiments. Although the conclusions depend somewhat on what experiment is done, we find most nuclei to be roughly spherical, with a volume roughly proportional to their mass—that is, with an approximately constant nuclear density.

Stability The properties of nuclei in their ground states depend on N and Z. Among light nuclei, those that are stable have $N \approx Z$, which we can understand on the basis of the Pauli exclusion principle; heavier nuclei that are stable have more neutrons than protons because of the Coulomb energy of repulsion. Nuclei are strongly bound; we can calculate nuclear binding energies from mass values (which can be measured very accurately). For most nuclei, the binding energy per nucleon is approximately constant, reflecting the very short range of the nuclear force.

Radioactivity Nuclides that are not stable are *radioactive:* they decay into different, more stable nuclear species by emitting various "radiations." Three kinds of radioactivity are observed: α decay, β decay, and γ emission. Radioactive decay is a statistical process—that is, the decay of individual nuclei is random—so the decay rate of a given sample declines exponentially. The rate of decay is usually characterized by a *half-life*, which is the time required for the decay rate of a sample to diminish by half. A very wide range of half-life values is observed.

Alpha Decay An α *particle* is a nuclear fragment that is equivalent to a He nucleus, consisting of two neutrons and two protons bound together. All heavy nuclides are unstable to α decay. (A nucleus is unstable if the total mass of the decay products is less than that of the parent nucleus.) The nuclide that results from the decay may itself be unstable; thus, there are whole series or chains of radioactive nuclides starting with the long-lived U and Th isotopes and continuing until a stable isotope of Pb is reached.

Beta Decay *Beta decay* is the emission by the nucleus of an electron or positron; its effect is to change the number of protons Z in the nucleus by ± 1 without changing the mass number A (the mass of an electron is very small compared to 1 mass unit). A β emission is accompanied by the emission of a *neutrino v*, which shares the available momentum and energy. As the neutrino has no charge, it is very difficult to observe directly.

Gamma Emission The γ *ray* is just a high-energy photon. It is produced when a nucleus in an excited state decays, emitting its extra excitation energy as a γ photon.

Nuclear Reactions Information about nuclear structure is obtained by bombarding nuclei with particles and observing the scattering and reactions that result. We build ever larger particle accelerators to do this. A given reaction is characterized by a *Q value*, which is the net amount of energy released; reaction probability is usually expressed as an effective area, the cross section σ.

Thermal Neutrons Reactions with neutrons are of particular importance in nuclear reactors. Neutrons that are scattered many times in a material without being absorbed lose kinetic energy until, eventually, they reach equilibrium with the thermal motion of the material. Neutrons of this energy are called *thermal neutrons*. For many elements, the cross section for neutron capture or fission reactions is very high for thermal neutrons.

Nuclear Energy Either the fusion of two light nuclei into a heavier one or the fission of a heavy nucleus into two lighter fragments releases energy, because in both cases the nuclei that result are more tightly bound. The energy release can be very large, many orders of magnitude larger than that of chemical reactions, because of the very strong nuclear binding. Both fission and fusion have been applied to make weapons, and fission (thus far) has been used as a civilian energy resource.

Fission For two nuclides—^{235}U and ^{239}Pu—enough energy is gained by the capture of a thermal neutron to cause fission. Since the fission process in turn releases free neutrons, a chain reaction is possible. To exploit the large cross section for

thermal neutrons, a low-mass *moderating material* must be present. Scattering in the moderator reduces the energy of fission neutrons to the thermal range without absorbing them. Water is the most common moderating material, although "ordinary" water will not work with natural uranium because there is too much neutron absorption. The reaction rate is controlled by changing the reproduction factor by inserting or extracting neutron-absorbing materials; this is practical only because some one percent of the fission neutrons are delayed by several seconds.

Breeder Reactor A *breeder* is a fast-neutron fission reactor that can make as much fissionable material (by neutron capture in ^{238}U, leading to ^{239}Pu) as it uses. Breeder reactors have the potential to provide a major long-range energy resource. Because fast (not thermal) neutrons must be used, however, control is more difficult, and the large-scale production of Pu poses safety problems. Currently, the real limitation on the exploitation of fission resources is the problem of storage and disposal of radioactive waste.

Fusion A self-sustaining controlled fusion reaction has yet to be achieved. The fundamental problem is that of containing the reacting material for an extended time at temperatures above 10^7 K. The plasma can be confined magnetically or "inertially." As a large-scale energy resource, fusion would pose many of the same problems as fission, but the "fuel" would be plentiful.

Interaction of Charged Particles with Matter The effects of charged nuclear particles are those caused by the ionization produced by the high-energy particles as they pass through matter. Charged particles passing through a material traverse a well-defined range, leaving a "wake" of ionization, and then stop. The energy loss rate is highest at low energies and is roughly proportional to the density of electrons in the material. At very high energies, there is also some energy loss directly to radiation (bremsstrahlung).

Neutrons and γ Rays The energy loss mechanisms for uncharged particles are quite different from those for charged particles. Neutrons are unaffected by electrons, and they interact with matter only via nuclear reactions. Gamma rays interact with electrons by such processes as the photoelectric affect or Compton scattering, which remove the photon altogether rather than gradually reducing its energy. The "stopping" of uncharged particles is a statistical process, and we observe that the intensity of a beam decreases exponentially as the thickness of matter traversed increases.

Radiation Dose The main determinant of radiation damage (mechanical or biological) to a material is the density with which energy is deposited in the material; correspondingly, radiation dose is defined as energy delivered per unit mass. For biological effects, the type of radiation also matters, and the relative biological effectiveness (RBE) is factored into calculation of *dose equivalent*.

Biological Effects No acute effects are observed for doses below 25 rem. At doses over 100 rem, damage to the gastrointestinal and blood-forming systems are observed. Doses over 500 rem are usually fatal. The effects of doses too small to produce acute effects are long term, mainly an enhanced probability of contracting cancer. At very low doses the effects, if any, cannot be measured; it is assumed, to be conservative, that effects are proportional to dose all the way to zero.

II.   Numbers and Key Equations

Numbers

$$1 \text{ unified mass unit (u)} = 1.66 \times 10^{-27} \text{ kg}$$
$$= 931.5 \text{ MeV}/c^2$$

1 becquerel (Bq) = 1 decay/s

$$1 \text{ curie (Ci)} = 3.7 \times 10^{10} \text{ Bq}$$

$$1 \text{ barn} = 10^{-28} \text{ m}^2$$

1 gray (Gy) = 1 J/kg

1 rad = 0.01 Gy

Key Equations

Nuclear radius

$$R = R_0 A^{1/3}$$

Binding energy

$$E_b = (ZM_H + Nm_n - M_A)c^2$$

Decay of radioactive nuclide

$$\Delta N = -\lambda N \, \Delta t \qquad N = N_0 e^{-\lambda t}$$

Half-life

$$t_{1/2} = \frac{0.693}{\lambda}$$

Cross section

$$\sigma = \frac{R}{I}$$

Q value

$$Q = \Sigma_i \, Mc^2 - \Sigma_f \, Mc^2$$

Lawson's criterion

$$n\tau > 10^{14} \text{ s·particles/cm}^3$$

III.   Possible Pitfalls

Since all energy has mass, you can get the binding energy of a nucleus or the Q value of a reaction simply from mass differences. Spontaneous processes must go in the direction of lower rest mass; any process involving increased rest mass must have outside energy input.

The nuclear charge actually changes in an α or β emission since these particles have charge; thus, the chemical character of the element changes. When a uranium nucleus emits an α particle, it isn't uranium any more (it's thorium).

The half-life and the mean lifetime of a radioactive nuclide are not the same thing.

Charged particles moving through matter stop smoothly over a well-defined range. The same is not true of uncharged particles, which disappear or are removed from the beam at random. The *intensity* of a beam of uncharged particles decreases exponentially passing through matter.

The nuclear properties of a given nuclide depend on both N and Z, whereas the chemical properties depend only on Z. (The atomic number Z determines what the element is.)

IV.   Questions and Answers

Questions

1.   One of the principal differences between the strong nuclear force and the "classical" forces of electromagnetism and gravitation is that the nuclear force acts only over a very short range. How do we know this?

2.   When a radioactive decay by α emission is followed by a β decay, it is almost always a β^- rather than a β^+ decay. Why?

3.   The radioactive properties of a particular nuclide are independent of the chemical situation of the atom. That is, the decay of ^{14}C is the same in pure carbon, in a CO_2 molecule, or in a large biological molecule. Why?

4.   In the last 80 years, industrialized nations have been dumping millions of tons of CO_2 from the burning of coal and petroleum products into the air. There is essentially no ^{14}C in these "fossil fuels." Why not? What effect will great amounts of ^{14}C-free carbon added to the atmosphere have on the future use of carbon dating?

5.   The stable isotopes of neon are ^{20}Ne, ^{21}Ne, and ^{22}Ne. If ^{25}Ne existed, would it be a β^+ emitter, a β^- emitter, an α emitter, or stable? Explain.

6.   Complete each of the following nuclear reaction equations:
(*a*) $^1H + {}^7Li \rightarrow$ _____ $+ n$;   (*b*) $^1H + {}^{40}Ar \rightarrow {}^{40}K +$ _____ ;
(*c*) $^1H + {}^{40}Ar \rightarrow {}^{37}Cl +$ _____ ;   (*d*) $^2H +$ _____ $\rightarrow {}^{29}Si + {}^1H$;
(*e*) $^{24}Mg + {}^1H \rightarrow$ _____ $+ \gamma$

7.   All nuclides having a mass number higher than around 200 are unstable. Why?

8.   In calculating the radiation dose you received from an external radiation source, you would ordinarily consider only the γ radiation that the source emits. Why?

9. Give a simple physical reason why the cross section for neutron-induced reactions at low neutron energy is proportional, in many cases, to $1/v$.

10. What is the moderator for in a fission reactor?

11. Sustained nuclear fusion reactions, as an energy source, occur continually in the sun but are awfully hard to arrange on earth. Why?

12. In the carbon dating of archeological samples, what exactly is it that we are dating?

Answers

1. Because the strong nuclear force, although the strongest, has no observable effects on any scale as large as the size of an atom or larger, and because the strong nuclear force is observed to "saturate" in that a nucleon in the nucleus interacts only with its few nearest neighbors.

2. An α particle consists of two neutrons and two protons. Since the proportion of neutrons in the nucleus tends to be higher at higher mass numbers, emission of an α particle tends to result in a daughter nuclide with somewhat too high a proportion of neutrons. Thus, if the daughter is β-unstable it will most likely emit a β^- particle, thereby changing a neutron (in the nucleus) into a proton.

3. Because the chemical situation of the atom is the result of the interactions of the few outermost electrons in the atom, whereas radioactive decay is a nuclear property.

4. There's no ^{14}C in these fuels because they are the products of the biological decay of living organisms millions or even billions of years dead; all their ^{14}C has long since decayed away. If enough ^{14}C-free carbon is dumped into the atmosphere to dilute the proportion of ^{14}C significantly, then currently living organisms will acquire a correspondingly lower proportion of ^{14}C; unless accounted for, this will bias carbon dating measurements in the future, making carbon samples seem older than they actually are.

5. It would be a β^- emitter since it has (apparently) too many neutrons for stability and the effect of β^- decay is to transform a neutron into a proton.

6. (a) ^{1}H + ^{7}Li → ^{7}Be + n;   (b) ^{1}H + ^{40}Ar → ^{40}K + n;
(c) ^{1}H + ^{40}Ar → ^{37}Cl + $\underline{\alpha}$;   (d) ^{2}H + $^{28}\underline{\text{Si}}$ → ^{29}Si + ^{1}H;
(e) ^{24}Mg + ^{1}H → $^{25}\underline{\text{Al}}$ + γ

7. The underlying reason for their instabililty is the Coulomb repulsion between protons, which more and more at higher Z acts to drive the nucleus apart. The primary mechanism by which they decay is α emission.

8. Because α and β radiation have very short ranges in matter and so are likely to be absorbed in the intervening air or in your clothing. Only γ rays have a long enough range in matter to dose your innards from the outside.

9. The strong nuclear force has a very short range; the amount of time that a neutron spends "within range" will be of the order of this range divided by the velocity of the neutron.

10. Its function is to slow the fission neutrons down by exchanging energy in scattering without absorbing the neutrons, so that the fission reactions that keep the reactor going have the advantage of the very large cross sections for thermal neutrons.

11. Because the basic requirement for a sustained reaction is that the material in which fusion is taking place be confined at reasonable density for a "long" time at temperatures approaching 10^8 K. These conditions are met in the interior of a star like the sun but are awfully hard to arrange in the laboratory.

12. What the technique measures is the length of time since the living tissue, of which the carbon being tested was once part, died.

V. Problems and Solutions

Problems

1. Consider the fission reaction

$$n + {}^{235}\text{U} \rightarrow {}^{140}\text{Xe} + {}^{93}\text{Sr} + \text{free neutrons}$$

(a) How many free neutrons are given off? (b) The atomic masses of ^{93}Sr, ^{140}Xe, and ^{235}U are 92.91417 u, 139.92088 u, and 235.04394 u, respectively; the mass of a neutron is 1.008665 u. How much energy is released in this fission reaction? (c) How much energy (in kilocalories) would be produced if 1 g of ^{235}U were fissioned entirely by this reaction?

How to Solve It

- The total number of nucleons is not changed in the fission reaction.

- There is a net decrease of rest mass in the process. A corresponding amount of energy is released, most of it as kinetic energy of the fragments.

- Look at the last part simply as a unit conversion—MeV/nucleus to kcal/g.

2. From the mass values in Table 32-1 on page 838 in the text, calculate the binding energy per nucleon for (*a*) iron (^{56}Fe) and (*b*) lead (^{208}Pb).

How to Solve It

- The binding energy is the energy corresponding to the difference in mass between the nucleus and its constituent nucleons.

- Divide the binding energy by the mass number to get the binding energy per particle.

- The binding energy per nucleon is greatest for medium-weight nuclides, like iron, and less for both heavy and light nuclides.

3. If I have 0.1 g of ^{85}Kr (half-life 10.7 y), what is its decay rate in curies?

How to Solve It

- The curie, like the becquerel, is a measure of decay rate.

- The fraction of the nuclei present that decay per unit time is the decay constant λ. You can calculate this from the half-life.

- Calculate the number of krypton atoms in the sample and, from that, the decay rate.

4. The initial counting rate from a radioactive source is 1000 counts/s. Exactly one day (24 h) later, the counting rate is 125 counts/s. (*a*) What is the half-life of this source? (*b*) What is the counting rate after an additional 30 h?

How to Solve It

- How many half-lives have passed in 24 h?

- Calculate the decay constant from the half-life.

- Because the 30 h given in (*b*) is not a whole number of half-lives, you will have to use the exponential decay formula to solve that part.

5.   Assume that the amount of ^{14}C in living matter results in 15 decays/min for each gram of carbon. In a sample of charcoal from an archeological site, I find 260 decays/min from the ^{14}C in 55 g of carbon. How old is this sample?

How to Solve It

- You know the present decay rate; the "initial" rate is that given for living matter.

- Calculate the decay constant from the known half-life of ^{14}C (5700 y).

- Use the exponential decay equation to find the time for which the sample has been decaying.

6.   A certain sample of wood contains 4.0 g of carbon and shows a ^{14}C decay rate of 12.4 counts/min. What is the age of this sample?

How to Solve It

- The decay constant for ^{14}C was calculated in Problem 5.

- You know the present decay rate; the "initial" rate is that for living matter.

- Use the exponential decay equation to find the time for which the sample has been decaying.

7.   If the reproduction factor in a reactor is 1.003 and the mean time between fission generations is 1.0 ms (note that this value ignores the delayed fission neutrons), how long does it take for the power level in the reactor to double?

How to Solve It

- From the given reproduction factor, how many generations are required for the power level in the reactor to double?

- To what time does this number of generations correspond?

8.   The atomic masses of ^{14}C and ^{14}N are 14.003242 u and 14.003074 u, respectively. Calculate the Q value of the reaction

$$^{14}N + n \rightarrow {}^{14}C + {}^1H$$

(This reaction between cosmic-ray neutrons and nitrogen in the atmosphere is the main source of radioactive carbon on earth.)

How to Solve It

- Calculate how much rest mass disappears in this reaction.

- The Q value is the energy equivalent of this much mass; multiply by c^2 to get the corresponding energy. (In particularly appropriate units, c^2 = 931.5 MeV/u).

9. Lead 1 cm thick reduces the intensity of γ rays from a certain radioactive source by 30 percent. (*a*) By how much will 3 cm of lead reduce the intensity of the same radiation? (*b*) What thickness of lead is needed to reduce the intensity by a factor of 1000?

How to Solve It
- The reactions that remove photons from the beam occur at random, so each centimetre of lead removes the same fraction of photons from the beam.

- How many times must the intensity be reduced by 30 percent to reduce it by a factor of 1000?

10. The range of 10-MeV α particles is about 11 cm in air and about 130 μm in water. On the average, one ion pair (a positive ion and a free electron) is produced in either air or water for each 34 eV of energy lost by the α particle. What is the average number of ion pairs per micrometre of track length produced by 10-MeV α particles (*a*) in air and (*b*) in water?

How to Solve It
- This is very straightforward; the α particles lose so many electron-volts per micrometre, on the average, and each ion pair corresponds to so many electronvolts.

- The result of part (*b*) shows why α particles are so damaging to living tissues (which are more or less water): they can cause a lot of disruption in a space the size of a living cell (a few micrometres).

11. (*a*) Estimate the size of a deuteron (^{2}H nucleus) from $R \approx$ (1.4 fm)$A^{1/3}$, and use it to calculate the electrostatic potential energy of two deuterons that are just touching, assuming them to be spherical. (*b*) If the two deuterons are to react, they must be "thrown together" hard enough to overcome this repulsion; this is the reason that fusion reactions require enormous temperatures. At what temperature of deuterium gas would this amount of energy equal 10 kT (where k is Boltzmann's constant)? At this temperature, we'd expect a few deuterons to be energetic enough to interact.

How to Solve It
- Find the radius of a deuteron; their effective separation is just equal to one diameter.

- Since we assume the deuterons are spherical, we can use the point-charge formula for their electrostatic potential energy.

Solutions

1.   (*a*) The total number of nucleons must be the same before and after the decay. If n free neutrons are emitted,

$$1 + 235 = 140 + 93 + n$$

$$n = \underline{3}$$

(*b*) The Q value is the energy corresponding to the difference between the total mass before the reaction and the total mass after it. Thus, here

$$Q = [1.008665 \text{ u} + 235.04394 \text{ u} - 92.91417 \text{ u} - 139.92088 \text{ u}$$

$$- (3)(1.008665 \text{ u})](931.5 \text{ MeV/u})$$

$$= \underline{178.4 \text{ MeV}}$$

(*c*) If 1 g of ^{235}U were fissioned entirely by this reaction, the total energy release would be

$$E = (1\text{g})\, \frac{(178 \text{ MeV})(1.602 \times 10^{-13} \text{ J/MeV})}{(235 \text{ u})(1.661 \times 10^{-24} \text{ g/u})(4184 \text{ J/kcal})}$$

$$= \underline{1.57 \times 10^7 \text{ kcal}}$$

2.   (*a*) $\underline{8.72 \text{ MeV}}$ (*b*) $\underline{7.87 \text{ MeV}}$

3.   From the given half-life, the decay constant is

$$\lambda = \frac{0.693}{\tau_{1/2}} = \frac{0.693}{10.7 \text{ y}}$$

$$= 6.48 \times 10^{-2} \text{ y}^{-1} = 2.05 \times 10^{-9} \text{ s}^{-1} \text{ or decays/s}$$

In 0.1 g of ^{85}Kr, there are

$$N = \frac{0.1 \text{ g}}{85 \text{ g/mol}} (6.022 \times 10^{23} \text{ atoms/mol}) = 7.08 \times 10^{20} \text{ atoms}$$

so the decay rate is

$$R = \lambda N$$

$$= (2.05 \times 10^{-9} \text{ decays/s})(7.08 \times 10^{20})$$

$$= 1.45 \times 10^{12} \text{ decays/s} = 1.45 \times 10^{12} \text{ Bq}$$

$$= \frac{1.45 \times 10^{12} \text{ Bq}}{3.70 \times 10^{10} \text{ Bq/Ci}} = \underline{39.2 \text{ Ci}}$$

4. (a) $\underline{8 \text{ h}}$ (b) $\underline{9.3 \text{ counts/s}}$

5. The present decay rate per gram of carbon in the sample (the proper name for this is *specific gravity*) is

$$S = \frac{R}{m} = \frac{260 \text{ decays/min}}{55 \text{ g}} = 4.7 \text{ decays/min·g}$$

If the ^{14}C has been decaying for a time t, then

$$S = S_0 e^{-\lambda t}$$

$$4.7 \text{ decays/min·g} = (15 \text{ decays/min·g}) e^{-\lambda t}$$

$$e^{-\lambda t} = 4.7/15 = 0.313$$

$$\lambda t = -1.16$$

Now

$$\lambda = \frac{\ln(2)}{t_{1/2}} = \frac{0.693}{5730 \text{ y}} = 1.21 \times 10^{-4} \text{ y}^{-1}$$

so

$$t = \frac{1.16}{1.21 \times 10^{-4} \text{ y}^{-1}} = \underline{9600 \text{ y}}$$

6. $\underline{13{,}000 \text{ y}}$

7. Let R_0 be the initial reaction rate. If $k = 1.003$ and the rate has doubled after N generations, then

$$R_N = (1.003)^N R_0 = 2R_0$$

Thus,

$$1.003^N = 2$$

$$N \ln(1.003) = \ln(2) = 0.693$$

$$N = \frac{0.693}{0.00300} = 231$$

This corresponds to a time of $(231)(1 \text{ ms}) = \underline{0.231 \text{ s}}$.

8. $\underline{0.626 \text{ MeV}}$

9. (*a*) One 1-cm layer reduces the intensity by 30 percent, that is, to 0.70 of its initial value, so three layers reduce it by $(0.70)^3 = 0.343$. Thus, a 3-cm thickness of lead reduces the initial intensity by 66 percent.
(*b*) To reduce the intensity to $\frac{1}{1000}$th of its initial value will require N layers, where

$$(0.70)^N = 0.001$$

$$N \log(0.70) = \log(0.001)$$

$$-0.155 N = -3$$

$$N = 19.4$$

The required thickness is thus $(19.4)(1 \text{ cm}) = \underline{19.4 \text{ cm}}$

10. (*a*) $\underline{2.7 \text{ ion pairs}/\mu\text{m}}$ (*b*) $\underline{2300 \text{ ion pairs}/\mu\text{m}}$

11. (*a*) The radius of a deuteron is

$$R = R_0 A^{1/3} = (1.4 \text{ fm})2^{1/3} = 1.76 \text{ fm}$$

so the effective separation of the two deuterons is 3.52 fm, and their electrostatic potential energy is

$$U = \frac{ke^2}{r} = \frac{(8.99 \times 10^9 \text{ N·m}^2/\text{C}^2)(1.602 \times 10^{-19} \text{ C})^2}{3.52 \times 10^{-15} \text{ m}}$$

$$= 6.55 \times 10^{-14} \text{ J} = \underline{0.409 \text{ MeV}}$$

(*b*) At temperature T,

$$10 \, kT = 6.55 \times 10^{-14} \text{ J}$$

$$T = \frac{(6.55 \times 10^{-14} \text{ J})}{(10)(1.38 \times 10^{-23} \text{ J/K})} = \underline{4.75 \times 10^7 \text{ K}}$$

This overestimates considerably the amount of energy needed to get some fusion going, but it at least gives you an idea of why the temperature has to be so high.

CHAPTER 33

Elementary Particles

I. Key Ideas

Elementary Particles We have long sought the fundamental particles of which matter is constituted; our ideas change as we learn more and more about matter on the ultimate scale. Atoms were once regarded as "fundamental," then nucleons and electrons. Since then, hundreds of subatomic particles have been discovered or created. According to our present understanding, everything can be considered to be made up of just two families of particles, the leptons and the quarks.

Classifications Particles with spin $\frac{1}{2}$ obey the Pauli exclusion principle and are described by the Dirac equation; those with integer spin do not obey the Pauli exclusion principle. The Dirac equation views an antiparticle, such as a positron, as a "hole" in an infinite "sea" of negative-energy electron states. The position was observed in 1932; the antiproton, in 1955. Particles and their antiparticles can be created from or can annihilate to electromagnetic radiation of the corresponding energy.

The Four Forces Four basic forces are known to act in nature: the strong nuclear or *hadronic* interaction, the electromagnetic force, the weak nuclear interaction, and gravity. Fundamental particles that are affected by the strong interaction are *hadrons*. The hadrons include baryons—nucleons and other "heavy" particles—and mesons. Particles that decay via the strong interaction have lifetimes of the order of 10^{-23} s or less, whereas those that do not have much longer lifetimes—perhaps 10^{-10} s. The leptons—electrons, muons, and neutrinos—decay by means of the weak interaction but are unaffected by the strong force.

Conservation Laws The attitude of particle physicists is that whatever can happen, happens. That a reaction that otherwise ought to occur does not suggests that it would violate some conservation law. This has led physicists to postulate various conserved properties of elementary particles besides energy and momentum. These include baryon and lepton number and strangeness (which is conserved in hadronic decays but not in those caused by the weak interaction).

Quarks Leptons, so far as we can tell, are point particles without internal structure. This is not true of the hadrons. It is now thought that all the hadrons are combinations of *quarks*. Baryons are systems of three quarks or antiquarks; mesons, of one quark and one antiquark. Quarks are fractionally charged. In the original model, there were three kinds (*flavors*) of quarks; subsequently, a fourth had to be introduced. It carries another conserved quantity called *charm*. Like strangeness, charm is conserved in hadronic (strong) interactions and is changed by ± 1 by the weak interaction. There are now thought to be six flavors of quarks. The strength of the quark model is that all possible combinations of three quarks can be identified with known hadrons. Isolated quarks have never been observed, and according to current theory, they cannot be.

Field Particles In addition to the quarks and leptons, there are also the particles that are the means by which forces are exerted by one elementary particle on another. For example, the electromagnetic force between charged particles is described as an exchange of (virtual) photons. The other basic interactions are described in the same way. The weak interaction is carried by *vector bosons,* the strong nuclear interaction by *gluons,* and the gravitational force by *gravitons.*

Unified Theories Attempts to describe all the elementary forces of nature by a single unified theory began with Einstein. At present a successful theory unites the electromagnetic and weak nuclear interactions, but a grand unification theory incorporating the strong nuclear force still does not exist.

II.   Key Equations

Key Equations

Positron annihilation

$$e^- + e^+ \rightarrow \gamma + \gamma$$

Neutron decay

$$n \rightarrow p^+ + e^- + \bar{\nu}_e$$

III.   Possible Pitfalls

There are a lot of different ways we distinguish and characterize the "elementary" particles. Note that they overlap a lot—for instance, the hadrons include both baryons and mesons.

The quarks violate quantization of charge as we usually observe it, as their charges are fractions of e. However, the quarks have not been observed as isolated particles, and it is thought that they cannot be.

IV.   Questions and Answers

Questions

1. Could a fast proton decay into heavier particles in flight if it had enough kinetic energy? What prevents a process like $p \rightarrow \pi^+ + n$ from happening?

2. The free antineutron (if it could live long enough!) would β-decay into an antiproton and a positron. Is it a neutrino or an antineutrino that would be emitted in the process?

3. The uncharged π^0 is much shorter lived than the π^+ or π^-. Why is this?

4. Reactions induced by proton beams of sufficient energy on hydrogen targets can produce "strange" particles. At least two strange particles are always produced, never just one. Why?

5. Doubly charged baryons are known to exist. Is this consistent with the quark model? If doubly charged mesons were found, would this be consistent with the model?

Answers

1. No, because for this to happen would violate conservation of momentum and energy. This is easiest to see in a frame of reference attached to the original proton, which is also the center-of-mass frame. Rest energy would increase in the given decay since the pion and neutron together have more mass than the proton; but in the center-of-mass frame it is clear that there is nowhere for this extra energy to come from.

2. By the conservation of lepton number, since the positron is an antiparticle, this would have to be a neutrino.

3. The charged pions must decay by the weak interaction since their decays produce leptons. The neutral π^0 decays via the (much stronger) electromagnetic force into two photons.

4. The initial state is one of two protons, so the strangeness is zero. Thus, by the conservation of strangeness (for hadronic interactions), if one particle of nonzero strangeness is produced, another of opposite strangeness must be produced also to keep the total equal to zero.

5. Three quarks (in the original three-quark model, at least, they would have to be uuu) can produce a baryon of charge $+2e$, but a quark and an antiquark (this is what mesons are) could not.

V. Problems and Solutions

Problems

1. A neutral pion at rest decays into two photons:

$$\pi^0 \rightarrow \gamma + \gamma$$

What are the energies of the photons?

How to Solve It
- The two photons must have the same momentum by the conservation of momentum and, therefore, the same energy.
- The only initial energy is the rest energy of a pion—the two photons divide this.

2. What minimum photon energy is needed for the pair production of a positive and a negative muon?

How to Solve It

- The mass values you need are in Table 33-1 on page 868 in the text.

- The minimum energy corresponds to the rest energy of the two created muons.

3. Each of the following processes violates at least one conservation law. Which one(s)? (a) $n + p \rightarrow \Sigma^0 + K^0 + \pi^+$; (b) $\Sigma^+ \rightarrow p^+ + \pi^+ + \pi^-$; and (c) $\Sigma^+ \rightarrow \pi^+ + \pi^- + \pi^-$.

How to Solve It

- Compare total charge, baryon number, strangeness, and the like on both sides of each reaction.

Solutions

1. The rest mass of a neutral pion is

$$m_\pi = 135 \text{ MeV}/c^2$$

Thus, the photon energy is

$$E_\gamma = \frac{1}{2} (135 \text{ MeV}/c^2)c^2 = \underline{67.5 \text{ MeV}}$$

2. $\underline{211 \text{ MeV}}$

3. (a) The K^0 and π^+ are mesons; the neutron, proton, and Σ^0 are baryons. Thus, there are two baryons on the left and only one on the right, so this process would not conserve baryon number. It also violates the conservation of strangeness, as the K^0 has nonzero strangeness.

(b) This violates the conservation of strangeness, as there are only zero-strangeness particles on the right-hand side of the equation.

(c) This violates the conservation of baryon number—there are only mesons on the right—and also the conservation of charge.